INTRODUCTION TO *Electromagnetic Engineering*

Roger F. Harrington

Professor Emeritus
Syracuse University

DOVER PUBLICATIONS, INC.
Mineola, New York

Bibliographical Note

This Dover edition, first published in 2003, is a corrected republication of the work originally published by McGraw-Hill Book Company, Inc., New York, Toronto, and London, in 1958.

Library of Congress Cataloging-in-Publication Data

Harrington, Roger F.
Introduction to electromagnetic engineering / Roger F. Harrington.
p. cm.
"This Dover edition...is a corrected republication of the work originally published by McGraw-Hill Book Company, Inc., New York...in 1958"–T.p. verso.
Includes bibliographical references and index.
ISBN-13: 978-0-486-43241-0
ISBN-10: 0-486-43241-6
1. Electromagnetic theory. I. Title: Electromagnetic engineering. II. Title.

QC670.H37 2003
537'.1—dc22

2003055389

Manufactured in the United States by LSC Communications
4500057132
www.doverpublications.com

PREFACE

A knowledge of electromagnetic theory is becoming more and more a necessity in the practice of electrical engineering. This need has been reflected in several books on this topic recently written for and by electrical engineers. These books have, however, largely retained the physicist's viewpoint, neglecting the background already acquired by engineering students in electric circuit theory. To be learned most efficiently, a new subject should draw upon the background of the students as much as possible. The development in this text is therefore based on circuit theory rather than on the classical force-relationship approach. In other words, the theory of electric circuits is used to provide a system of "experiments" already familiar to the electrical engineer. The field concepts are then introduced as a logical extension of circuit concepts already known.

This text is written for introductory study of electromagnetic theory at the junior, senior, or first-year graduate level. The only prerequisite is knowledge of calculus and circuit theory. Vector analysis is introduced and used as needed. The mksc system of units, standard among electrical engineers, is employed throughout. As far as practicable, current engineering notation and nomenclature have been used. The few departures have been made in an attempt to reduce the already unwieldy aggregate of quantities. (For example, q, q_l, q_s, and q_v are used to denote the various types of charge distributions, rather than q, λ, σ, and ρ.) When a quantity has more than one name in common usage, the name most descriptive in terms of circuit concepts has been chosen (for example, *magnetic flux density* rather than *magnetic induction*). Orientation and nomenclature for coordinate systems also have been standardized in the text. In other words, the author has spent considerable time attempting to organize and standardize the notation throughout. It is hoped that this will result in more efficient learning.

An analytical approach to the theory has been emphasized. This is in contrast to the qualitative approach stressed in college physics or in introductory electrical engineering courses. The two approaches are, of course, complementary. and it is hoped that the student already has some familiarity with the qualitative aspects of the subject. The mathematics of the text are approached from an engineer's viewpoint rather than a mathematician's. Physical concepts are used in place of abstract mathematics wherever possible. This is the reason for using the concept of "equivalent magnetic charge and current" in preference to "Green's function theory." Also, so far as practical, proofs of the important theorems are given. It is felt that such proofs help the student both in understanding and in using the theorems.

The purpose of the text is threefold: (*a*) to lead the student to familiarity with and understanding of the various field quantities and concepts, (*b*) to teach necessary vector analysis and associated mathematics, and (*c*) to teach the student to solve a number of elementary field problems. Wherever there is such a multiplicity of objectives, material must be presented more slowly than if only one objective were held. It is hoped that delaying calculation of resistance, capacitance, and inductance until Chapter 9 will cause no undue concern, for many other concepts should be introduced first in order to provide a firm foundation. Any teacher desiring to introduce these calculations earlier can do so without destroying the usefulness of the text.

Primary emphasis is on static fields, but the relationship of static to time-varying fields is established early. Only an introductory treatment of the solution to time-varying field problems is given in Chapter 10. Readers already familiar with field theory will note that several mathematical methods of solving static field problems are omitted. The author feels that inclusion of such methods as separation of variables and complex functions, if they were presented as completely as the other material, would create too voluminous a text. It is recommended that a course based on this text be followed by one on boundary-value problems. For the minimum sequence of fields courses that an interested graduate student should complete, a course on harmonic-time-varying fields is recommended. An undergraduate student might omit the mathematically oriented boundary-value course and proceed profitably to a lower-level course on harmonic-time-varying fields. Another course that could be profitably pursued after study of this text would be a course given from the physicist's viewpoint, emphasizing the microscopic (atomic) and force aspects of electromagnetic theory.

At the end of each chapter are a number of questions for discussion. These emphasize some of the more important concepts in the text and suggest extensions of these concepts and other concepts not covered in the text. They may be profitably used to stimulate classroom discussion or to suggest lecture material to the teacher. Also, a reader studying this text by himself can use the questions to test his understanding of the material. It should be emphasized, however, that not all questions have well-defined answers.

Sets of problems are also included at the end of each chapter. The student can solve most of these using only the concepts of the preceding text. A few represent extensions of the text material, and, for these, hints are given to lead the student to the solution. Some of the problem results are of general interest and are listed in the Index.

Material that requires continual reference throughout the text (for example, vector analysis) has been summarized in the appendixes. Also at the end of the text is a bibliography of electromagnetic theory. This list is not complete, but represents the literature that has come to the attention of the author during his writing. It is hoped that the reader will be able to refer to some of these books, for the author has by no means completely covered the vast theory of electromagnetic fields.

Information concerning errors and omissions as well as constructive comments from users of this text will be appreciated. The author wishes to thank his colleagues, in particular Dr. W. R. LePage and Dr. C. R. Cahn, for their helpful discussions and encouragement. The expert typing and secretarial help of Gladys McDowell and Irma Gellen is also gratefully acknowledged.

Roger F. Harrington

CONTENTS

CHAPTER 1

BASIC CONCEPTS

1-1. Units and Dimensions. The measurement of a quantity involves a *unit* and a *number.* The unit is a reference amount of the quantity, and the number expresses the ratio of the magnitude of that quantity to the magnitude of the unit. When the unit of a quantity is assigned an arbitrary value, that unit is called a *fundamental unit* and the quantity a *fundamental quantity.* When the unit of a quantity is defined in terms of other units, that unit is called a *secondary unit* and the quantity a *secondary quantity.*

In mechanics length, mass, and time are usually chosen as fundamental quantities. In the mks (meter-kilogram-second) system the meter is the unit of length, the kilogram is the unit of mass, and the second is the unit of time. To describe electrical phenomena, it is convenient to consider an additional quantity to be fundamental. The quantity taken is usually electric charge. In the mksc (meter-kilogram-second-coulomb) system the coulomb is the unit of charge, the other fundamental units being those of the mks system. Throughout this book, we shall use the mksc system of units. It is especially convenient because the units of electrical quantities in this system are identical with those used in engineering practice.

The equations of physics express the relationships among numbers associated with quantities. These equations are valid for any choice of units for the fundamental quantities. If the fundamental units are changed, the secondary units change in accordance with their defining equations. A *dimensional equation* is an equation which expresses the manner in which the unit of a quantity changes as the fundamental units are changed. The symbols L, M, T, and Q denote ratios between possible new units of length, mass, time, and charge, and their respective units in the mksc system. Such ratios have been given the name *dimensions.* The terminology "mass has the dimension M," symbolically written [mass] $= M$, means that the unit of mass can be chosen independently. The quantities length, mass, time, and charge in the mksc system are said to have *fundamental dimensions* because their units are arbitrary. All other quantities are said to have *secondary dimensions* because their units depend upon the fundamental units. For example, the dimensional equation for voltage is [voltage] = ML^2/T^2Q. This equation specifies the change in the unit of voltage which must be made when the fundamental units are changed if the same system of equations is to be used. For example, if the unit of mass is doubled, the unit of voltage must be doubled; if the unit of time is doubled, the unit of voltage must be quartered. The dimensions of a secondary quantity can be determined from any equation relating this secondary quantity to quantities of known dimensions.

Because equations are written so as to be valid for any choice of fundamental units, they must "check dimensionally." This means that quantities equal to one another must have the same dimensions and that quantities added together in the same equation must have the same dimensions. The argument of any function expressible as a power series must be dimensionless if successive terms of the series are to have the same dimensions. Examination of an equation to verify that these conditions are met is called a *dimensional check*. We can be sure that an equation that fails to check dimensionally is not generally valid, but a successful check is no assurance that an equation is correct.

A table of quantities considered in this book, their symbolic representation, their units, and their dimensions, is given in Appendix A.

1-2. Linearity and Superposition. Suppose two quantities, f and g, are interrelated such that for each "stimulus" g there corresponds a "response" f. Let f_i be the response to the specific stimulus g_i. If the application of stimulus $g_i + g_j$ produces the response $f_i + f_j$, i and j arbitrary, then the system is *linear* and the *principle of superposition* applies. In other words, a system is linear if the total response is the sum of the partial responses.

The linearity of a system is reflected in the linearity of the equations representing it. Consider the equation

$$\mathcal{L}(f) = g$$

where $\mathcal{L}$ is an *operator*. $\mathcal{L}(f)$ may be as simple as constant times f, or it may involve integration, differentiation, powers, and roots of functions of f. Basically, the above equation states merely that f and g are interrelated. If

$$\mathcal{L}(f_i + f_j) = \mathcal{L}(f_i) + \mathcal{L}(f_j) \qquad \mathcal{L}(kf) = k\mathcal{L}(f)$$

then $\mathcal{L}$ is a *linear operator*, the preceding equation is a *linear equation*, and the principle of superposition applies. Applying this test to possible operators, we can construct a list of common linear operations. For example, successive differentiation of f and successive integration of f are linear operations.

1-3. Circuit Quantities. In the theory of electric circuits we are usually interested in *voltage* and *current,* occasionally in *electric charge* and *magnetic flux.* The basic concepts of these four quantities are summarized in this section.

Electric charge is the quantity by which "amount" of electricity is measured. There are two kinds of electric charge, called *positive* and *negative,* possessed by protons and electrons, respectively. An electrically neutral body contains equal amounts of positive and negative charge. A body is positively charged if it contains an excess of positive charge over negative charge. It is negatively charged if the reverse relation holds. The unit of electric charge is the *coulomb* in the mksc system of units. The number of coulombs on a body equals the number of coulombs of positive charge minus the number of coulombs of negative charge. Thus, electric charge is an algebraic quantity. The algebraic number of coulombs of a quantity of charge is denoted by the symbol q.

The physical manifestation of electric charge is a force exerted on a charged body placed in the vicinity of other charged bodies. Two bodies whose charges have the same sign repel each other. Two bodies attract each other if their charges have opposite sign. Small amounts of charge (small numbers of coulombs) give rise to large amounts of force (large numbers of newtons). For example, two small bodies each charged to 10^{-3} coulomb and separated by 1 meter in vacuum exhibit an attractive or repulsive force of 9×10^3 newtons (2020 pounds).

Another example indicating the order of magnitude of a coulomb involves the circuit concept of capacitance. A 1-microfarad capacitor charged to 100 volts maintains a charge of 10^{-4} coulomb on the plate connected to the positive terminal of a battery. A charge of -10^{-4} coulomb is on the other plate. Thus, for purposes of expressing the amount of stored charge, the coulomb is a large unit.

The electric current through a surface is the *net time rate* at which charge passes through the surface in the *positive reference direction.* Let us illustrate what is meant by this statement. Whenever we speak of a current we should designate the direction in which we consider the charge to flow. This is called the *reference direction* of the current. Positive charge crossing the surface in the reference direction gives rise to a positive current. Negative charge crossing the surface in opposition to the reference direction also gives rise to a positive current. Positive charge crossing against the reference direction and negative charge crossing in the reference direction both give rise to negative current. The term *net time rate* refers to the algebraic sum of the four possibilities mentioned above.

In the mksc system, current is considered to be a secondary quantity; consequently, the unit of current is based on the units of the fundamental quantities. The net rate of flow of one coulomb per second, called the *ampere,* is taken as the unit of current. The algebraic number of amperes in a quantity of current we denote by the symbol i. In engineering practice we are concerned with currents ranging from thousands of amperes down to microamperes.

It is possible, in fact usual, to find a current (flow of charge) in a body and still have the body electrically neutral (uncharged). Even though the positive and negative charges are in motion, there can still be equal amounts of positive and negative charge within the body. A neutral wire carrying current experiences a force when placed in the vicinity of other current-carrying wires. This force is called *magnetic force.* It is not so large an effect as electric force between charged bodies. For example, two long parallel wires carrying 1 ampere of current and separated by 1 meter in vacuum experience a force of attraction or repulsion of only 2×10^{-7} newton (4.5×10^{-8} pound) per meter of length. Many of the ammeters used in circuit measurements utilize this magnetic force to measure current.

Since charges in the vicinity of other charges are subjected to a force, there must be a change of energy from one form (or place) to another associated with the motion of charge. This change in energy per unit positive charge associated with the transportation of charge is called *voltage.* In general, the voltage depends upon the path over which the charge is transported. However, in the special case in which currents do not vary with respect to time, the voltage is independent of the path over which the charge is transported, so long as we remain outside of "sources." In this case the voltage is called *potential difference* and it has a unique value between any two points. This is the basis of direct-current circuit analysis. When the currents vary "slowly" with respect to time, the voltage is "almost" independent of the path over which the charge is transported, so long as we remain outside of sources and inductors. The voltage is then, for most practical purposes, uniquely defined between two points, and again the term potential difference is often used. This is the basis of alternating-current circuit analysis.

The unit of voltage in the mksc system of units is the *volt.* One volt is the voltage associated with a change of energy of one joule per coulomb of charge transported. Given a set of terminals, we arbitrarily designate one to be the *positive reference terminal,* and identify it by a + mark. The voltage having reference polarity at the positive reference terminal is then the change in energy per unit charge transported over the given path, with its sign determined as follows: If the transportation of a positive charge from the positive reference terminal to the other terminal is accompanied by a change in energy from electrical to some other form (thermal, mechanical, etc.), then the voltage is algebraically positive.

If either the sign of the charge or the direction of transport is reversed, and if the energy change is again from electrical to other forms, then the voltage is algebraically negative. If both the sign of the charge and the direction of transport are reversed, the voltage is algebraically positive. The algebraic number of volts in a quantity of voltage is denoted by the symbol v. The range of voltages considered in engineering work is large, extending from megavolts down to microvolts.

When current varies with respect to time or when the physical geometry of a system changes with respect to time, a force is exerted upon charge in addition to the coulomb (static) force. Thus, the change in energy associated with the transportation of charge, and therefore the voltage, is a function of the changing current or the changing geometry. To account for this "induced" voltage, the concept of *magnetic flux* is introduced. The unit of magnetic flux in the mksc system of units is the *weber*. A rate of change of one weber per second induces one volt in the circuit which surrounds the flux. Although we talk of flux "lines" as if they existed in reality, it appears that there is no physical entity associated with them. Thus, we view magnetic flux as a mathematical quantity; physical pictures of it are merely an aid to our thinking. Magnetic flux is an algebraic quantity, requiring, like electric current, a reference direction. The algebraic number of webers of magnetic flux through a surface we shall denote by the symbol ψ. The weber is a fairly large unit of magnetic flux; most engineering problems involve less than a weber of flux for flat surfaces. We can, however, get the effect of large quantities of flux by having a circuit encircle the flux many times or by having the flux intersect a surface many times. This is what is done in transformers. The number of turns in a winding times the flux enclosed is called the *flux linkage*. Flux linkages of the order of thousands of webers are readily obtainable in transformers.

1-4. Linear Circuit Elements. The fundamental elements of circuit theory are *resistors*, *inductors*, and *capacitors*. These are two-terminal configurations of matter for which the terminal voltages and currents obey (approximately) simple mathematical laws. The mathematical laws are considered to define *ideal* circuit elements.

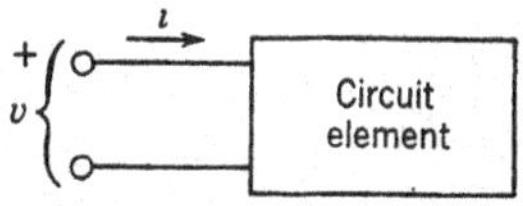

FIG. 1-1. Reference conditions for a two-terminal circuit element.

Before we can write the basic equations for circuit elements, reference conditions for voltage and current must be established. These we take as shown in Fig. 1-1. The voltage-current relationship for an *ideal linear resistor* is then given by

$$v = Ri \tag{1-1}$$

where R is a constant. This equation defines the mathematical quantity of resistance R. The unit of resistance is the *ohm*, that resistance for which one ampere of current is accompanied by one volt across the terminals. The voltage-current relationship for an *ideal linear inductor* is

$$v = L\frac{di}{dt} \tag{1-2}$$

where L is a constant. This defines the mathematical quantity of inductance L. The unit of inductance is the *henry*, that inductance for which a rate of change of current of one ampere per second is accompanied by one volt across the terminals. The voltage-current relationship for an *ideal linear capacitor* is

$$i = C\frac{dv}{dt} \tag{1-3}$$

where C is a constant. This defines the mathematical quantity of capacitance C. The unit of capacitance is the *farad*, that capacitance for which one ampere through it is associated with a rate of change of voltage of one volt per second across the terminals.

We can also express the linear element relationships for L and C in terms of magnetic flux ψ and electric charge q. First, consider a simple inductor consisting of a single loop of wire, as shown in Fig. 1-2. The loop is pictured as being closed by a short path between the terminals, along which the voltage v has the reference polarity shown. The reference direction of the flux ψ is also given in the figure. Faraday's law of induction for this case is then given by

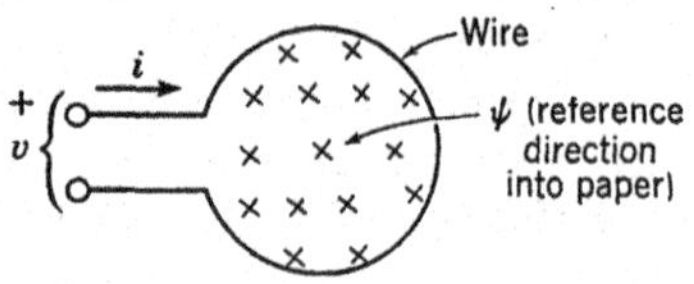

FIG. 1-2. A single-turn inductor.

$$v = \frac{d\psi}{dt} \tag{1-4}$$

For our purposes, this can be viewed as a definition of ψ, the magnetic flux linking the circuit. The element law for the ideal linear inductor can now be written as

$$\psi = Li \tag{1-5}$$

where i has the reference direction shown in Fig. 1-2. Equations (1-4) and (1-5) are equivalent to Eq. (1-2). We have used the single-turn inductor of Fig. 1-2 to illustrate Eq. (1-5), but this equation is valid for any inductor if the quantity ψ is interpreted correctly. Suppose we have an inductor of several turns. The ψ of Eq. (1-5) must then be considered as the flux intersecting the total surface bounded by the wire. A single line of flux may intersect this surface several times. Thus, the ψ of Eq. (1-5) is often referred to as the flux linkage, and should not be confused with the flux intersecting a simple flat surface. This point will be clarified further when we take up the concept of magnetic flux density.

Now consider a simple capacitor consisting of two parallel plates, as shown in Fig. 1-3. With the reference conditions of i and q as shown in the figure, the equation for conservation of charge is

$$i = \frac{dq}{dt} \tag{1-6}$$

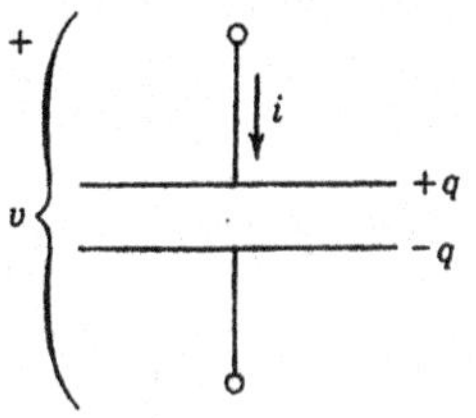

FIG. 1-3. A parallel-plate capacitor.

The element law for the ideal linear capacitor can now be written as

$$q = Cv \tag{1-7}$$

where v has the reference polarity as shown in Fig. 1-3 and q is the charge on the top plate of the capacitor. Equation (1-7) is valid for any capacitor so long as q is interpreted as the algebraic charge on the plates connected to the terminal of positive reference voltage. Equations (1-6) and (1-7) are equivalent to Eq. (1-3).

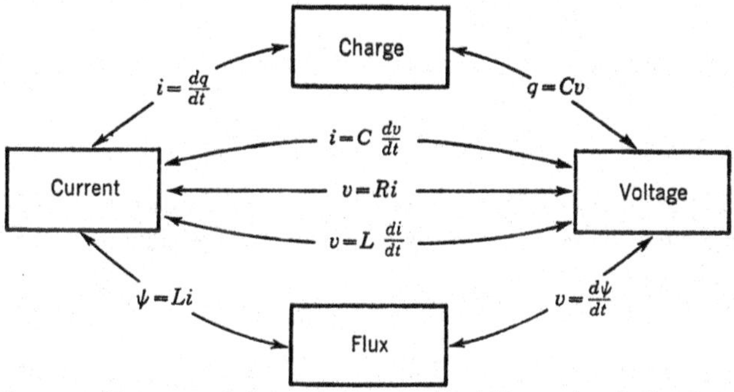

FIG. 1-4. Summary of the equations for ideal linear circuit elements.

The mathematical relationships for ideal linear circuit elements are summarized by Fig. 1-4. The reference conditions and interpretation of the symbols are those used in the above discussions. If reference conditions are changed, the equations may be altered by minus signs.

1-5. Kirchhoff's Laws Generalized. The voltage and current laws of Kirchhoff, in conjunction with the circuit element laws of the preceding section, form the basis of electric circuit analysis. We shall now review the interpretation of Kirchhoff's laws, and modify their formulation to show explicitly the roles of electric charge and magnetic flux.

The usual expression of Kirchhoff's voltage law for circuits is

$$\sum_n v_n = 0 \tag{1-8}$$

where the v_n all have positive reference polarities in (or opposite to) the "direction of travel" around the circuit. To illustrate Eq. (1-8), consider the series *RLC* circuit of Fig. 1-5. For this circuit, Eq. (1-8) becomes

$$v_1 + v_2 + v_3 + v_4 = 0$$

if the wires are perfect conductors and the "stray inductance" of the loop is negligible. If the wires connecting the circuit elements have resistance, we can consider them as resistors and include the voltage dropped across them in the application of Kirchhoff's voltage law. If the inductive effect of the loop is significant, we can add equivalent stray inductances to the circuit, and still apply Eq. (1-8). However, we can also account for this inductive effect of the loop in terms of magnetic flux. Viewing the loop of Fig. 1-5 as a single-turn inductor, we have

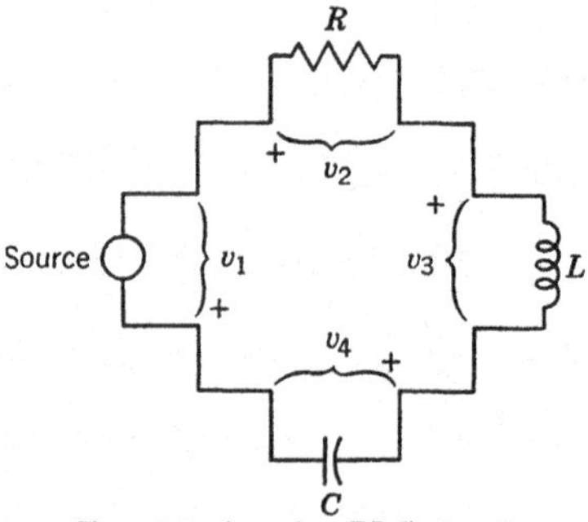

Fig. 1-5. A series *RLC* circuit.

$$v_1 + v_2 + v_3 + v_4 = -\frac{d\psi}{dt}$$

where ψ is the magnetic flux enclosed by the loop with reference direction *into* the paper. The general form of the voltage law for circuits can evidently be written as

$$\sum_n v_n = -\frac{d\psi}{dt} \tag{1-9}$$

where the reference directions of the v_n's and ψ can be summarized as follows. Let the fingers of the right hand point round the loop in the direction of the + to − terminals of the v_n. The reference direction of ψ is then the direction in which the thumb points. It is to be emphasized that Eq. (1-9) is not "more correct" than Eq. (1-8); it merely shows explicitly the "induced" voltage in the loop in terms of the flux enclosed by the loop. In Eq. (1-8), one of the v_n's can be considered to be the voltage due to the stray inductance of the loop, giving the effect of the right-hand term of Eq. (1-9). However, for our purposes, it is more convenient to view Eq. (1-9) as the basic voltage law for circuits.

The usual expression of Kirchhoff's current law for circuits is

$$\sum_n i_n = 0 \tag{1-10}$$

where the i_n all have reference directions away from (or toward) the junction under consideration. To illustrate Eq. (1-10), consider the parallel *RLC* circuit of Fig. 1-6. For this circuit, Eq. (1-10) becomes

$$i_1 + i_2 + i_3 + i_4 = 0$$

if the surrounding medium is a perfect insulator and the "stray capacitance" of the junction is negligible. If there is a leakage current from the top junction to the bottom junction of Fig. 1-6, we can add resistance between the two junctions to account for it. If the capacitance between the two junctions is significant, we can add equivalent stray capacitance to the circuit, and still apply Eq. (1-10). However, this capacitive effect of the junctions can also be accounted for in terms of the charge which collects on the junctions. Viewing the two junctions of Fig. 1-6 as forming a simple capacitor, we have

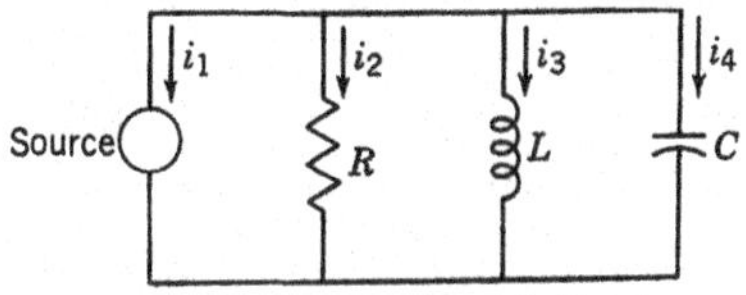

FIG. 1-6. A parallel *RLC* circuit.

$$i_1 + i_2 + i_3 + i_4 = -\frac{dq}{dt}$$

where q is the charge on the top junction. The general form of the current law for circuits can evidently be written as

$$\sum_n i_n = -\frac{dq}{dt} \tag{1-11}$$

where the i_n's have reference direction *away* from the junction and q is the algebraic positive charge collecting on the junction. Again, Eq. (1-11) is not "more correct" than Eq. (1-10), but merely shows explicitly the "capacitive" current from the junction in terms of the charge on the junction. In Eq. (1-10), one of the i_n's can be considered to be this current, giving the effect of the right-hand term of Eq. (1-11). However, for our purposes, it is more convenient to view Eq. (1-11) as the basic current law for circuits.

The extended voltage and current laws for circuits, Eqs. (1-9) and (1-11), also apply to inductors and capacitors. Consider Eq. (1-9) as applied to the single-turn inductor of Fig. 1-2. Taking the wire to be a perfect conductor, we have only the voltage across its terminals to consider. If we apply the rule for determining reference conditions in applying Eq. (1-9) to the single-turn inductor, we find that the reference direction for ψ in Fig. 1-2 is opposite to that required for Eq. (1-9). Therefore, the ψ of Fig. 1-2 is minus the ψ of Eq. (1-9), giving $v = d\psi/dt$, which is identical with Eq. (1-4). Now consider Eq. (1-11) as applied to the parallel-plate capacitor of Fig. 1-3. Taking the surrounding medium to be a perfect insulator, we have only the current in the wire to consider. We note that in Fig. 1-3 the reference direction for i is toward the top plate of the capacitor, whereas for Eq. (1-11) the reference direction for current is away from the junction. Considering the top plate of the capacitor and the connecting wire as a junction, we have the i of Fig. 1-3 equal to minus i as used in Eq. (1-11). Thus, Eq. (1-11) reduces to $i = dq/dt$ when applied to Fig. 1-3, which is identical with Eq. (1-6). The two examples above illustrate how the extended voltage and current equations apply not only outside the circuit elements, as do the usual Kirchhoff equations, but also inside the elements.

1-6. Power and Energy. By definition, voltage is work per unit charge transferred, and current is rate at which charge is transferred. It therefore follows that *power* P is given by

$$P = vi \tag{1-12}$$

Since both v and i are algebraic quantities, requiring reference conditions, P must also be an algebraic quantity requiring a reference direction. Given two wires, the reference direction for P is the same as the direction for i in the wire having positive reference voltage. This is illustrated by Fig. 1-7, the network being called an *algebraic load.* If the reference condition of either v or i (but not both) were reversed, the reference direction for P would be reversed, and the network would be called an *algebraic source.* The network is called an *actual load* or an *actual source* according as the *average power* is toward or away from the network. The unit of power is the *watt,* it being the power transferred by the action of one volt and one ampere. The watt is the rate of flow of energy of one joule per second, the *joule* being the unit of energy in the mksc system of units.

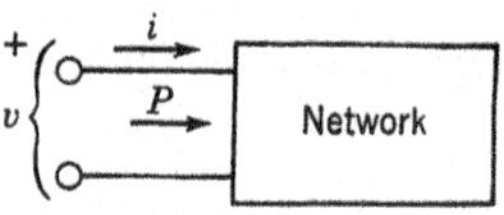

FIG. 1-7. Reference direction for power.

The *energy* W transferred to any algebraic load in the time interval $t_1 < t < t_2$ is

$$W = \int_{t_1}^{t_2} P(t)\,dt = \int_{t_1}^{t_2} v(t)i(t)\,dt \tag{1-13}$$

where the reference conditions are as shown in Fig. 1-7. The algebraic power flow toward a linear resistor P_R is given by

$$P_R = i^2R = v^2/R \tag{1-14}$$

where we have made use of Eq. (1-1). Since R, i^2, and v^2 are always positive, the power flow is always toward the resistor. Thus, the resistor is at all times an actual load, and we say that energy is being *dissipated* in the resistor. The amount of energy dissipated in the resistor W_R for the time interval $t_1 < t < t_2$ is given by

$$W_R = R \int_{t_1}^{t_2} i^2(t)\, dt = \frac{1}{R} \int_{t_1}^{t_2} v^2(t)\, dt \tag{1-15}$$

This energy is dissipated as heat, and is often called the *Joule heat loss.*

The algebraic power flow toward a linear inductor P_L is given by

$$P_L = Li \frac{di}{dt} \tag{1-16}$$

where we have made use of Eq. (1-2). Since i and di/dt can be independently positive or negative, the actual power flow at any instant may be either toward or away from the inductor. The algebraic energy transferred to the inductor W_L during a time interval $t_1 < t < t_2$ is given by

$$W_L = L \int_{t_1}^{t_2} i(t) \frac{di(t)}{dt}\, dt = L \int_{i_1}^{i_2} i\, di$$

where i_1 and i_2 are the values of i at the times t_1 and t_2, respectively. Upon integrating the above equation, we obtain

$$W_L = \tfrac{1}{2}L(i_2^2 - i_1^2)$$

If we choose t_1 to be a time for which $i_1 = 0$, we have

$$W_L = \tfrac{1}{2}Li^2 \tag{1-17}$$

where $i = i_2$ is the value of current at any instant of time $t = t_2$ which may be of interest. Since L and i^2 are always positive, the energy transferred to the inductor in changing the current from zero to some value is always positive. This energy is said to be *stored* in the inductor, and is returned to the source when the current is reduced to zero. We assume that there is zero energy stored in an inductor when the current through it is zero, and therefore Eq. (1-17) gives the total *magnetic energy* stored in the inductor. An alternative form for magnetic energy can be obtained by using Eq. (1-5). Substituting this into Eq. (1-17), we have

$$W_L = \tfrac{1}{2}\psi i \tag{1-18}$$

where ψ is the magnetic flux passing through a surface bounded by the wire of the inductor (the flux linkage).

The algebraic power flow toward a linear capacitor P_C is given by

$$P_C = Cv \frac{dv}{dt} \tag{1-19}$$

where we have used Eq. (1-3). Since v and dv/dt can be independently positive or negative, the actual power flow at any instant may be either toward or away from the capacitor. The algebraic energy transferred to the capacitor W_C during a time interval $t_1 < t < t_2$ is given by

$$W_C = C \int_{t_1}^{t_2} v(t) \frac{dv(t)}{dt} dt = C \int_{v_1}^{v_2} v \, dv$$

where v_1 and v_2 are the values of v at the times t_1 and t_2, respectively. Integration of the above equation gives

$$W_C = \tfrac{1}{2}C(v_2^2 - v_1^2)$$

If we choose t_1 to be a time for which $v_1 = 0$, we have

$$W_C = \tfrac{1}{2}Cv^2 \qquad (1\text{-}20)$$

where $v = v_2$ is the voltage at any instant of time $t = t_2$ which may be of interest. Since C and v^2 are always positive, the energy transferred to the capacitor in changing the voltage from zero to some value is always positive. This energy is said to be *stored* in the capacitor, and is returned to the source when the voltage is reduced to zero. We assume that there is zero energy stored in a capacitor when the voltage across it is zero, and therefore Eq. (1-20) gives the total *electric energy* stored in the capacitor. An alternative form for electric energy can be obtained by using Eq. (1-7). Substituting this into Eq. (1-20), we have

$$W_C = \tfrac{1}{2}qv \qquad (1\text{-}21)$$

where q is the charge on the capacitor plate for which the voltage is reference positive.

1-7. Circuit Sources. In the mathematics of circuit analysis, we represent the effect of sources in terms of (ideal) *voltage sources* and *current sources*. By definition, a voltage source is a source for which the voltage across its terminals follows a prescribed law of variation with respect to time, regardless of the terminal current. The symbolic representation of a voltage source is shown in Fig. 1-8*a*. By definition, a current source is a source for which the current in its terminals follows a prescribed law of variation with respect to time, regardless of the terminal voltage. The symbolic representation of a current source is shown in Fig. 1-8*b*.

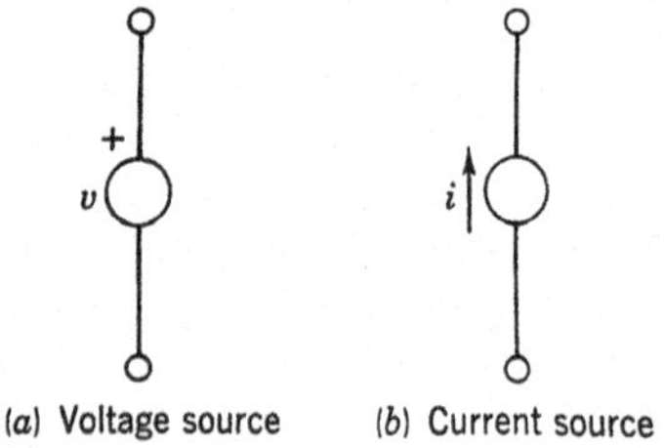

FIG. 1-8. Representation of ideal circuit sources.

The above-postulated ideal sources are useful (1) to replace actual sources for purposes of analysis and (2) to show explicitly on circuit diagrams those voltages and currents which are known or treated as independent parameters. Some physical sources can be represented (approximately) by ideal sources. Others can be represented by networks containing ideal sources and linear circuit elements; such networks are called *linear sources*. Any physical source can be replaced by a voltage source if we know the output voltage for the "load" under consideration. Similarly, any physical source can be replaced by a current source if we know the output current for the load under consideration. Thus, we use ideal sources to replace actual sources; analysis of actual sources is not treated in this text.

Let us now look at the power and energy relationships for ideal circuit sources. For sources, we choose the reference conditions shown in Fig. 1-9. Then Eq. (1-12) gives the instantaneous power transferred *from* the source and Eq. (1-13) gives the energy transferred *from* the source in a time interval $t_1 < t < t_2$. If the source is a *constant* (d-c) voltage source, the energy output becomes

$$W = v \int_{t_1}^{t_2} i(t)\, dt \tag{1-22}$$

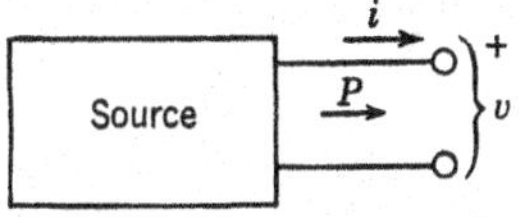

FIG. 1-9. Reference conditions for an algebraic source.

since v is constant. This integral of current has significance in terms of a collection of charge only when there is an ideal capacitor in series with the source (see Prob. 1-11). If the source is a *constant* (d-c) current source, the energy output becomes

$$W = i \int_{t_1}^{t_2} v(t)\, dt \tag{1-23}$$

since i is constant. This integral of voltage has significance in terms of magnetic flux only when there is an ideal inductor in parallel with the source (see Prob. 1-12).

1-8. Mutual-inductance. When the electric circuit contains several coils, part of the magnetic flux associated with the current in one inductor may link another inductor. In this case the two inductors are said to be *mutually coupled*, and they possess a *mutual-inductance* as well as self-inductances. The magnetic flux through the loop of each inductor is considered to be the sum of the partial fluxes associated with the individual currents. Figure 1-10 shows the general situation.

The first subscript on the flux component refers to the current with which the flux is associated; the second subscript refers to the inductor linked by the flux. We assume that each component of the magnetic flux is proportional to its current, so that we may write

$$\begin{aligned} \psi_{1,1} &= L_1 i_1 \qquad & \psi_{1,2} &= M_{1,2} i_1 \\ \psi_{2,2} &= L_2 i_2 \qquad & \psi_{2,1} &= M_{2,1} i_2 \end{aligned} \tag{1-24}$$

According to Eq. (1-4), the terminal voltages of the inductors are

$$v_1 = \frac{d\psi_1}{dt} = \frac{d\psi_{1,1}}{dt} + \frac{d\psi_{2,1}}{dt}$$
$$v_2 = \frac{d\psi_2}{dt} = \frac{d\psi_{2,2}}{dt} + \frac{d\psi_{1,2}}{dt}$$

Using Eqs. (1-24), we may write these as

$$v_1 = L_1 \frac{di_1}{dt} + M_{2,1} \frac{di_2}{dt}$$
$$v_2 = M_{1,2} \frac{di_1}{dt} + L_2 \frac{di_2}{dt}$$

This is the usual voltage-current relationship for linear coupled inductors.

The two mutual-inductances are not independent of each other and, in fact, are always equal for linear inductors. This may be shown on the

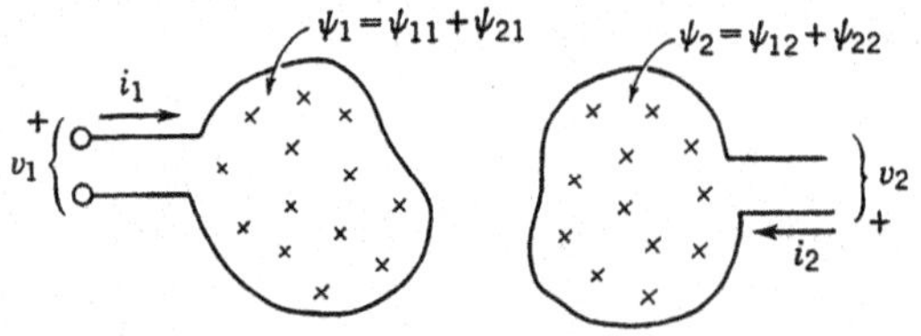

FIG. 1-10. Mutually coupled coils (flux reference direction into paper).

hypothesis that the magnetic energy stored in the inductors depends only on the final values of the currents. Let us hold $i_2 = 0$ and bring i_1 from zero up to its final value, $i_{1,f}$. The energy delivered to the inductors is

$$W_{\text{I}} = \int_{t_a}^{t_b} v_1 i_1 \, dt = \int_0^{i_{1,f}} L_1 i_1 \, di_1 = \tfrac{1}{2} L_1 i_{1,f}^2$$

Next, hold i_1 constant at its final value and bring i_2 from zero up to its final value, $i_{2,f}$. The energy delivered to the inductors is

$$W_{\text{II}} = \int_{t_b}^{t_c} (v_1 i_1 + v_2 i_2) \, dt = \int_0^{i_{2,f}} (M_{2,1} i_{1,f} + L_2 i_2) \, di_2$$
$$= M_{2,1} i_{1,f} i_{2,f} + \tfrac{1}{2} L_2 i_{2,f}^2$$

The total magnetic energy is the sum of W_{I} and W_{II}, or

$$W = \tfrac{1}{2} L_1 i_1^2 + \tfrac{1}{2} L_2 i_2^2 + M_{2,1} i_1 i_2$$

where the subscripts f have been dropped, since the final values are arbitrary. If the order in which the currents are built up is reversed, a similar procedure shows that the energy is given by

$$W = \tfrac{1}{2} L_1 i_1^2 + \tfrac{1}{2} L_2 i_2^2 + M_{1,2} i_1 i_2$$

These two expressions for W will be identical for all values of i_1 and i_2 only if

$$M_{1,2} = M_{2,1} = M \tag{1-25}$$

The voltage-current and energy relations for a pair of linear coupled inductors thus become

$$v_1 = L_1 \frac{di_1}{dt} + M \frac{di_2}{dt}$$
$$v_2 = M \frac{di_1}{dt} + L_2 \frac{di_2}{dt} \tag{1-26}$$
$$W = \tfrac{1}{2}L_1 i_1^2 + \tfrac{1}{2}L_2 i_2^2 + M i_1 i_2$$

The mutual-inductance M is an algebraic quantity, and may be either positive or negative.

If there are more than two coupled inductors, the above discussion applies to each pair, with the others open-circuited. When all inductors are carrying current, the total voltage across any one inductor is the sum of a self-inductance term plus terms involving mutual-inductance between this and every other inductor.

QUESTIONS FOR DISCUSSION

1-1. What determines the number of fundamental quantities to be used in a given subject?

1-2. What determines the choice of specific fundamental quantities?

1-3. What determines the size of the fundamental units?

1-4. What determines whether a quantity should have dimensions or be dimensionless?

1-5. Should all equations check dimensionally?

1-6. Given $f = ma$, determine the dimensions of force f in the mks system.

1-7. Which equations (by number) of this chapter are nonlinear?

1-8. Discuss the concept of charge.

1-9. Discuss the concept of current.

1-10. Discuss the concept of voltage.

1-11. Discuss the concept of magnetic flux.

1-12. What is the difference between voltage in general and potential difference?

1-13. Why are reference conditions important?

1-14. What is the relationship of the "mathematical" circuit elements to the actual circuit elements?

1-15. If the reference polarity of v in Fig. 1-1 were reversed and the reference direction for i kept the same, what effect would this have on Eqs. (1-1) to (1-3)?

1-16. If both the reference polarity of v and the reference direction of i in Fig. 1-1 were reversed, what effect would this have on Eqs. (1-1) to (1-3)?

1-17. Is there a reference condition for q in Eq. (1-7)?

1-18. An inductor constructed of a perfect conductor would have no voltage drop along the wire. How can this be reconciled with the existence of a terminal voltage?

1-19. In an ideal capacitor, no charge flows between the two plates. How can this be reconciled with the existence of a current in the capacitor leads?

1-20. Discuss the difference between the two forms of Kirchhoff's voltage law, Eqs. (1-8) and (1-9).

1-21. Discuss the difference between the two forms of Kirchhoff's current law, Eqs. (1-10) and (1-11).

1-22. Under what conditions could we write the generalized Kirchhoff equations $\Sigma v_n = d\psi/dt$ and $\Sigma i_n = dq/dt$?

1-23. Where does the ψ of Eq. (1-9) exist in Fig. 1-5?

1-24. Where does the q of Eq. (1-11) exist in Fig. 1-6?

1-25. Discuss the concepts of power and energy.

1-26. How can power and energy be measured?

1-27. What is the "internal impedance" of a voltage source?

1-28. What is the "internal impedance" of a current source?

1-29. Discuss the use of ideal sources to represent "independent parameters."

1-30. What work must be done to transport q coulombs across a potential difference of v volts?

1-31. Discuss the concept of mutual-inductance.

1-32. When there is more than one coil, what does "the flux due to the current in coil 1" mean?

1-33. Given two coils, how can the sign of M be determined?

PROBLEMS

1-1. Given [energy] $= ML^2/T^2$, determine the dimensions of v, i, q, ψ, R, L, and C in terms of M, L, T, and Q. [*Hint:* Start with Eqs. (1-6) and (1-13).]

1-2. Keeping our system of equations, if we take the unit of charge to be the "statcoulomb," where 1 coulomb $= 3 \times 10^9$ statcoulombs, and if we keep the units kilogram, meter, and second, what will be the magnitude of the new units of current, voltage, and magnetic flux?

1-3. List as many linear operators as you can.

1-4. A 10-microfarad capacitor is charged to 300 volts. How much charge is stored on its plates? How much energy is stored in it?

1-5. If continuous transport of 10 coulombs of positive charge per second and 5 coulombs of negative charge per second takes place from terminals (1) to (2), and 3 coulombs of positive charge per second and 5 coulombs of negative charge per second takes place from terminals (2) to (1), what is the net current from terminal (1) to (2)?

1-6. If 10 joules of mechanical energy must be expended to transport 2 coulombs of positive charge from terminal (1) to terminal (2), what is the potential difference between terminals (1) and (2), reference positive at terminal (1)?

1-7. If a 10-turn closely-wound coil maintains 10 volts across its terminals, what is the rate of change of flux linkage? What is the rate of change of flux through a simple cross section of the coil?

1-8. It is found that when the voltage between the top and bottom wires of Fig. 1-6 is linearly increasing at the rate of 1 volt per second, the sum of the currents $i_1 + i_2 + i_3 + i_4$ is equal to -10^{-10} ampere. What is the stray capacitance between the two wires?

1-9. How much energy is dissipated in a 10-ohm resistor in 1 minute if 10 volts are maintained across its terminals? What happens to this energy?

1-10. How much energy is stored in a 1-henry coil carrying 0.1 ampere? What happens to this energy if the current is switched off?

1-11. Consider the series circuit of a switch, resistor, capacitor, and constant (d-c) voltage source, as shown in Fig. 1-11. The switch is closed and the circuit is allowed to reach steady-state conditions. What is the energy delivered by the source, the energy stored in the capacitor, and the energy dissipated in the resistor in terms of v, C, and R?

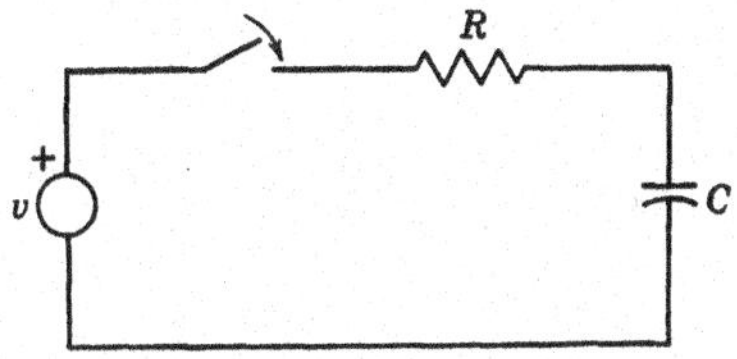

FIG. 1-11. Circuit for Prob. 1-11.

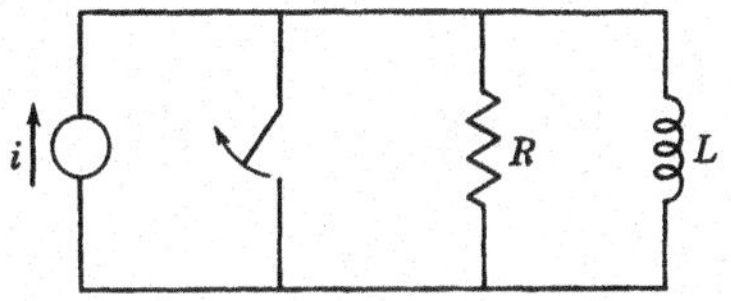

FIG. 1-12. Circuit for Prob. 1-12.

1-12. Consider the parallel circuit of a switch, resistor, inductor, and constant (d-c) current source, as shown in Fig. 1-12. The switch is opened and the circuit is allowed to reach steady-state conditions. What is the energy delivered by the source, the energy stored in the inductor, and the energy dissipated in the resistor, in terms of i, L, and R?

1-13. Consider two coils as shown in Fig. 1-10. It is known that a current of 1 ampere in coil 1 produces 0.1 weber of flux through coil 1 and 0.01 weber of flux through coil 2. A current of 1 ampere in coil 2 produces 0.05 weber of flux through itself. What flux through coil 1 does 1 ampere of current in coil 2 produce? Determine the values of L_1 and L_2, and the magnitude of M. For the orientation of the coils shown in Fig. 1-10, would you expect M as defined by Eq. (1-24) to be positive or negative?

1-14. When an electron moves freely through a potential difference the work done on the system is converted into kinetic energy of the electron. What is the kinetic energy of an electron starting from rest and passing through a potential difference of 1 volt? This energy is called the *electron volt*. The charge of an electron is -1.602×10^{-19} coulomb.

1-15. Assuming that the current in a No. 12 copper wire is due to the motion of 1 electron per atom, what is the average velocity of the electrons when the current is 1 ampere? The mass of No. 12 wire is 29.4 grams per meter length. The number of atoms per gram is N/w, where $N = 6.02 \times 10^{23}$ is Avogadro's number and $w = 63.6$ is the atomic weight of copper.

CHAPTER 2

CHARGE DENSITY AND CURRENT DENSITY

2-1. Charge Density. In circuit theory, we speak of a charge on a conductor or in a region. If we have a charged conductor and divide it into two parts, we find that there is a portion of the original charge on each part. This leads us to the concept that charge is distributed over the conductor with a certain charge density. From atomic theory, however, we know that electric charge is composed of aggregations of charged particles (electrons, protons, etc.). We might therefore question the validity of introducing a continuous charge density. But the charges with which we are concerned are composed of such a large number of atomic particles that this "atomicity" of charge is unimportant. Our procedure is analogous to the use of mass density in mechanics, even though mass is composed of atomic particles.

We define the *volume charge density* q_v at a point in a region of continuous charge distribution as the limit of the charge per volume contained within an element of volume surrounding the point as the element volume goes to zero.* In mathematical form, this is

$$q_v = \lim_{\Delta\tau \to 0} \frac{\Delta q}{\Delta \tau} \qquad (2\text{-}1)$$

where Δq denotes the charge contained in the infinitesimal volume element $\Delta\tau$. The volume charge density is a scalar point function and is measured in *coulombs per cubic meter*. The total charge contained in a region of continuous charge distribution is the volume integral of this charge density. Thus

$$q = \iiint q_v \, d\tau \qquad (2\text{-}2)$$

gives the charge within the region over which the volume integral is taken.

Sometimes the charge appears to be contained within a thin surface layer. We define the *surface charge density* q_s as the limit of charge per area on an element of surface as the element area goes to zero. In

* The symbol ρ is also commonly used to denote volume charge density. To avoid confusion with the cylindrical coordinate ρ and to keep the introduction of new symbols to a minimum, we use the notation q_v.

mathematical form, this is

$$q_s = \lim_{\Delta s \to 0} \frac{\Delta q}{\Delta s} \tag{2-3}$$

where Δq denotes the charge contained on the infinitesimal surface element Δs. As implied by the defining equation, surface charge density is measured in *coulombs per square meter*. The total charge on a surface having a surface density of charge is the surface integral of q_s. Thus

$$q = \iint q_s \, ds \tag{2-4}$$

gives the charge on the surface over which the integral is taken.

Still another possibility is that the charge may be distributed as a filament of charge. In this case we define the *linear charge density* q_l as the limit of charge per length on an element of the filament as the element length goes to zero. Thus

$$q_l = \lim_{\Delta l \to 0} \frac{\Delta q}{\Delta l} \tag{2-5}$$

where Δq is the charge on the infinitesimal line element Δl. The linear charge density is evidently measured in *coulombs per meter*. The total charge contained on a filament is the line integral of q_l. We therefore have

$$q = \int q_l \, dl \tag{2-6}$$

giving the charge on the filament over which the integral is taken.

For purposes of exposition, it is convenient to consider surface and linear charge densities to be special cases of volume charge density. A surface charge density can be thought of as being a volume charge density which is infinite on a surface such that $q_v \, d\tau = q_s \, ds$ on that surface. Similarly, a linear charge density can be thought of as a volume charge density which is infinite along a filament such that $q_v \, d\tau = q_l \, dl$ on that filament. Thus, we shall view Eq. (2-2) as the general form, with Eqs. (2-4) and (2-6) obtainable from it by means of suitable limiting operations.

2-2. Vectors. A *vector* is a quantity having both magnitude and direction in space. Examples of vector quantities in mechanics are force, velocity, and acceleration. Symbols representing vector quantities will be in boldface type. In this section, we shall consider some of the fundamental vector concepts.

A vector can be represented graphically by an arrow, the tail of which usually rests at the point under consideration. The length of the arrow is proportional to the magnitude of the vector, and the arrow points in the direction of the vector. Two vectors (or the same vector function at two different points) are considered equal if both their magnitudes and their directions are the same. Two vectors can be added by placing the tail of one vector at the head of the other, the sum being represented by an arrow drawn from the tail of the first to the head of the second. This is illustrated by Figs. 2-1*a*, *b*, and *c* for the two vectors labeled **A** and **B**. From Figs. 2-1*b* and *c*, it is evident that

$$\mathbf{A} + \mathbf{B} = \mathbf{B} + \mathbf{A} \tag{2-7}$$

The negative of a vector is a vector of the same magnitude pointing in the opposite direction. Subtraction of one vector from another is the same

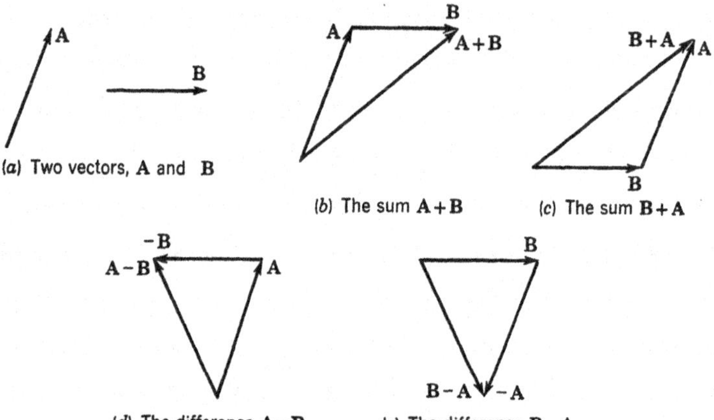

(*a*) Two vectors, **A** and **B**

(*b*) The sum **A**+**B**

(*c*) The sum **B**+**A**

(*d*) The difference **A**−**B**

(*e*) The difference **B**−**A**

FIG. 2-1. Addition and subtraction of vectors.

as addition of the first vector to the negative of the second vector. This is illustrated by Figs. 2-1*d* and *e*. Addition (or subtraction) of several vectors is fundamentally the same as addition (or subtraction) of two vectors, since the sum (or difference) of the first two can be found, this new vector added to the next (or to the negative of the next), and so on. The rule for adding and subtracting vectors is sometimes called the *parallelogram rule*, for the sum and difference of two vectors are the diagonals of the parallelogram having the two vectors as adjacent sides.

The magnitude of a vector is designated by the same letter in italics as that used in boldface type for the vector. For example, the magnitude of **A** is denoted by A. The *scalar product* or *dot product* of two vectors is defined as the product of their magnitudes times the cosine of the angle between them. Thus

$$\mathbf{A} \cdot \mathbf{B} = AB \cos \eta \tag{2-8}$$

where η is the angle between **A** and **B**. In terms of the vector picture, Fig. 2-2*a*, the scalar product can be viewed as the magnitude of **A** times the "projection" of **B** on **A**. The projection of **B** on **A** is called the *component* of **B** in the direction of **A**. We denote a component of a vector by the same letter in italics as that used in boldface for the vector, with a subscript to indicate which component is being considered. For example, we could denote the component of **B** in the direction of **A** by B_a, and Eq. (2-8) could then be written $\mathbf{A} \cdot \mathbf{B} = AB_a$. Two vectors are *orthogonal*, or perpendicular to each other, if their scalar product is zero,

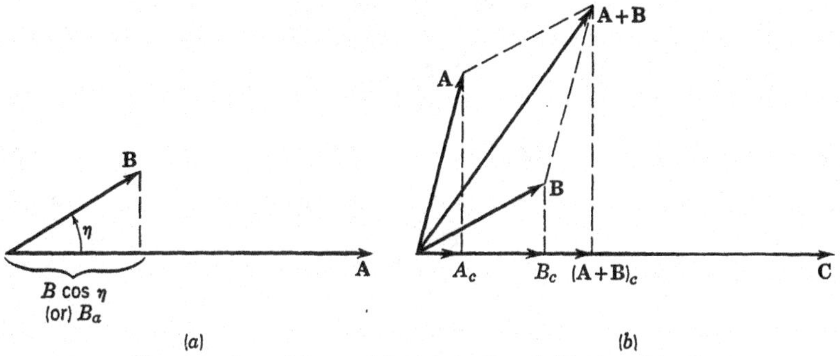

FIG. 2-2. Quantities used in (*a*) $\mathbf{A} \cdot \mathbf{B}$ and (*b*) $(\mathbf{A} + \mathbf{B}) \cdot \mathbf{C}$.

for η is then $\pm 90°$. The scalar product of a vector with itself is its magnitude squared. That is,

$$\mathbf{A} \cdot \mathbf{A} = A^2 \tag{2-9}$$

which follows from Eq. (2-8) since η is now zero. It also follows from the definition of the scalar product that

$$\mathbf{A} \cdot \mathbf{B} = \mathbf{B} \cdot \mathbf{A} \tag{2-10}$$

which is called the *commutative law*. It is evident from the geometry of Fig. 2-2*b* that

$$(\mathbf{A} + \mathbf{B}) \cdot \mathbf{C} = \mathbf{A} \cdot \mathbf{C} + \mathbf{B} \cdot \mathbf{C} \tag{2-11}$$

which is the *distributive law*.

A *unit vector* is defined as a vector with unit magnitude. We shall use the notation **u** to denote a unit vector, with a subscript used when possible to indicate direction. The scalar product of a vector with a unit vector is the component of that vector in the direction of the unit vector. For the vectors of Fig. 2-2*a*, if $\mathbf{u}_a$ is a unit vector pointing in the direction of **A**, then

$$B_a = \mathbf{B} \cdot \mathbf{u}_a \tag{2-12}$$

Multiplication of a vector by a scalar quantity gives a vector in the original direction with magnitude equal to the product of the scalar times the magnitude of the original vector. Thus, the vector **A** of Fig. 2-2*a* could be written

$$\mathbf{A} = \mathbf{u}_a A \tag{2-13}$$

where $\mathbf{u}_a$ is a unit vector in the **A** direction. A *constant vector* is a vector whose magnitude and direction are everywhere the same. Unit vectors are not necessarily constant vectors, since their direction may be a function of their position.

The *vector product* or *cross product* of two vectors is defined as a vector of magnitude equal to the product of the magnitudes of the two vectors times the sine of the angle between them. The direction of the vector product is perpendicular to the two vectors, pointing in the direction of the thumb of the right hand when the fingers point in the direction of rotation of the first vector into the second. This we write

$$\mathbf{A} \times \mathbf{B} = \mathbf{u}_n AB \sin \eta \tag{2-14}$$

where η is the angle between **A** and **B**, and where $\mathbf{u}_n$ is a unit vector perpendicular (or normal) to both **A** and **B** and pointing in the direction indicated by the above *right-hand rule.* The vector picture of Eq. (2-14) is given in Fig. 2-3. Note that the vector **A** × **B** has a magnitude equal to the area of the parallelogram with two adjacent sides **A** and **B**. The unit normal $\mathbf{u}_n$ is perpendicular to the surface in which **A** and **B** (or the parallelogram) lie. Two vectors are parallel or antiparallel if their vector product is zero, since η is then 0° or 180°. Vector products do not obey the commutative law, but

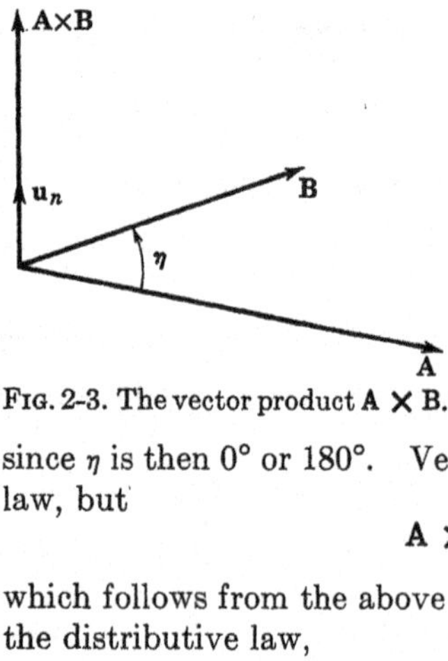

FIG. 2-3. The vector product **A** × **B**.

$$\mathbf{A} \times \mathbf{B} = -\mathbf{B} \times \mathbf{A} \tag{2-15}$$

which follows from the above rules. Vector products do, however, obey the distributive law,

$$(\mathbf{A} + \mathbf{B}) \times \mathbf{C} = \mathbf{A} \times \mathbf{C} + \mathbf{B} \times \mathbf{C} \tag{2-16}$$

The proof of Eq. (2-16) is left to the reader (see Prob. 2-5).

We shall be primarily interested in *vector fields,* which are vector functions defined throughout a region. It is, of course, impossible to represent a vector field completely by means of arrows, for this would require an arrow at every point. We must, therefore, make some sort of compromise when we picture vector functions. If a function is continuous and has continuous derivatives, we call it a *well-behaved* vector function. We can represent well-behaved vector fields by arrows at a discrete number of points, as shown in Fig. 2-4*a*. Another common way of picturing a vector field is by *field lines,* defined as lines everywhere tangent to the tails of the arrows which represent the vector. Such a picture would give us information about the direction of the vector at all points. To give a qualitative picture of the magnitude of the vector, it is often possible to make the density of field lines approximately proportional to the magnitude of the vector function. This sort of picture is shown in Fig. 2-4*b*. Other methods of representing vector fields can be devised and may have advantages in special cases. A vector function is said to be *singular* at those points where it is not well-behaved, that is, where it is discontinuous or has discontinuous derivatives. Many of the

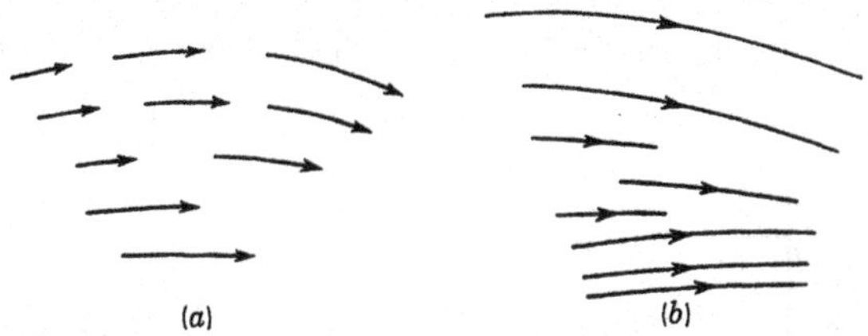

FIG. 2-4. Graphic representation of vector fields (*a*) by arrows at discrete points and (*b*) by field lines, the density of which indicates magnitude.

vector functions that we shall have occasion to use are singular at specific points, lines, or surfaces.

To deal analytically with vectors, it is usually convenient to refer them to coordinate systems and to use the "coordinate components" of the vectors. We shall restrict consideration to rectangular, cylindrical, and spherical coordinate systems. In these systems, we use the coordinate quantities (x,y,z), (ρ,ϕ,z), and (r,θ,ϕ), respectively, as shown in Fig. 2-5. *Coordinate surfaces* are those defined by setting some coordi-

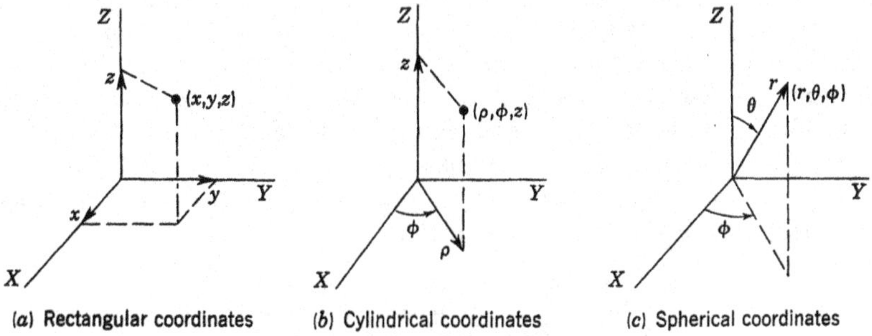

FIG. 2-5. Coordinate systems.

nate equal to a constant. For example, the ρ-coordinate surfaces are cylinders ρ = constant. A *coordinate line* is the intersection of two coordinate surfaces. For example, the ϕ-coordinate lines are circles formed by intersecting ρ = constant and z = constant coordinate surfaces.

Coordinate unit vectors are defined as vectors of unit length, tangent to the coordinate lines and pointing in the positive coordinate directions.

In rectangular coordinates, the coordinate unit vectors are designated $\mathbf{u}_x$, $\mathbf{u}_y$, $\mathbf{u}_z$, and point in the x, y, z directions, respectively.* This is shown in Fig. 2-6*a*. The coordinate unit vectors are vector functions, or fields, for they are defined everywhere throughout space. A constant vector is a vector function whose magnitude *and* direction are everywhere the same. The rectangular-coordinate unit vectors are constant vectors, since they point in the same direction everywhere in space. In cylindrical coordinates, we use the unit vectors $\mathbf{u}_\rho$, $\mathbf{u}_\phi$, $\mathbf{u}_z$, which point in the ρ, ϕ, z directions, respectively. These are shown in Fig. 2-6*b*. Note

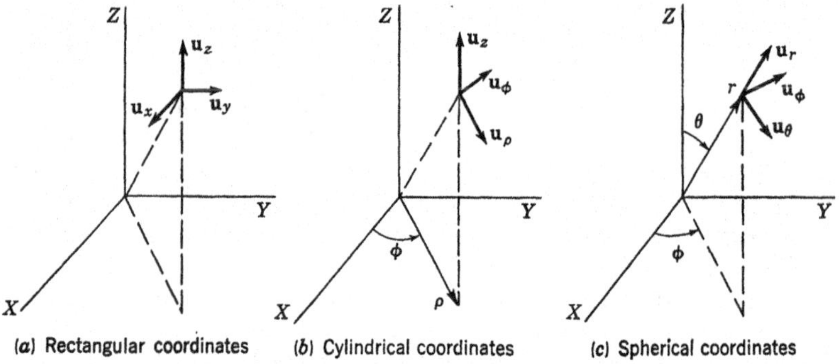

FIG. 2-6. Coordinate unit vectors.

that $\mathbf{u}_\rho$ and $\mathbf{u}_\phi$ are *not* constant vectors, since their direction in space changes from point to point. For example, at $\phi = 0°$, $\mathbf{u}_\rho$ points in the x direction, while at $\phi = 90°$, $\mathbf{u}_\rho$ points in the y direction. This illustrates the fact that a unit vector is not necessarily a constant vector. In spherical coordinates, $\mathbf{u}_r$, $\mathbf{u}_\theta$, $\mathbf{u}_\phi$, pointing in the r, θ, ϕ directions, respectively, are the coordinate unit vectors. These are illustrated in Fig. 2-6*c*. None of the spherical-coordinate unit vectors are constant vectors. Since the above coordinate systems are orthogonal, the unit vectors in each system are at every point orthogonal. It follows from the definition of the scalar product that

$$\begin{aligned} \mathbf{u}_x \cdot \mathbf{u}_y &= \mathbf{u}_y \cdot \mathbf{u}_z = \mathbf{u}_z \cdot \mathbf{u}_x = 0 \\ \mathbf{u}_\rho \cdot \mathbf{u}_\phi &= \mathbf{u}_\phi \cdot \mathbf{u}_z = \mathbf{u}_z \cdot \mathbf{u}_\rho = 0 \\ \mathbf{u}_r \cdot \mathbf{u}_\theta &= \mathbf{u}_\theta \cdot \mathbf{u}_\phi = \mathbf{u}_\phi \cdot \mathbf{u}_r = 0 \end{aligned} \tag{2-17}$$

The scalar product of a unit vector with itself is, of course, unity. In view of the definition of the vector product, we get

$$\begin{array}{lll} \mathbf{u}_x \times \mathbf{u}_y = \mathbf{u}_z & \mathbf{u}_y \times \mathbf{u}_z = \mathbf{u}_x & \mathbf{u}_z \times \mathbf{u}_x = \mathbf{u}_y \\ \mathbf{u}_\rho \times \mathbf{u}_\phi = \mathbf{u}_z & \mathbf{u}_\phi \times \mathbf{u}_z = \mathbf{u}_\rho & \mathbf{u}_z \times \mathbf{u}_\rho = \mathbf{u}_\phi \\ \mathbf{u}_r \times \mathbf{u}_\theta = \mathbf{u}_\phi & \mathbf{u}_\theta \times \mathbf{u}_\phi = \mathbf{u}_r & \mathbf{u}_\phi \times \mathbf{u}_r = \mathbf{u}_\theta \end{array} \tag{2-18}$$

* The symbols **i**, **j**, **k** are also commonly used for $\mathbf{u}_x$, $\mathbf{u}_y$, $\mathbf{u}_z$.

The vector product of a unit vector with itself is, of course, zero. Equations (2-17) and (2-18) assume that all vectors are at the same point.

Once we establish a coordinate system, we can represent vector quantities in terms of the known unit vectors. A knowledge of three noncoplanar components of a vector is sufficient to determine both the magnitude and the direction of the vector. We can therefore represent vectors in terms of their components in the directions of the unit vectors. For example, in rectangular coordinates, we have, according to Eq. (2-12), $A_x = \mathbf{A} \cdot \mathbf{u}_x$, which is called the x component of $\mathbf{A}$. The y and z components are determined in the same manner. We then form three new vectors, $\mathbf{u}_x A_x$, $\mathbf{u}_y A_y$, and $\mathbf{u}_z A_z$, and the original vector is simply the sum of these three *component vectors.* Thus, the vector $\mathbf{A}$ can be written

$$\mathbf{A} = \mathbf{u}_x A_x + \mathbf{u}_y A_y + \mathbf{u}_z A_z \qquad (2\text{-}19)$$

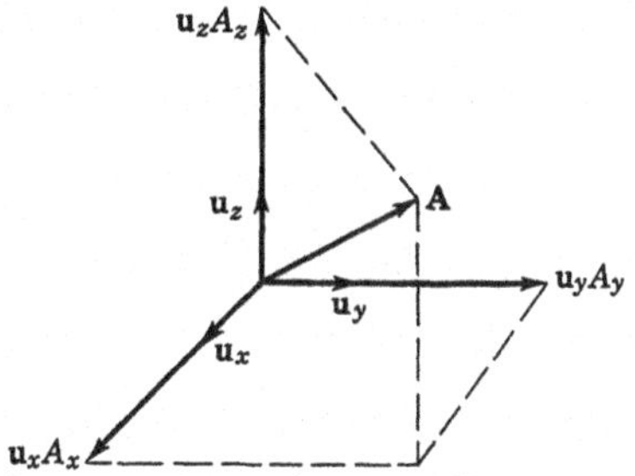

FIG. 2-7. Vector picture for Eq. (2-19).

The vector picture for this equation is shown in Fig. 2-7. In a similar manner we might choose to represent $\mathbf{A}$ in *cylindrical components* A_ρ, A_ϕ, A_z, obtaining

$$\mathbf{A} = \mathbf{u}_\rho A_\rho + \mathbf{u}_\phi A_\phi + \mathbf{u}_z A_z \qquad (2\text{-}20)$$

The vector picture for this equation would be similar to Fig. 2-6 except that the appropriate $\mathbf{u}_\rho$, $\mathbf{u}_\phi$, $\mathbf{u}_z$ would be shown. Finally, we might choose to represent $\mathbf{A}$ in terms of *spherical components* A_r, A_θ, A_ϕ, obtaining

$$\mathbf{A} = \mathbf{u}_r A_r + \mathbf{u}_\theta A_\theta + \mathbf{u}_\phi A_\phi \qquad (2\text{-}21)$$

In general, a vector function can be represented in terms of three components, each component being a scalar function.

Now that we have established a method for representing vectors in specific coordinate systems, we can formalize the basic vector manipulations. If $\mathbf{A}$ and $\mathbf{B}$ are two vectors at the same point in space, we can represent them in terms of component vectors, as in Eqs. (2-19) to (2-21), and we can add corresponding component vectors. This gives

$$\begin{aligned}\mathbf{A} + \mathbf{B} &= \mathbf{u}_x(A_x + B_x) + \mathbf{u}_y(A_y + B_y) + \mathbf{u}_z(A_z + B_z) \\ &= \mathbf{u}_\rho(A_\rho + B_\rho) + \mathbf{u}_\phi(A_\phi + B_\phi) + \mathbf{u}_z(A_z + B_z) \\ &= \mathbf{u}_r(A_r + B_r) + \mathbf{u}_\theta(A_\theta + B_\theta) + \mathbf{u}_\phi(A_\phi + B_\phi) \qquad (2\text{-}22)\end{aligned}$$

Since $\mathbf{u}_x$, $\mathbf{u}_y$, and $\mathbf{u}_z$ are constant vectors, the first of Eqs. (2-22) holds even if $\mathbf{A}$ and $\mathbf{B}$ are at different points in space. The above equations can be readily extended for the addition of three or more vectors. Performing the scalar-product operations on $\mathbf{A}$ and $\mathbf{B}$ when these are represented in terms of components, and making use of Eqs. (2-17), we obtain

$$\begin{aligned}\mathbf{A} \cdot \mathbf{B} &= A_xB_x + A_yB_y + A_zB_z \\ &= A_\rho B_\rho + A_\phi B_\phi + A_zB_z \\ &= A_rB_r + A_\theta B_\theta + A_\phi B_\phi\end{aligned} \tag{2-23}$$

From the above equations and Eq. (2-9), it follows that

$$\begin{aligned}A = \sqrt{A_x^2 + A_y^2 + A_z^2} &= \sqrt{A_\rho^2 + A_\phi^2 + A_z^2} \\ &= \sqrt{A_r^2 + A_\theta^2 + A_\phi^2}\end{aligned} \tag{2-24}$$

Expressing **A** and **B** in terms of rectangular vector components, performing the vector-product operation, and making use of Eqs. (2-18), we obtain

$$\begin{aligned}\mathbf{A} \times \mathbf{B} &= \mathbf{u}_x(A_yB_z - A_zB_y) + \mathbf{u}_y(A_zB_x - A_xB_z) + \mathbf{u}_z(A_xB_y - A_yB_x) \\ &= \begin{vmatrix} \mathbf{u}_x & \mathbf{u}_y & \mathbf{u}_z \\ A_x & A_y & A_z \\ B_x & B_y & B_z \end{vmatrix}\end{aligned} \tag{2-25}$$

The last form of Eq. (2-25) is the determinant form, to be expanded in the usual manner. Similarly, we obtain for cylindrical and spherical coordinates,

$$\mathbf{A} \times \mathbf{B} = \begin{vmatrix} \mathbf{u}_\rho & \mathbf{u}_\phi & \mathbf{u}_z \\ A_\rho & A_\phi & A_z \\ B_\rho & B_\phi & B_z \end{vmatrix} = \begin{vmatrix} \mathbf{u}_r & \mathbf{u}_\theta & \mathbf{u}_\phi \\ A_r & A_\theta & A_\phi \\ B_r & B_\theta & B_\phi \end{vmatrix} \tag{2-26}$$

It should be apparent by now that the operations $\mathbf{A} + \mathbf{B}$, $\mathbf{A} \cdot \mathbf{B}$, and $\mathbf{A} \times \mathbf{B}$ in terms of components have the same form for all right-hand orthogonal coordinate systems, *so long as the two vectors are at the same point in space.*

Formalization of vector manipulations in terms of coordinate components gives us a useful tool for proving vector identities. The vectors exist independently of the coordinate system, so a vector identity proven in one coordinate system is valid in any other coordinate system and is, in general, true. Vector identities are usually most easily proved in rectangular components. For example, suppose we wish to prove the identity

$$\mathbf{A} \times (\mathbf{B} \times \mathbf{C}) = (\mathbf{A} \cdot \mathbf{C})\mathbf{B} - (\mathbf{A} \cdot \mathbf{B})\mathbf{C} \tag{2-27}$$

Expanding the left-hand side in rectangular components, we obtain

$$\begin{aligned}\mathbf{A} \times (\mathbf{B} \times \mathbf{C}) = \mathbf{u}_x[A_y(B_xC_y - B_yC_x) - A_z(B_zC_x - B_xC_z)] \\ + \mathbf{u}_y[A_z(B_yC_z - B_zC_y) - A_x(B_xC_y - B_yC_x)] \\ + \mathbf{u}_z[A_x(B_zC_x - B_xC_z) - A_y(B_yC_z - B_zC_y)]\end{aligned}$$

The two terms on the right-hand side of Eq. (2-27) are individually

$$(\mathbf{A}\cdot\mathbf{C})\mathbf{B} = (\mathbf{u}_xB_x + \mathbf{u}_yB_y + \mathbf{u}_zB_z)(A_xC_x + A_yC_y + A_zC_z)$$
$$(\mathbf{A}\cdot\mathbf{B})\mathbf{C} = (\mathbf{u}_xC_x + \mathbf{u}_yC_y + \mathbf{u}_zC_z)(A_xB_x + A_yB_y + A_zB_z)$$

Subtracting the second of the above equations from the first and rearranging, we can obtain the preceding equation. This proves Eq. (2-27) in general.

The vector relationships derived in this section, as well as those given in the problems and derived later in the text, are summarized for easy reference in Appendix B.

2-3. Surface Integral of a Vector. The rate of flow of a liquid can be represented by a vector, the magnitude of which represents the density

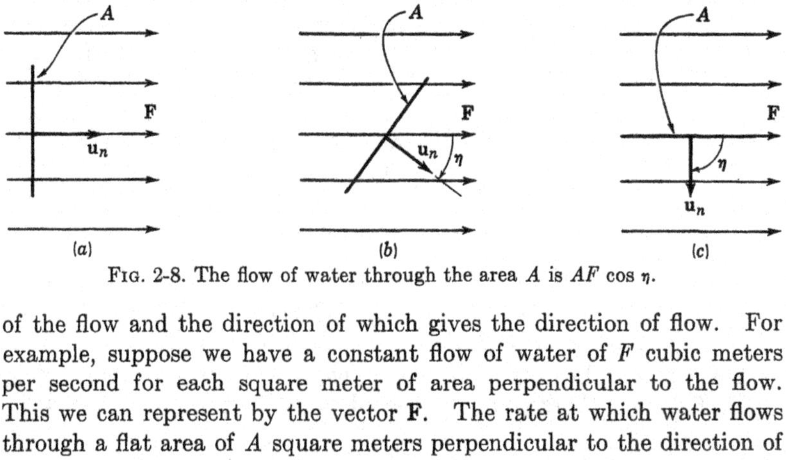

FIG. 2-8. The flow of water through the area A is $AF \cos \eta$.

of the flow and the direction of which gives the direction of flow. For example, suppose we have a constant flow of water of F cubic meters per second for each square meter of area perpendicular to the flow. This we can represent by the vector $\mathbf{F}$. The rate at which water flows through a flat area of A square meters perpendicular to the direction of flow is evidently FA. This is illustrated by Fig. 2-8*a*. If the area A is turned so that it is no longer perpendicular to the direction of flow, the rate at which water flows through it is then $FA \cos \eta$, where η is the angle through which it is turned. This is illustrated in Fig. 2-8*b*. The form of this equation is that of the scalar product. If we define a unit vector $\mathbf{u}_n$ perpendicular to the area,* we can write

$$FA \cos \eta = A\mathbf{F}\cdot\mathbf{u}_n = AF_n$$

Alternatively, we could define a vector $\mathbf{A}$ such that its direction is that of $\mathbf{u}_n$, and write $FA \cos \eta = \mathbf{F}\cdot\mathbf{A}$. If the area is parallel to the direction of flow, as shown in Fig. 2-8*c*, there is no water flowing through it. In this case $\eta = 90°$ and the above scalar product is zero. This example illustrates the use of "flow" vectors and "area" vectors.

Suppose we now consider a well-behaved vector function $\mathbf{B}$ and a

* The symbol $\mathbf{n}$ is also commonly used to denote the unit normal to a surface.

"smooth" surface S, as shown in Fig. 2-9. For a differential element of surface ds we have essentially a plane surface, and the vector **B** is essentially constant over ds. Thus, the "flow" of **B** through ds is $B \cos \eta \, ds$, where η is the angle between the perpendicular to ds and **B**. We could define a unit vector $\mathbf{u}_n$ normal to ds, and write

$$B \cos \eta \, ds = \mathbf{B} \cdot \mathbf{u}_n \, ds = B_n \, ds$$

Alternatively, we could define a differential normal vector $d\mathbf{s}$, with magnitude ds and direction $\mathbf{u}_n$, and write $B \cos \eta \, ds = \mathbf{B} \cdot d\mathbf{s}$.

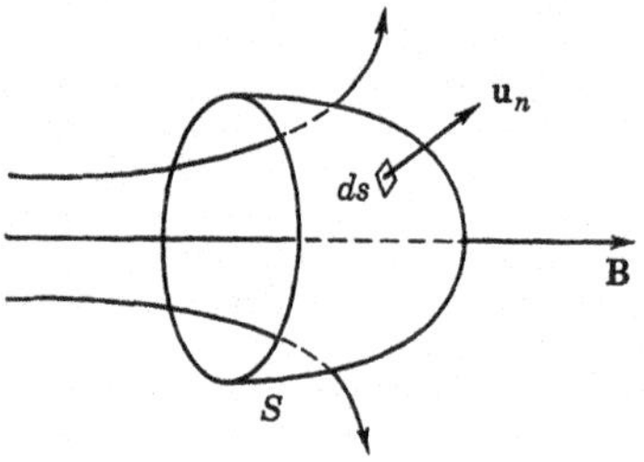

FIG. 2-9. The flow of **B** through S is given by Eq. (2-28).

Note that there is an indeterminancy of 180° in the direction of $\mathbf{u}_n$; that is, $\mathbf{u}_n$ could point in either direction from the surface. When we choose a direction for $\mathbf{u}_n$ we are determining the flow of **B** in that direction. In other words, the choice of $\mathbf{u}_n$ determines the reference direction for the flow of **B**. The flow of **B** through the entire surface S can now be found by integrating over the contributions from each differential element of surface. Thus*

$$\text{flow of } \mathbf{B} = \iint \mathbf{B} \cdot d\mathbf{s} = \iint \mathbf{B} \cdot \mathbf{u}_n \, ds = \iint B_n \, ds \tag{2-28}$$

This integral is called the *surface integral of a vector*. More precisely, it is the surface integral of the normal component of **B** over S.

2-4. Current Density. The basic concept of electric current is the rate of flow of charge, and we should therefore expect to have a flow

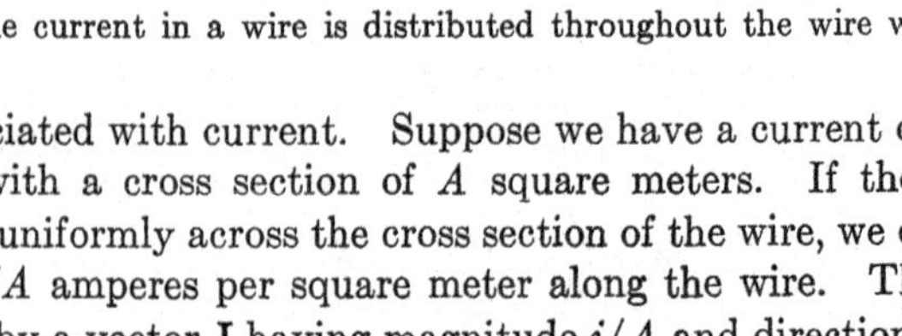

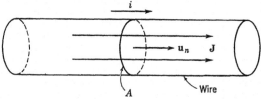

FIG. 2-10. The current in a wire is distributed throughout the wire with a current density **J**.

vector associated with current. Suppose we have a current of i amperes in a wire with a cross section of A square meters. If the current is distributed uniformly across the cross section of the wire, we can say that there are i/A amperes per square meter along the wire. This state we can denote by a vector **J** having magnitude i/A and direction that of the current. This is shown in Fig. 2-10. The vector **J** is called the *electric*

* The flow of a vector is also called the *flux* of a vector in mathematical language.

current density or, more precisely, the *volume density of current elements.* The latter name will be justified in the next paragraph when we take up the concept of a current element. For the case shown in Fig. 2-10, the current in the wire in the reference direction shown is evidently $i = \mathbf{J} \cdot \mathbf{u}_n A$, where $\mathbf{u}_n$ is the unit normal to A. Alternatively, we could write $i = \mathbf{J} \cdot \mathbf{A}$, where the vector $\mathbf{A}$ has magnitude A and direction that of $\mathbf{u}_n$. If we extend the concept to a situation for which $\mathbf{J}$ is not constant and the surface S is not flat, we must use a surface integral as developed in the preceding section. We then have

$$i = \iint \mathbf{J} \cdot d\mathbf{s} \tag{2-29}$$

where i is the current through the surface of integration in the reference direction in which $d\mathbf{s}$ points and where $\mathbf{J}$ is the electric current density. The vector picture of Eq. (2-29) would be Fig. 2-9 with $\mathbf{B}$ replaced by $\mathbf{J}$.

Let us now look a little more closely at the definition of vector current density. The mathematical definition of $\mathbf{J}$ is

$$\mathbf{J} = \mathbf{u}_i \lim_{\Delta s \to 0} \frac{\Delta i}{\Delta s} \tag{2-30}$$

interpreted as follows: Δi is the current passing through the infinitesimal surface element Δs, which is oriented perpendicular to the direction of flow of Δi and thus intercepts the maximum value of current; $\mathbf{u}_i$ is a unit vector pointing in the direction of the current and is therefore also the unit vector normal to Δs. Note that this definition is essentially the same as the example represented by Fig. 2-10, except that it is applied to an infinitesimal area of surface. We call $\mathbf{J}$ a "volume distribution" of current, for it specifies the current density at all points within a region. However, as implied by the defining equation, $\mathbf{J}$ is measured in *amperes per square meter.* Thus, $\mathbf{J}$ appears to be some sort of "surface density." This paradox can be dispelled by the concept of a *current element,* defined as the magnitude of a current times the length over which it extends. An element of volume $\Delta\tau$ can be expressed as a cross section Δs perpendicular to the current times a length Δl in the direction of the current; that is, $\Delta\tau = \Delta s\, \Delta l$. The quantity $J\, \Delta\tau = J\, \Delta s\, \Delta l = \Delta i\, \Delta l$ is, therefore, an incremental current element. We can rewrite Eq. (2-30)

$$\mathbf{J} = \mathbf{u}_i \lim_{\Delta\tau \to 0} \frac{\Delta i\, \Delta l}{\Delta\tau} \tag{2-31}$$

making it evident that $\mathbf{J}$ is a volume density of current elements.

For some applications, it is convenient to introduce a surface distribution of electric current. This is visualized as a sheet of electric current of infinitesimal thickness on a surface. The definition of *electric surface current* $\mathbf{J}_s$, or *surface density of current elements*, is

$$\mathbf{J}_s = \mathbf{u}_i \lim_{\Delta l \to 0} \frac{\Delta i}{\Delta l} \tag{2-32}$$

interpreted as follows: Consider a surface on which we have a current, as shown in Fig. 2-11. We take an infinitesimal element of length Δl along the surface and perpendicular to the current lines. The current crossing Δl is denoted Δi. The unit vector $\mathbf{u}_i$ points in the direction of the current and therefore lies along the surface perpendicular to Δl. $\mathbf{J}_s$ is usually called a surface distribution of current because it specifies the current over a surface, although it is more precise to call it a surface density of current elements. It is measured in *amperes per meter*, as implied by Eq. (2-32). The total current crossing a specified path on a given surface is evidently

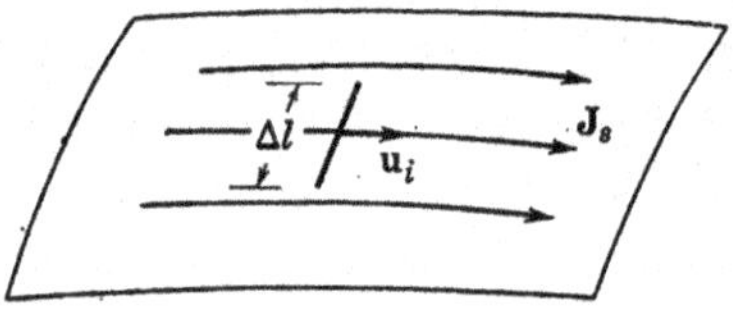

FIG. 2-11. A surface distribution of current.

$$i = \int \mathbf{J}_s \cdot \mathbf{u}_p \, dl \tag{2-33}$$

where $\mathbf{u}_p$ is the unit vector perpendicular to the path and lying along the surface, *not* perpendicular to the surface. The reference direction of i in Eq. (2-33) is the direction in which $\mathbf{u}_p$ points.

A linear distribution of electric current is viewed as a finite value of current confined to a line. This is a filament of current, which we denote by i, and call simply a current. It is in just this way that we view currents in thin wires; the filament of current could correctly be called a linear density of current elements.

For purposes of exposition, it is often convenient to regard surface and line distributions of current as special cases of volume distribution of current. A surface distribution can be pictured as the limit of a volume distribution as its thickness goes to zero. Similarly, a filament of current can be viewed as the limit of a volume distribution as its transverse (to the direction of i) dimensions go to zero. Thus, we can view Eq. (2-29) as being general and Eq. (2-33) as well as the trivial case $i = i$ as being obtainable from it by suitable limiting processes.

2-5. Conservation of Charge. We are now prepared to express the circuit law for charge and current, Eq. (1-11), in terms of charge density and current density. To summarize, we have

$$\sum_n i_n = -\frac{dq}{dt} \tag{2-34}$$

where the i_n are the currents leaving a junction (reference direction away from the junction) and q is the charge collecting on the junction (reference positive on the junction). Consider a junction as shown in Fig. 2-12. From a "field" standpoint, we consider the current leaving the junction to be distributed with a certain current density **J** and the charge on the junction to be distributed with a certain charge density q_v. We take a *closed* surface S and express the electric current crossing S in the outward direction as the surface integral of the normal component of **J**. Thus, we have

$$\sum_n i_n = \oint\!\!\!\oint \mathbf{J} \cdot d\mathbf{s} \tag{2-35}$$

where the positive normal to S is taken *outward* and the circle on the integral sign emphasizes that S is a closed surface. For example, if in Fig. 2-12 the region surrounding the wires is a perfect insulator, we have the current confined to filaments, and Eq. (2-35) becomes

$$\oint\!\!\!\oint \mathbf{J} \cdot d\mathbf{s} = i_1 - i_2 + i_3 + i_4 - i_5$$

The minus signs arise because we have chosen reference conditions for some of the i_n inward. If the surrounding medium is not a perfect insulator there will also be a "leakage" current which must be added to the right-hand side of the above equation. This leakage current is the current that crosses S outside of the wires and is therefore the surface integral of **J** over the part not covered by wires. The charge enclosed by S is the volume integral of q_v. Thus, we have

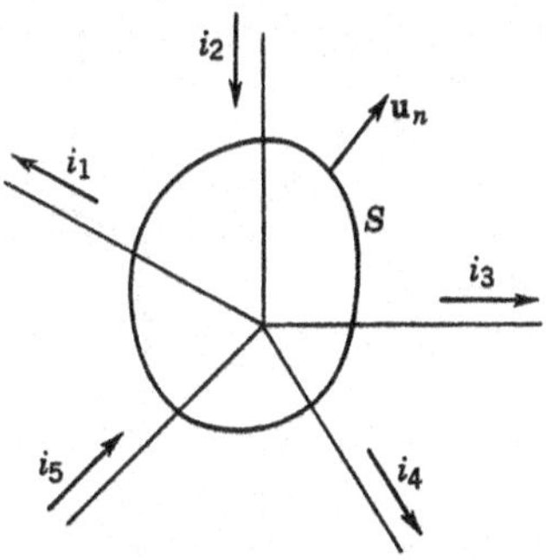

FIG. 2-12. A junction of wires.

$$q = \iiint q_v \, d\tau \tag{2-36}$$

where the integral is taken throughout the region enclosed by S. If the medium surrounding the wires is a perfect insulator, the charge will collect only along the wires, as line distributions of charge. Otherwise, there may be some charge distributed throughout the material external to the wires.

If we now combine Eqs. (2-34), (2-35), and (2-36), we obtain

$$\oint\!\!\!\oint \mathbf{J} \cdot d\mathbf{s} = -\frac{d}{dt} \iiint q_v \, d\tau \tag{2-37}$$

where the positive direction of $d\mathbf{s}$ is outward and the volume integral is taken throughout the region enclosed by the surface. This equation expresses the *conservation of charge*, for it states that any change of charge in a region must be accompanied by a flow of charge across the surrounding surface. In other words, net charge is neither created nor destroyed, but merely transported. We view Eq. (2-37) as the field equation relating charge and current and Eq. (2-34) as the circuit equation.

2-6. Divergence. When the surface integral of a vector over a closed surface is a positive quantity, the vector lines must, on the average, point outward from the surface. We then say that the vector "diverges" from the region enclosed. This is suggested by Fig. 2-13. The *average divergence* of **A** over the region τ is defined as the outward flow of **A** divided by the volume enclosed, or

$$\text{average divergence of } \mathbf{A} = \frac{1}{\tau} \oint\!\!\oint \mathbf{A} \cdot d\mathbf{s} \tag{2-38}$$

If we apply this definition to an infinitesimal element of volume, we obtain the mathematical concept of *divergence*. In other words, the divergence of a vector is the limit of the average divergence as the region becomes vanishingly small.

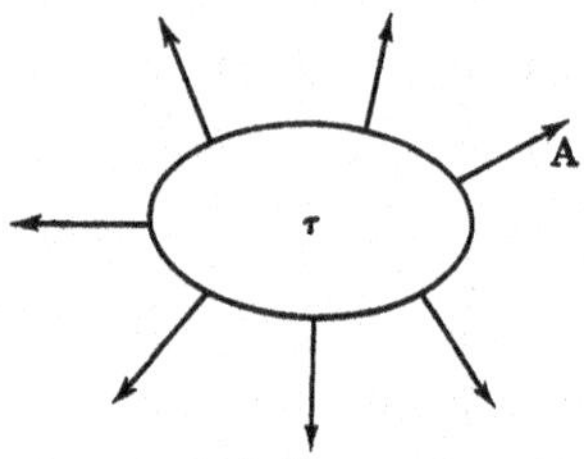

FIG. 2-13. A diverges from τ.

The mathematical definition of the divergence of a vector **A**, denoted by div **A** or $\nabla \cdot \mathbf{A}$ (read "del dot A"), is

$$\text{div } \mathbf{A} = \nabla \cdot \mathbf{A} = \lim_{\Delta\tau \to 0} \frac{1}{\Delta\tau} \oint\!\!\oint \mathbf{A} \cdot d\mathbf{s} \tag{2-39}$$

where the integration is taken over the infinitesimal surface surrounding $\Delta\tau$. In the limit, we have a property of **A** at a point, so div **A** is a scalar function of position. In applying Eq. (2-39), it is assumed that **A** is well-behaved; that is, it is continuous and has continuous derivatives in the vicinity of the point under consideration.

If a vector field having divergence is represented by field lines, as in Fig. 2-4*b*, the lines would appear to begin at points of positive divergence and to end at points of negative divergence. If the vector represents the flow of a quantity, points of positive divergence are "sources of flow" and points of negative divergence are "sinks of flow." For example, current begins at points of decreasing charge density and ends at points of increasing charge density. It should be noted that a vector with zero divergence need not be zero, since the flow may be continuous.

In view of the limiting operation in Eq. (2-39), we should expect divergence to be a type of derivative. Let us first obtain this derivative in rectangular coordinates and components. A representative differential volume element is shown in Fig. 2-14. The outward flow of **A** over the surface of this differential element is the sum of the surface integrals over the six sides of the cube. Consider the two sides perpendicular to the z direction. The outward-pointing unit normals are $\mathbf{u}_n = \mathbf{u}_z$ for the side at $z + dz$, and $\mathbf{u}_n = -\mathbf{u}_z$ for the side at z. **A** is well-behaved, so A_z may be considered to be approximately constant over each face of the cube. Over the "bottom" face we have

$$\mathbf{A} \cdot d\mathbf{s} = -A_z(z)\, dx\, dy \qquad \text{(bottom face)}$$

where the notation $A_z(z)$ means that A_z is evaluated at the position z. The rate of change of A_z with respect to z is, by definition, $\partial A_z/\partial z$ (partial derivative). The value of A_z over the "top" face is its value

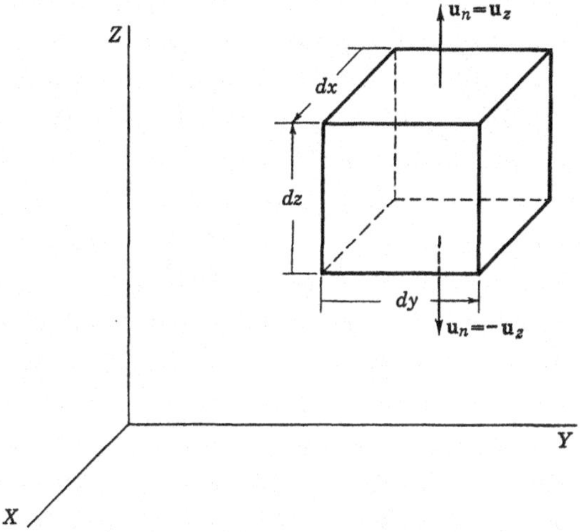

FIG. 2-14. A differential volume element in rectangular coordinates.

over the bottom face plus the rate of change times the distance between the two faces. That is

$$A_z(z + dz) = A_z(z) + \frac{\partial A_z}{\partial z}\, dz$$

so over the top face we have

$$\begin{aligned} \mathbf{A} \cdot d\mathbf{s} &= [A_z(z + dz)]\, dx\, dy \\ &= \left[A_z(z) + \frac{\partial A_z}{\partial z}\, dz \right] dx\, dy \qquad \text{(top face)} \end{aligned}$$

The contribution to the surface integral from both the top and bottom faces is therefore

$$\mathbf{A} \cdot d\mathbf{s} = \frac{\partial A_z}{\partial z}\, dx\, dy\, dz \qquad \text{(top and bottom faces)}$$

Following similar steps for the other two pairs of faces, we find the total surface integral over the differential cube to be

$$\oint\!\!\oint \mathbf{A} \cdot d\mathbf{s} = \left(\frac{\partial A_x}{\partial x} + \frac{\partial A_y}{\partial y} + \frac{\partial A_z}{\partial z} \right) dx\, dy\, dz$$

If we now divide this by the volume of the cube, $d\tau = dx\,dy\,dz$, according to Eq. (2-39) we obtain

$$\operatorname{div} \mathbf{A} = \nabla \cdot \mathbf{A} = \frac{\partial A_x}{\partial x} + \frac{\partial A_y}{\partial y} + \frac{\partial A_z}{\partial z} \tag{2-40}$$

This is the differential form of divergence in rectangular coordinates and components. It is a scalar, as called to mind by the dot-product symbolism of $\nabla \cdot \mathbf{A}$. We can view ∇ (del) as the "differential operator"

$$\nabla = \mathbf{u}_x \frac{\partial}{\partial x} + \mathbf{u}_y \frac{\partial}{\partial y} + \mathbf{u}_z \frac{\partial}{\partial z} \tag{2-41}$$

Equation (2-40) results from an application of our formal rules for scalar multiplication between the operator ∇ and a vector $\mathbf{A}$. However, ∇ is

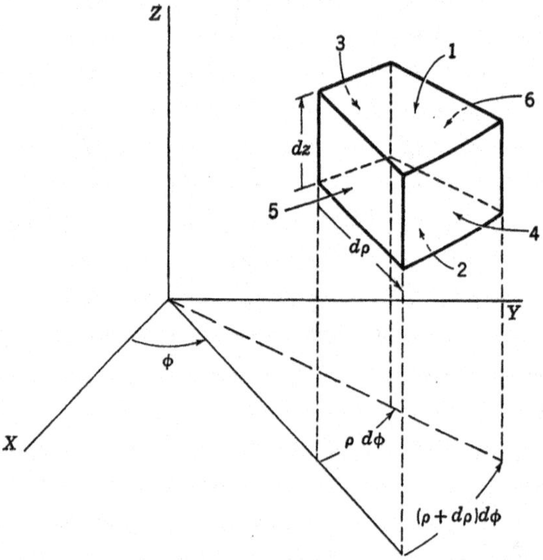

FIG. 2-15. A differential volume element in cylindrical coordinates.

not a vector. It has meaning only when it operates on a function according to Eq. (2-40) or according to other equations we shall consider later.

By applying Eq. (2-39) to a differential volume element in cylindrical coordinates, we can obtain the differential form of divergence in cylindrical coordinates and components. A representative volume element is shown in Fig. 2-15. For convenience, let us number the sides 1 to 6, as indicated in Fig. 2-15. For the top and bottom sides, that is, sides 1 and 2, we proceed as in rectangular coordinates, the only change being that the area is now $\rho\,d\phi\,d\rho$ instead of $dx\,dy$. This gives for the contribution from these two sides.

$$\mathbf{A}\cdot d\mathbf{s} = \frac{\partial A_z}{\partial z}\,dz\,\rho\,d\phi\,d\rho \qquad \text{(sides 1 and 2)}$$

We do not have the symmetry that we had in rectangular coordinates, so each pair of surfaces must be considered explicitly. An additional complication is introduced because of the difference in area between sides 3 and 4, that is, those perpendicular to the ρ direction. Side 3 has the area $\rho\,d\phi\,dz$, so for this side

$$\mathbf{A}\cdot d\mathbf{s} = -A_\rho(\rho)\rho\,d\phi\,dz \qquad \text{(side 3)}$$

Side 4 has the area $(\rho + d\rho)\,d\phi\,dz$, so for this side

$$\begin{aligned}\mathbf{A}\cdot d\mathbf{s} &= [A_\rho(\rho + d\rho)](\rho + d\rho)\,d\phi\,dz\\ &= \left[A_\rho(\rho) + \frac{\partial A_\rho}{\partial \rho}\,d\rho\right](\rho + d\rho)\,d\phi\,dz \qquad \text{(side 4)}\end{aligned}$$

Neglecting the term in $(d\rho)^2$, we have for both of these sides

$$\begin{aligned}\mathbf{A}\cdot d\mathbf{s} &= \left(A_\rho + \rho\frac{\partial A_\rho}{\partial \rho}\right)d\rho\,d\phi\,dz\\ &= \frac{\partial}{\partial \rho}(\rho A_\rho)\,d\rho\,d\phi\,dz \qquad \text{(sides 3 and 4)}\end{aligned}$$

Finally, for the remaining two sides, we have

$$\begin{aligned}\mathbf{A}\cdot d\mathbf{s} &= -A_\phi(\phi)\,d_\rho\,dz \qquad \text{(side 5)}\\ \mathbf{A}\cdot d\mathbf{s} &= [A_\phi(\phi + d\phi)]\,d\rho\,dz\\ &= \left[A_\phi(\phi) + \frac{\partial A_\phi}{\partial \phi}\,d\phi\right]d\rho\,dz \qquad \text{(side 6)}\end{aligned}$$

The contribution from these last two sides is thus

$$\mathbf{A}\cdot d\mathbf{s} = \frac{\partial A_\phi}{\partial \phi}\,d\phi\,d\rho\,dz \qquad \text{(sides 5 and 6)}$$

The total outward flow of **A** from the element is the sum of the contributions from all six sides, or

$$\oint\!\!\oint \mathbf{A}\cdot d\mathbf{s} = \left[\frac{\partial}{\partial \rho}(\rho A_\rho) + \frac{\partial A_\phi}{\partial \phi} + \rho\frac{\partial A_z}{\partial z}\right]d\rho\,d\phi\,dz$$

According to Eq. (2-39), to obtain the divergence we should divide by the volume $d\tau = \rho\,d\rho\,d\phi\,dz$, obtaining

$$\text{div}\,\mathbf{A} = \nabla\cdot\mathbf{A} = \frac{1}{\rho}\frac{\partial}{\partial \rho}(\rho A_\rho) + \frac{1}{\rho}\frac{\partial A_\phi}{\partial \phi} + \frac{\partial A_z}{\partial z} \qquad (2\text{-}42)$$

This is the differential form of divergence in cylindrical coordinates and components. We do not define any differential operator ∇ in cylindrical coordinates because it would not be the vector transformation of Eq. (2-41).

The divergence of a vector in spherical coordinates and components is obtained by applying the defining equation to a differential volume element of the spherical system. The details of the work are left to the student as an exercise. The final result is

$$\nabla \cdot \mathbf{A} = \frac{1}{r^2}\frac{\partial}{\partial r}(r^2 A_r) + \frac{1}{r \sin\theta}\frac{\partial}{\partial \theta}(\sin\theta\, A_\theta) + \frac{1}{r\sin\theta}\frac{\partial A_\phi}{\partial \phi} \tag{2-43}$$

We have no occasion to define a vector operator ∇ in spherical coordinates; the notation $\nabla \cdot \mathbf{A}$ merely symbolizes div $\mathbf{A}$.

The *divergence theorem*

$$\oint\!\!\!\oint \mathbf{A} \cdot d\mathbf{s} = \iiint \nabla \cdot \mathbf{A}\, d\tau \tag{2-44}$$

is a direct consequence of the fundamental definition of divergence. The volume integral is taken throughout the region enclosed by the surface. The validity of Eq. (2-44) can be demonstrated as follows. We take a surface S, as represented in Fig. 2-16, and divide the region enclosed into differential elements of volume. For each elemental volume and surrounding surface, according to Eq. (2-39) we have

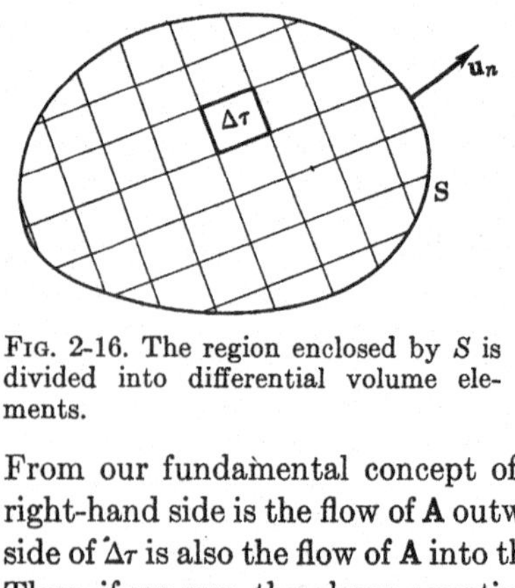

FIG. 2-16. The region enclosed by S is divided into differential volume elements.

$$\nabla \cdot \mathbf{A}\,\Delta\tau = \oint\!\!\!\oint_{\Delta s} \mathbf{A} \cdot d\mathbf{s}$$

From our fundamental concept of the surface integral of a vector, the right-hand side is the flow of $\mathbf{A}$ outward from $\Delta\tau$. The flow of $\mathbf{A}$ out of one side of $\Delta\tau$ is also the flow of $\mathbf{A}$ into the side of the adjacent volume element. Thus, if we sum the above equation over the entire region enclosed by S, all surface integrals of $\mathbf{A}$ cancel except those over S. The sum of the right-hand side is simply $\oint\!\!\!\oint \mathbf{A} \cdot d\mathbf{s}$ over S, which is the left-hand side of Eq. (2-44). The sum of the left-hand side over all volume elements enclosed is just the right-hand side of Eq. (2-44).

The vector formulas derived in this section are summarized for easy reference in Appendix B.

2-7. The Equation of Continuity. If we now apply Eq. (2-37), which expresses the conservation of charge, to a differential element of volume, we have

$$\oint\!\!\!\oint \mathbf{J} \cdot d\mathbf{s} = -\frac{\partial}{\partial t}(q_v\, \Delta\tau)$$

where we have replaced the ordinary derivative with a partial derivative to show that the coordinates of q_v and $\Delta\tau$ are treated as constants. Dividing through by $\Delta\tau$ and letting the volume element become vanishingly small, we find that the left-hand side becomes the definition of the divergence of **J**. Therefore

$$\nabla \cdot \mathbf{J} = -\frac{\partial q_v}{\partial t} \tag{2-45}$$

which, by analogy to the corresponding fluid-flow relationship, is called the *equation of continuity*. It is, of course, no more than Eq. (2-37) applied to a differential volume element and, therefore, expresses the conservation of charge. We say that Eq. (2-37) is the integral form and that Eq. (2-45) is the differential form of the equation of conservation of charge. Equation (2-45) states that points of changing charge density are flow sources of **J**.

The equivalence of Eqs. (2-37) and (2-45) is readily demonstrable by means of the divergence theorem, Eq. (2-44). If we integrate Eq. (2-45) throughout a region, we obtain

$$\iiint \nabla \cdot \mathbf{J}\, d\tau = -\iiint \frac{\partial q_v}{\partial t}\, d\tau$$

We can apply the divergence theorem to the left-hand side and remove the time differentiation from under the integral sign on the right-hand side. This gives Eq. (2-37). Actually, Eq. (2-37) is more general than Eq. (2-45), since in Eq. (2-37) it is not required that q_v and **J** be well-behaved at all points within the enclosed region.

QUESTIONS FOR DISCUSSION

2-1. Discuss the physical limitations on the use of a charge-density function.

2-2. What change could be made in Eq. (2-1) to account for the atomicity of charge?

2-3. Discuss the concept of visualizing surface, line, and point charge distributions as limiting cases of volume charge distribution.

2-4. Explain the parallelogram rule for addition and subtraction of vectors.

2-5. Why are two different products of vectors (scalar and vector) introduced?

2-6. Why do we determine the direction of the vector $\mathbf{A} \times \mathbf{B}$ by a right-hand rule instead of a left-hand rule?

2-7. Is a unit vector a constant vector?

2-8. Which of Eqs. (2-17) and (2-18) apply if the **u**'s are not at the same point in space?

2-9. Justify the formulas $A_x = \mathbf{A} \cdot \mathbf{u}_x$, $A_r = \mathbf{A} \cdot \mathbf{u}_r$, etc.

2-10. Why are the second and third equations of each set of Eqs. (2-22) and (2-23), and Eqs. (2-26), not valid in general if **A** and **B** are at different points in space?

2-11. Give some special cases for which **A** and **B** are at different points, although the formulas for **A** + **B**, **A** · **B**, and **A** × **B** in cylindrical components, Eqs. (2-22), (2-23), and (2-26), apply.

2-12. Repeat Ques. 2-11 for spherical components.

2-13. Discuss the concept of a well-behaved function.

2-14. Discuss the concept of a singular function.

2-15. Devise a method for picturing a vector field different from those of Fig. 2-4.

2-16. Why is it possible to prove vector identities in rectangular components only and then state that they are valid in general?

2-17. Discuss the concept of a vector area, and show that **A** × **B** is the vector area of a parallelogram having adjacent sides **A** and **B**.

2-18. Show that **C** · **A** × **B** is the volume of a parallelopiped having adjacent sides **A**, **B**, and **C**.

2-19. Discuss the concept of the flow (or flux) of a vector.

2-20. Discuss the three common ways of representing the flow of a vector, Eq. (2-28).

2-21. What difficulties arise if we attempt to determine the flow of a singular vector?

2-22. What change in Eq. (2-30) could be made to account for the atomicity of charge?

2-23. What is the reference direction for current in Eq. (2-37)?

2-24. Equation (2-37) can be written concisely as $i = -dq/dt$. Explain this notation.

2-25. Discuss the concept of divergence.

2-26. Give a definition of surface divergence applicable to a surface current distribution.

2-27. What difficulties would be encountered if the divergence theorem were applied to a singular vector function? (See Prob. 2-21.)

2-28. Discuss the relationship of the equation of continuity, Eq. (2-45), to the Kirchhoff current law, Eq. (1-11).

PROBLEMS

2-1. Given a sphere 1 meter in diameter, containing a total charge of 10^{-6} coulomb, what is the charge density if the charge is uniformly distributed throughout the sphere?

2-2. For the same size sphere and the same total charge as in Prob. 2-1, what is the charge density if charge density is proportional to distance from the center?

2-3. For the same size sphere and the same total charge as in Prob. 2-1, what is the surface charge density if the charge is uniformly distributed over the surface of the sphere?

2-4. If a filament 1 meter long supports a linear charge density, its magnitude varying as a half sine wave over the length (zero at the ends) and having peak value of 10^{-6} coulomb per meter, what is the total charge on the filament?

2-5. By geometrical methods prove Eq. (2-16).

2-6. Starting with Eqs. (2-19) to (2-21), derive Eqs. (2-22) to (2-26) using the fundamental concepts.

2-7. By expanding in rectangular components, prove the vector identity

$$\mathbf{A} \cdot \mathbf{B} \times \mathbf{C} = \mathbf{B} \cdot \mathbf{C} \times \mathbf{A} = \mathbf{C} \cdot \mathbf{A} \times \mathbf{B}$$

2-8. Given:

$$\mathbf{A} = \mathbf{u}_x 5 + \mathbf{u}_y 3 + \mathbf{u}_z 2$$
$$\mathbf{B} = \mathbf{u}_x 2 + \mathbf{u}_y 3 + \mathbf{u}_z 3$$

find A, B, $\mathbf{A} + \mathbf{B}$, $\mathbf{A} - \mathbf{B}$, $\mathbf{A} \cdot \mathbf{B}$, $\mathbf{A} \times \mathbf{B}$, η, $(\mathbf{A} \cdot \mathbf{B})\mathbf{A}$, $\mathbf{A} \cdot (\mathbf{A} \times \mathbf{B})$, $\mathbf{A} \times (\mathbf{A} \times \mathbf{B})$, $\mathbf{B} \cdot [\mathbf{A} \times (\mathbf{A} \times \mathbf{B})]$.

2-9. For A and B as given in Prob. 2-8, if $\mathbf{B} \times \mathbf{C} = 0$ and $\mathbf{A} \cdot \mathbf{C} = 45$, find C.

2-10. Derive the equations of Appendix B, Sec. 1*b*. (*Hint:* Apply geometrical and trigonometrical concepts to the coordinate pictures.)

2-11. Derive the equations of Appendix B, Sec. 6. (*Hint:* Use the results of Prob. 2-10.)

2-12. Given $\mathbf{B} = \mathbf{u}_r B_0/r^2$, where B_0 is a constant, determine the flow of B through a hemispherical surface of unit radius. What is the flow of B if the radius is 2?

2-13. Given that the electric current density is $\mathbf{J} = \mathbf{u}_\rho J_0/\rho$, where J_0 is a constant, determine the current out of a z-directed cylinder of unit height and radius ρ_0.

2-14. In a z-directed cylinder of radius ρ_0, the current density is $\mathbf{J} = \mathbf{u}_z J_0 \cos(\pi\rho/2\rho_0)$. Determine the current in the z direction.

2-15. On a spherical surface we have a surface current given by $\mathbf{J}_s = \mathbf{u}_\theta J_0/\sin\theta$. What is the current crossing the equator ($\theta = 90°$)?

2-16. There are z-directed line currents of $+1$ ampere at the coordinates ($x = n$, $y = m$), $n = \ldots -2, -1, 0, 1, 2, \ldots$, $m = \ldots -2, -1, 0, 1, 2, \ldots$. Determine the current through the rectangle enclosed by the lines ($z = 0$, $x = 5.5$), ($z = 0$, $x = -5.5$), ($z = 0$, $y = 3.5$), and ($z = 0$, $y = -3.5$).

2-17. For the situation of Prob. 2-13, what is the rate of change of the charge contained in the cylinder?

2-18. A metal sphere of radius $r = 0.1$ meter is immersed in a poorly conducting medium and fed by an insulated wire carrying current $i = 0.01$ ampere, as shown in

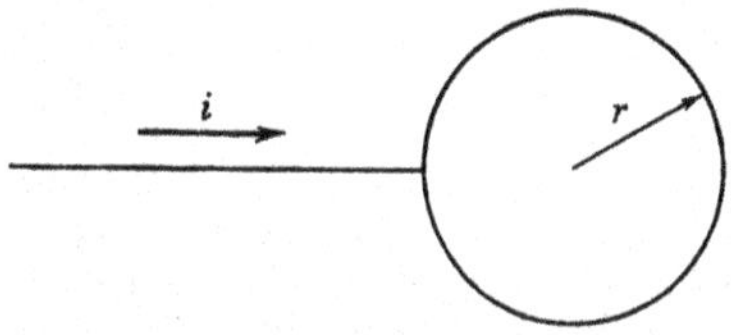

FIG. 2-17. Figure for Prob. 2-18.

Fig. 2-17. If the current density at the surface of the sphere (excluding the current i) is $\mathbf{J} = \mathbf{u}_r 0.07$, at what rate is charge collecting on the sphere?

2-19. Derive Eq. (2-43).

2-20. If $\mathbf{A} = \mathbf{u}_x x + \mathbf{u}_y y + \mathbf{u}_z z$, determine $\nabla \cdot \mathbf{A}$ using rectangular coordinates. Change A to cylindrical components and determine $\nabla \cdot \mathbf{A}$ using cylindrical coordinates. Change A to spherical components and determine $\nabla \cdot \mathbf{A}$. The vector A, as defined above, is called the radius vector from the origin (see Sec. 5-2).

2-21. Determine $\nabla \cdot \mathbf{B}$ for B as given in Prob. 2-12. Determine the flow of B out of a sphere of unit radius by evaluating first the left-hand side of Eq. (2-44) and then the right-hand side. How do you account for this apparent discrepancy?

2-22. If the current density is $\mathbf{J} = \mathbf{u}_x \sin x + \mathbf{u}_y e^{-2y} + \mathbf{u}_z z$, determine the time rate of change of the electric charge density.

2-23. Repeat Prob. 2-22 for $\mathbf{J} = \mathbf{u}_x e^{-3y} \sin 3x + \mathbf{u}_y e^{-3y} \cos 3x$.

2-24. Write Eq. (2-45) explicitly in rectangular coordinates and components. Repeat for the cylindrical and spherical cases.

2-25. Obtain Eqs. (2-4) and (2-6) from Eq. (2-2) using an appropriate limiting process.

2-26. Obtain Eq. (2-33) from Eq. (2-29) using an appropriate limiting process.

CHAPTER 3

ELECTRIC INTENSITY AND MAGNETIC FLUX DENSITY

3-1. Line Integral of a Vector. Force is a vector quantity; to specify it requires both a magnitude and a direction. Suppose we have a block on an inclined plane and we neglect friction. The block is subjected to the force of gravity, represented by the vector **F** in Fig. 3-1*a*. The component of **F** along the plane is the force that can produce motion. We

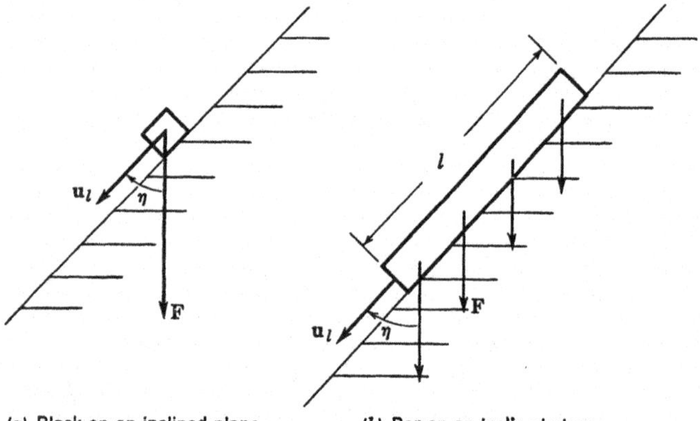

(*a*) Block on an inclined plane (*b*) Bar on an inclined plane

Fig. 3-1. Illustration of "force" vectors and "length" vectors.

define a unit vector along the plane to be $\mathbf{u}_l$, as shown in Fig. 3-1*a*. In vector notation, the component of force of interest is

$$F \cos \eta = \mathbf{F} \cdot \mathbf{u}_l = F_l$$

Now consider a bar of matter on a similar inclined plane, as shown in Fig. 3-1*b*. If we define the vector **F** to be the force per unit length of the bar, the total force is Fl, where l is the length of the bar. In this case, we have the force acting along the inclined plane given by

$$Fl \cos \eta = \mathbf{F} \cdot \mathbf{u}_l l = \mathbf{F} \cdot \mathbf{l}$$

where **l** is a vector of magnitude l and direction $\mathbf{u}_l$. The scalar product of a force vector and a line vector, as illustrated above, is called the *action* of the vector along the line.

Suppose we now consider a well-behaved vector function **A** and a smooth curve C, as suggested by Fig. 3-2. For a differential element of the curve dl we have essentially a straight line, and the vector **A** is essentially constant along dl. Thus, the action of **A** along dl is $A \cos \eta \, dl$, where η is the angle between the tangent to dl and A. Defining a unit vector $\mathbf{u}_l$ which is tangent to dl, we can write $A \cos \eta \, dl = \mathbf{A} \cdot \mathbf{u}_l \, dl = A_l \, dl$. Alternatively, we could define a differential tangential vector $d\mathbf{l}$, with magnitude dl and direction $\mathbf{u}_l$, and write $A \cos \eta \, dl = \mathbf{A} \cdot d\mathbf{l}$. Note that there is an indeterminancy in the direction in which $\mathbf{u}_l$ points; that is, it could point in either direction along the curve. When we choose a direction for $\mathbf{u}_l$, we are determining the action of **A** in that direction. In other words, the choice of $\mathbf{u}_l$ determines the reference direction for the action of **A**. The action of **A** along the entire length of the curve can now be found by integrating over the contributions from each differential element of length. Thus

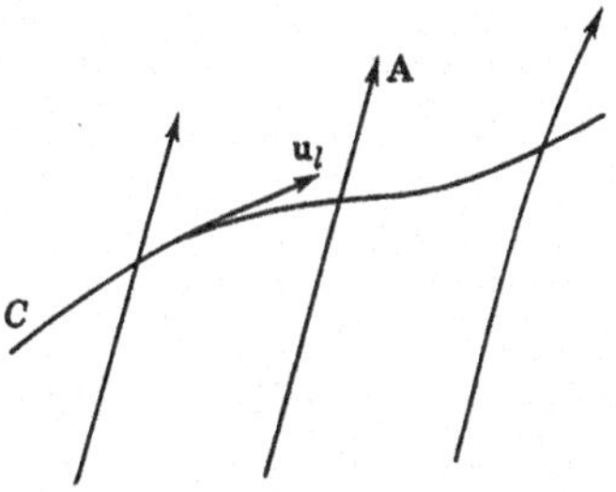

FIG. 3-2. The action of **A** along C is given by Eq. (3-1).

$$\text{action of } \mathbf{A} = \int \mathbf{A} \cdot d\mathbf{l} = \int \mathbf{A} \cdot \mathbf{u}_l \, dl = \int A_l \, dl \tag{3-1}$$

this integral being called the *line integral of a vector*. More precisely, it is the line integral of the tangential component of **A**.

3-2. Electric Intensity. It is experimentally observed that there is a force exerted on a charge in the vicinity of other charges. In vacuum, the force on a "test charge" is found to be a well-behaved function of the position of the test charge. It is proportional to the magnitude of the test charge, so long as the presence of the test charge does not alter the distribution of the other charges. In vacuum, we define the *electric intensity* **E** due to all charges excluding the test charge as the force per unit charge on the test charge. Thus, by definition

$$\mathbf{E} = \frac{\mathbf{F}}{q} \qquad \text{in vacuum} \tag{3-2}$$

where **F** is the force on the charge q. It is necessary that the test charge be sufficiently small in extent so that the **F** field (or **E** field) is essentially constant over the test charge. If a test charge is introduced into a system for the purpose of measuring **E**, it is desirable to make its magnitude small to minimize the redistribution of the original system of charges. However, regardless of the magnitude of the test charge, Eq. (3-2) defines the **E** field of all other charges as it exists with the test charge present.

Consider a system of stationary charges in vacuum. These set up an **E** field which can be "measured" according to Eq. (3-2). We recall that voltage is the work per unit charge associated with the transport of charge. Letting **F** denote the force on a charge being transported, we have the work expended given by

$$\text{Work} = \int \mathbf{F} \cdot d\mathbf{l}$$

When the field is due to stationary charges, that is, an *electrostatic* field, the force on a moving charge is the same as that on a stationary charge, so **F** is given by Eq. (3-2). The voltage associated with the transport of a charge in this case is therefore

$$v = \frac{\text{work}}{q} = \int \frac{\mathbf{F}}{q} \cdot d\mathbf{l} = \int \mathbf{E} \cdot d\mathbf{l}$$

Thus, in the electrostatic case, voltage is the action (line integral) of electric intensity as defined by Eq. (3-2).

If we attempted to use Eq. (3-2) to define **E** internal to matter, we would be beset by many difficulties. If the test charge were large compared to atomic dimensions, its presence would require displacement of some of the matter. If the matter were in contact with the test charge, mechanical force due to stresses in the body might be present in addition to the electrical force. If the test charge were made of atomic size to avoid the above-mentioned effects, the force would depend upon the position of the test charge with respect to the individual atomic charges of the matter. This would not only give us a wildly fluctuating field, but would actually fail to give us a unique value at a given point, because of the motion of the atomic particles.

To circumvent these and other difficulties, we can *redefine* **E** in such a way as to include the previous definition as a special case. We therefore make the general definition that *electric intensity is the vector whose action is voltage*, that is,

$$v = \int \mathbf{E} \cdot d\mathbf{l} \tag{3-3}$$

That such a vector can be found is a postulate, the validity of which is demonstrated by the consistent manner in which measurements can be predicted by its use. It is evident from Eq. (3-3) that the dimensions of **E** are voltage per length. The unit of electric intensity in the mksc system is therefore the *volt per meter*. The definition of **E** according to Eq. (3-3) results in a well-behaved, unique field internal to matter. Only for the special case of a stationary charge in vacuum can we unquestionably interpret **E** as a force per unit charge. The field that we have defined is often called a *macroscopic* field, for it obscures the atomic nature of matter. This is in contrast with the *microscopic* field obtained by considering the individual atomic charges of matter and applying Eq. (3-2) everywhere.

So far we have used the concepts of work and voltage without precisely establishing reference conditions. Referring to our definition of voltage in Sec. 1-3, we find that voltage is positive if energy is changed from electrical to mechanical when positive charge is transported from the positive reference terminal to the other terminal. For mechanical work to be gained, the electrical force must be in the direction of transport. The voltage specified by Eq. (3-3) must therefore have positive reference polarity at the *beginning* of the path of integration. This is shown in Fig. 3-3. The voltage between a pair of terminals may depend upon the path of integration; Eq. (3-3) gives the voltage for the path specified.

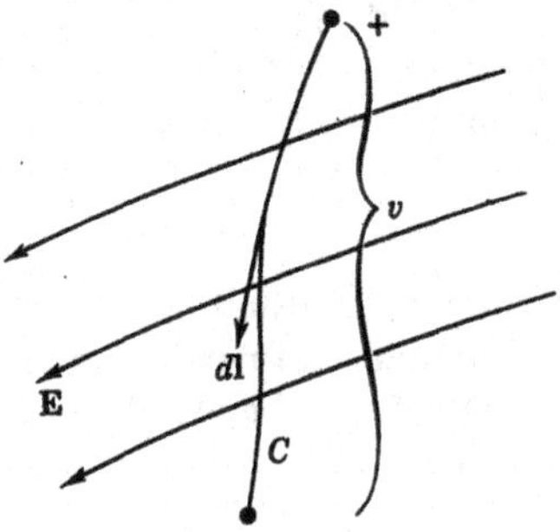

FIG. 3-3. The voltage for the path C is given by Eq. (3-3).

To illustrate the concept of electric intensity as voltage per length, consider the following "experiment." We take a sheet of conducting material and form a resistor by attaching perfectly conducting plates to each end. A d-c voltage is applied across the two end-plates. This is shown in Fig. 3-4. We now imagine a voltmeter connected to a pair of probes, the distance between which is very small and is called Δl. We

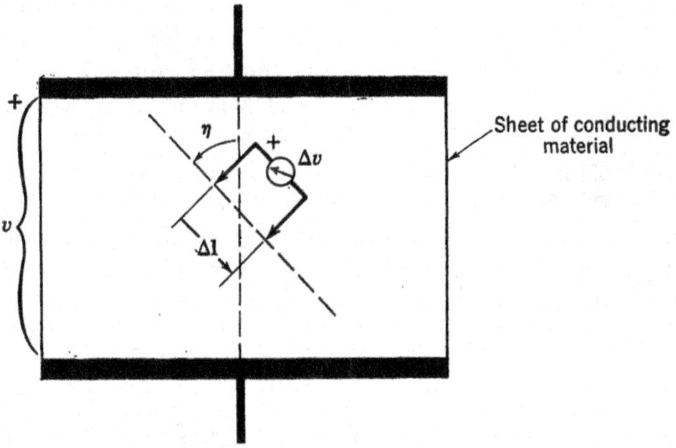

FIG. 3-4. The reading of the voltmeter Δv is proportional to the component of $\mathbf{E}$ in the direction of $\Delta\mathbf{l}$.

take the vector $\Delta\mathbf{l}$ to point from the positive voltmeter terminal to the negative one. Assuming that $\Delta\mathbf{l}$ is small enough so that $\mathbf{E}$ is essentially constant over the interval, from Eq. (3-3) we have $\Delta v = \mathbf{E} \cdot \Delta\mathbf{l} = E_l \, \Delta l$, where Δv is the voltmeter reading. Thus, the component of $\mathbf{E}$ in the direction of Δl is given by

$$E_l = \frac{\Delta v}{\Delta l}$$

providing Δl is sufficiently small. The maximum component of **E** is the magnitude of **E**, so we could rotate the voltmeter to the position of maximum positive reading and thereby measure **E**. Thus, an alternative definition of **E** would be

$$\mathbf{E} = \lim_{\Delta l \to 0} \frac{[\Delta v\, \mathbf{u}_l]_{\max}}{\Delta l}$$

where $\mathbf{u}_l$ is a unit vector pointing in the direction of $\Delta\mathbf{l}$ at the time of maximum voltmeter reading. The reading of the voltmeter rotated an angle η from the orientation of maximum reading would be

$$\Delta v = E\,\Delta l \cos \eta = \Delta v_{\max} \cos \eta$$

as evident from elementary vector concepts. We can visualize the voltage between a pair of terminals a large distance apart, according to Eq. (3-3), as the sum of the readings of a series of differential voltmeters strung along the path of integration.

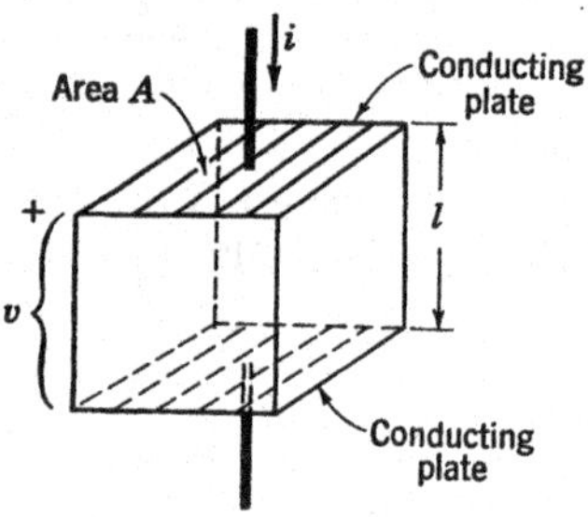

FIG. 3-5. A "block" resistor.

3-3. Conduction Current. In circuit theory, an ideal resistor is an element that obeys Ohm's law, $v = Ri$. The current is visualized as the motion of free electrons and is called a *conduction current* $\mathbf{J}^c$. Suppose we consider an ideal resistor constructed of a homogeneous block of conducting matter with perfectly conducting plates covering opposite faces, as shown in Fig. 3-5. The distance between the two plates we denote by l and the cross-sectional area of the block by A. We assume that the current flows with uniform density throughout the block, in a straight path perpendicular to the two plates. Defining a unit vector $\mathbf{u}_i$ pointing in the direction of current, we can write for the conduction current density

$$\mathbf{J}^c = \mathbf{u}_i J^c = \mathbf{u}_i \frac{i}{A}$$

A perfect conductor can be thought of as a conductor for which no voltage drop exists along its surface, so the electric intensity **E** must be perpendicular to the two plates. We assume that **E** points in this direction throughout the block and that it is of constant magnitude. It then follows that

$$\mathbf{E} = \mathbf{u}_i E = \mathbf{u}_i \frac{v}{l}$$

Substituting for v and i into Ohm's law, $i = v/R$, we have

$$J^c = \frac{l}{RA} E$$

and, since J^c and E have the same direction, we can write

$$\mathbf{J}^c = \frac{l}{RA} \mathbf{E} \tag{3-4}$$

We define the *conductivity* σ of the material within the block as

$$\sigma = \frac{l}{RA} \tag{3-5}$$

Thus, *the conductivity of a homogeneous (the same at all points) and isotropic (no preferential direction for current) conducting medium is equal to the conductance (reciprocal of resistance) between opposite faces of a unit cube (1 meter on a side).* The dimensions of conductivity are evidently conductance per length, and the unit in the mksc system is the *mho per meter.* A medium is *linear* if the conductivity is independent of the magnitude of the current.

Using Eq. (3-5), we can now write Eq. (3-4) as

$$\mathbf{J}^c = \sigma \mathbf{E} \tag{3-6}$$

which is called *Ohm's law for fields.* Equation (3-6) is the mathematical definition of linear, isotropic conducting media, and a variety of materials is found to obey it quite closely. If a medium is not homogeneous, we can consider σ to be a function of position. In this case, we view σ as related to a cube of differential size Δl on a side, according to

$$\sigma = \lim_{\Delta l \to 0} \frac{1}{R\,\Delta l} \tag{3-7}$$

Alternatively, σ can be viewed as the conductance of a unit cube constructed of a material having the same conductivity as that at the point under consideration.

Ohm's law for fields, Eq. (3-6), is quite logical according to the atomic picture of matter. The conduction current consists of the motion of charged particles, and the electric intensity is related to some sort of average force per unit charge. It is to be expected that the average motion of the charges will be in the direction of the force, or in the direction of $\mathbf{E}$, and will probably be proportional to it. This is what is stated by Eq. (3-6).

All matter has some conductivity; materials having very large values of σ are called *conductors,* and materials having very small values of σ are called *insulators* or *dielectrics.* The range of σ for materials encountered in practice is extremely large; σ is of the order of 10^7 for good conductors and 10^{-17} for good insulators. A table of conductivities for some common materials is given in Appendix C. In many cases, good conductors may be approximated, for purposes of analysis, by *perfect conductors* $\sigma = \infty$, and good insulators by *perfect insulators* $\sigma = 0$. The only perfect insulator in nature is a vacuum, which is unattainable in practice.

The flow of charge through a resistance is characterized by a loss of energy in the form of heat. From circuit concepts, we recall that the rate of energy dissipation in a resistor is $P = vi$. Substituting for $v = El$ and $i = J^c A$, we can write this

$$P = J^c E A l$$

for the ideal resistor of Fig. 3-3. However, Al is simply the volume τ of the block, and $J^c = \sigma E$; so we can express the above equation as

$$P = \sigma E^2 \tau$$

The electric intensity is constant throughout the block; so we can consider the power dissipation to be distributed with a *density of power dissipation* p_d given by

$$p_d = \sigma E^2 = \sigma \mathbf{E} \cdot \mathbf{E} = \mathbf{J}^c \cdot \mathbf{E} \tag{3-8}$$

The total power dissipation in the resistor is then the volume integral of the density of power dissipation taken over the block, that is

$$P_d = \iiint p_d \, d\tau = \iiint \sigma E^2 \, d\tau \tag{3-9}$$

We have attached the subscript d to the symbol P to denote the power dissipation due to conduction current. Equations (3-8) and (3-9) are valid for any field in linear, isotropic conducting media. This can be shown by considering each differential element of volume to be a resistor of the type of Fig. 3-5, and then integrating over all volume elements.

3-4. Magnetic Flux Density. From circuit theory, we have a concept of magnetic flux analogous to that of electric current. At every point in space, we picture a definite density of flux oriented in a definite direction. Letting Δs be an incremental element of area, $\mathbf{u}_n$ the unit normal to Δs, and $\Delta\psi$ the flux through Δs, we have *magnetic flux density* $\mathbf{B}$ at a point specified by*

$$\mathbf{B} = \lim_{\Delta s \to 0} \frac{(\Delta\psi \, \mathbf{u}_n)_{\max}}{\Delta s} \tag{3-10}$$

The subscript "max" implies that both $\Delta\psi$ and $\mathbf{u}_n$ are associated with Δs oriented to intercept maximum flux. For any other orientation of the incremental area, the magnetic flux intercepted is evidently

$$\Delta\psi = \mathbf{B} \cdot \Delta\mathbf{s} = \mathbf{B} \cdot \mathbf{u}_n \, \Delta s = B_n \, \Delta s$$

* $\mathbf{B}$ is also called "magnetic induction."

This is represented by Fig. 3-6. From Eq. (3-10), it follows that magnetic flux density **B** has the dimensions of magnetic flux per area. The unit of **B** in the mksc system is the *weber per square meter.*

The magnetic flux ψ through a smooth surface is given by summing the contributions from all differential elements of area. Thus, ψ is the flow (or flux) of **B**,

$$\psi = \iint \mathbf{B} \cdot d\mathbf{s} \tag{3-11}$$

the reference direction of ψ being that in which $d\mathbf{s}$ points. In the sense in which both flux density **B** and current density **J** are flow vectors, they are mathematically analogous. **B** is conceptually different from **J** in that magnetic flux is purely a mathematical concept whereas electric current represents an actual flow of something, i.e., charge.

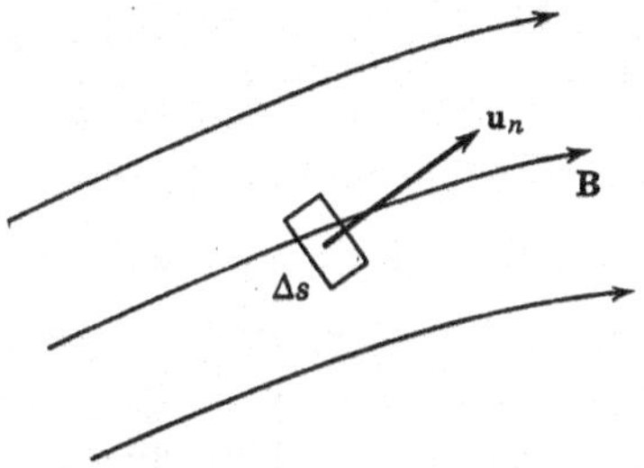

FIG. 3-6. The flux through Δs is proportional to the component of **B** in the direction of $\mathbf{u}_n$.

Magnetic flux density differs from current density in one important respect. It has been found that **B** *has no sources of flow in nature.* This means that lines **B** have no beginning or end and that they must either close on themselves or continue indefinitely. Mathematically, this property is expressed by

$$\nabla \cdot \mathbf{B} = 0 \tag{3-12}$$

In contrast, **J** satisfies Eq. (2-45) and therefore has sources of flow at points where charge density is changing. The integral form of Eq. (3-12) is

$$\oint\!\!\oint \mathbf{B} \cdot d\mathbf{s} = 0 \tag{3-13}$$

which is obtained by integrating Eq. (3-12) throughout a region and applying the divergence theorem. In words, Eq. (3-13) states that the magnetic flux intercepted by a closed surface is always zero. The validity of Eqs. (3-12) and (3-13) is a postulate borne out by experimental evidence.

So far as we are concerned, the magnetic flux density **B** is the vector whose flow is the magnetic flux ψ. There is, however, an alternative interpretation of **B** in certain special cases. This is in terms of force on a current element, which is in some respects analogous to the interpretation of **E** in terms of force on a point charge. It is observed experimentally that the force **F** on an uncharged current element $i\mathbf{l}$ is given by

$$\mathbf{F} = i\mathbf{l} \times \mathbf{B} \qquad \text{in vacuum} \tag{3-14}$$

The *moment* of a current element is defined as the magnitude of the current times the length over which it flows, which is il in Eq. (3-14).

Thus, in vacuum, we may define B from all sources other than the "test current element" as the maximum force per unit current moment on the test current element. The direction of **B** is then perpendicular both to $i\mathbf{l}$ and to **F** when $i\mathbf{l}$ is oriented for maximum force. It is determined according to the right-hand rule of Fig. 3-7. As shown by Eq. (3-14), the force **F** is always perpendicular both to $i\mathbf{l}$ and to **B**.

If we attempt to extend Eq. (3-14) as a defining equation for **B** internal to matter, we encounter difficulties similar to those in the case of electric intensity. Any test current element by its very nature must displace matter. If the matter were in contact with the current element, mechanical force in addition to magnetic force might be present. If we attempt to make the test current element of atomic size, we arrive at the concept of a single moving charge. This would respond to the "local" or "microscopic" force field about the individual atomic particles of matter. This concept can be used to define a *microscopic* **B** field, but this is not the same as the *macroscopic* **B** field we are considering. We must therefore accept Eq. (3-10) as our general defining equation, with Eq. (3-14) valid in the special case of a current element in vacuum. The equivalence of Eqs. (3-10) and (3-14) can be demonstrated through the use of Faraday's law, Eq. (1-4). However, this demonstration requires the use of concepts not available to us at this time, so we shall not attempt to present it.

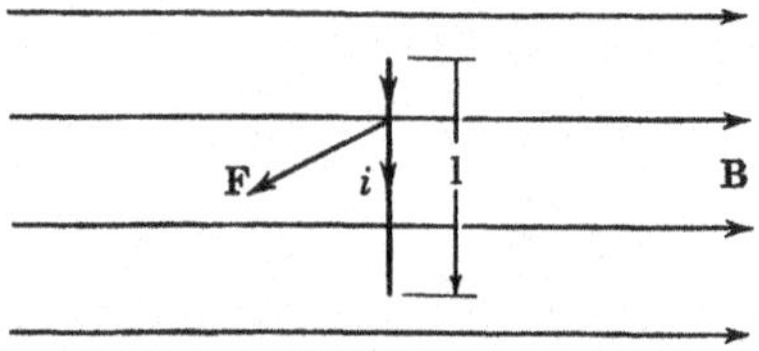

FIG. 3-7. The force on a current element is perpendicular both to $i\mathbf{l}$ and to **B**.

We mentioned above that the limiting case of a current element was a single moving charged particle. In this case there would be a net charge on the current element, and the total force on it would be a combination of both electric and magnetic forces. Given a charge q moving at a velocity $\mathbf{v}$, the current moment would be $q\mathbf{v}$, so the total force should be the sum of Eqs. (3-2) and (3-14), or

$$\mathbf{F} = q(\mathbf{E} + \mathbf{v} \times \mathbf{B}) \tag{3-15}$$

Thus, in the reference frame for which the charge is stationary, $\mathbf{v} \times \mathbf{B}$ appears to the charge to be an electric field. This is an example of the interrelationship between **E** and **B**, which is a property of time-varying fields. Equation (3-15), extended to apply internal to matter when q is of atomic size, is the basis of the cited microscopic field theory. When used in conjunction with the macroscopic field theory that we are treating, the application of Eq. (3-15) must be restricted to moving charge in vacuum.

3-5. Faraday's Law of Induction. In Sec. 1-4, we wrote Kirchhoff's voltage law for circuits in a form showing explicitly the role played by magnetic flux. This law is

$$\sum_n v_n = -\frac{d\psi}{dt} \tag{3-16}$$

where the v_n are the voltages around a closed circuit and ψ is the magnetic flux linking the circuit. In Sec. 3-2, we postulated that voltage is the line integral of electric intensity. Therefore, the sum of the voltages v_n around a closed loop is simply the line integral of $\mathbf{E}$ about the loop. In equation form, this is

$$\sum_n v_n = \oint \mathbf{E} \cdot d\mathbf{l}$$

where the circle on the integral sign emphasizes that a closed contour is taken. In Sec. 3-4, we postulated that magnetic flux is the surface integral of magnetic flux density. Therefore, the magnetic flux ψ is the flow of $\mathbf{B}$ over a surface bounded by the circuit, that is

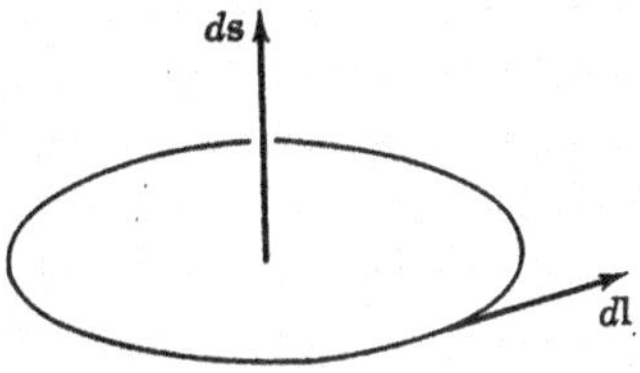

FIG. 3-8. Convention relating $d\mathbf{s}$ and $d\mathbf{l}$.

$$\psi = \iint \mathbf{B} \cdot d\mathbf{s}$$

Substituting into Eq. (3-16), we have the field theory form of *Faraday's law*,

$$\oint \mathbf{E} \cdot d\mathbf{l} = -\frac{d}{dt}\iint \mathbf{B} \cdot d\mathbf{s} \tag{3-17}$$

where the directions of $d\mathbf{l}$ and $d\mathbf{s}$ are as shown in Fig. 3-8. Throughout this text we adhere to the convention that $d\mathbf{l}$ encircles $d\mathbf{s}$ in the direction shown in Fig. 3-8. This can be remembered as a right-hand rule, fingers pointing in the direction of $d\mathbf{l}$ and thumb pointing in the direction of $d\mathbf{s}$.

In applying Eq. (3-17), we note that any number of surfaces may be visualized, all of which are bounded by the given contour. We might therefore wonder which surface should be taken. Because the divergence of $\mathbf{B}$ is zero, it follows that Eq. (3-17) applies to *any* surface bounded by the contour. This can be seen as follows: Suppose we have two surfaces S_1 and S_2, both of which are bounded by the given contour C, as shown in Fig. 3-9. We define normals $d\mathbf{s}_1$ and $d\mathbf{s}_2$ to S_1 and S_2 according to Fig. 3-8. The magnetic flux which appears in Eq. (3-17) for the two surfaces S_1 and S_2, respectively, is

$$\psi_1 = \iint_{S_1} \mathbf{B} \cdot d\mathbf{s}_1 \qquad \psi_2 = \iint_{S_2} \mathbf{B} \cdot d\mathbf{s}_2$$

We now view S_1 and S_2 of Fig. 3-9 as forming a single closed surface and we apply Eq. (3-13). In the application of this equation, ds is defined as pointing outward, so $ds = ds_1$ for S_1 and $ds = -ds_2$ for S_2. Therefore

$$\oiint_{S_1+S_2} \mathbf{B} \cdot d\mathbf{s} = \iint_{S_1} \mathbf{B} \cdot d\mathbf{s}_1 - \iint_{S_2} \mathbf{B} \cdot d\mathbf{s}_2 = \psi_1 - \psi_2 = 0$$

showing that $\psi_1 = \psi_2$. Thus, the right-hand side of Eq. (3-17) applies to any surface bounded by the contour.

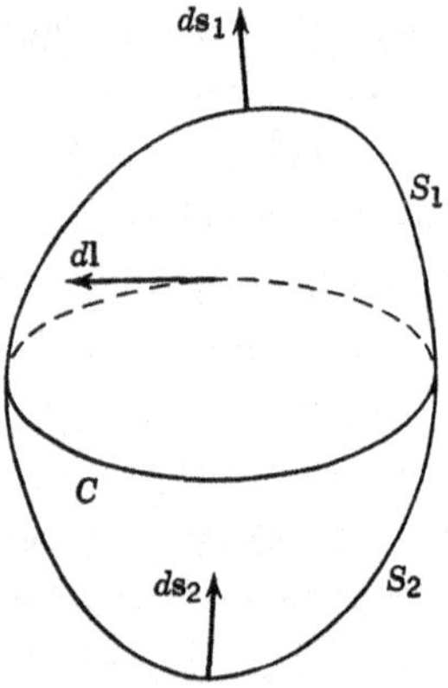

FIG. 3-9. The contour C is bounded by the two surfaces S_1 and S_2.

3-6. Circulation and Curl. In Faraday's law, Eq. (3-17), we have the line integral of **E** once around a closed path. The action of a vector once around a closed path is called the *circulation* of that vector about the path, that is

$$\text{circulation of } \mathbf{A} = \oint \mathbf{A} \cdot d\mathbf{l} \tag{3-18}$$

If the vector represents a force acting on an object, the circulation is equal to the work associated with moving the object once around the closed path.

The *average circulation* of a vector over a surface is defined as the circulation of the vector about the bounding contour divided by the area of the surface, that is

$$\text{average circulation of } \mathbf{A} = \frac{1}{s} \oint \mathbf{A} \cdot d\mathbf{l} \tag{3-19}$$

If we now take the limit of the average circulation as the area vanishes and, furthermore, orient the surface so that the average circulation is a maximum, we obtain the mathematical concept of *curl*. Considering an incremental element of area Δs and denoting the unit normal to it by $\mathbf{u}_n$, we define the vector curl **A**, represented symbolically by $\nabla \times \mathbf{A}$, to be

$$\text{curl } \mathbf{A} = \nabla \times \mathbf{A} = \lim_{\Delta s \to 0} \frac{1}{\Delta s} \left[\mathbf{u}_n \oint \mathbf{A} \cdot d\mathbf{l} \right]_{\max} \tag{3-20}$$

The subscript "max" implies that Δs is oriented so that $\oint \mathbf{A} \cdot d\mathbf{l}$ is maximum and that $\mathbf{u}_n$ is associated with this orientation. Then the direction of $\mathbf{u}_n$ is determined according to the right-hand rule, Fig. 3-8. The components of curl **A** may be found by evaluating the limit of the average circulation about loops of arbitrary orientation. Thus, the component of curl **A** in the direction of $\mathbf{u}_n$ is given by

$$\mathbf{u}_n \cdot \nabla \times \mathbf{A} = (\nabla \times \mathbf{A})_n = \lim_{\Delta s \to 0} \frac{1}{\Delta s} \oint \mathbf{A} \cdot d\mathbf{l} \tag{3-21}$$

The geometry of this situation is illustrated by Fig. 3-10. The magnitude of curl **A** is, of course, its maximum component, as indicated by Eq. (3-20). We shall not attempt to give a rigorous mathematical proof of the uniqueness of the vector curl **A** or of the general validity of Eq. (3-21). It may be found in the reference below.*

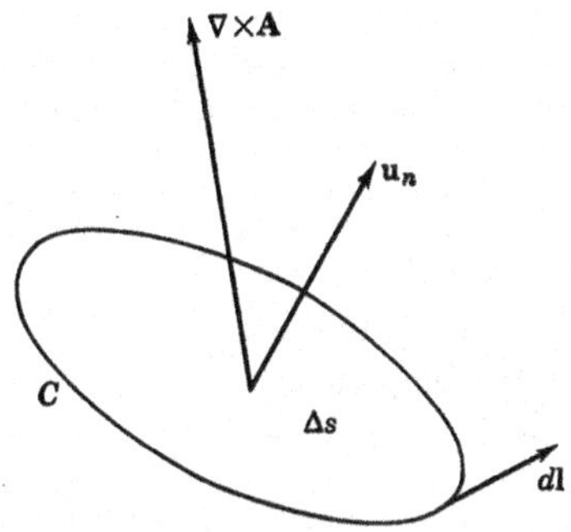

FIG. 3-10. The component of curl **A** in the direction of $\mathbf{u}_n$ is given by Eq. (3-21).

In any given coordinate system, we can obtain the coordinate components of curl **A** by applying Eq. (3-21) to differential loops whose sides are coordinate lines and whose normals are coordinate unit vectors. We shall first consider rectangular coordinates. A differential loop with z-directed normal is shown in Fig. 3-11. The line integral of **A** over the left side (as it appears in Fig. 3-11) of the loop we denote by $A_x\,dx$. Over the right side, A_x differs from A_x over the left side by the rate of change of A_x in the y direction times the separation of the two

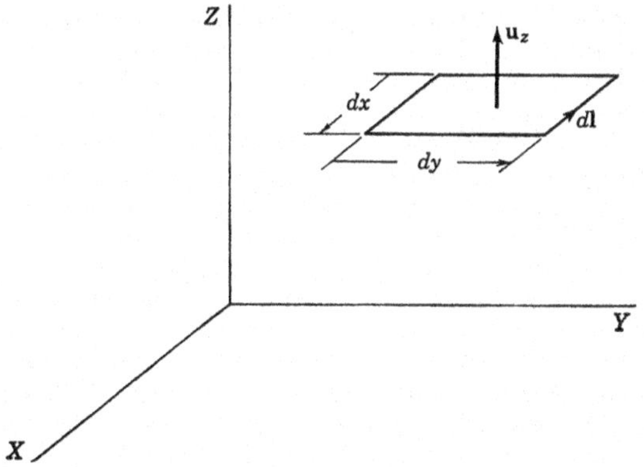

FIG. 3-11. Differential loop for determining $(\nabla \times \mathbf{A})_z$ in rectangular coordinates.

sides. Also, $d\mathbf{l}$ is in the opposite direction, giving for the line integral of **A** over the right side $-[A_x + (\partial A_x/\partial y)\,dy]\,dx$. The contribution to the line integral for both right and left sides of the loop is therefore $-(\partial A_x/\partial y)\,dy\,dx$.

* A. Abraham and R. Becker, "Classical Electricity and Magnetism," Hafner Publishing Company, New York, 1951, pp. 32–36.

In a similar fashion, the contribution for both back and front sides is found to be $(\partial A_y/\partial x)\,dx\,dy$. We thus have

$$\oint \mathbf{A} \cdot d\mathbf{l} = \left(\frac{\partial A_y}{\partial x} - \frac{\partial A_x}{\partial y}\right) dx\,dy$$

for the infinitesimal loop of Fig. 3-11. Dividing by $\Delta s = dx\,dy$ according to Eq. (3-21), we have the z-component of curl **A** given by

$$\mathbf{u}_z \cdot \nabla \times \mathbf{A} = (\nabla \times \mathbf{A})_z = \frac{\partial A_y}{\partial x} - \frac{\partial A_x}{\partial y}$$

Because of the symmetry of rectangular coordinates, the other two coordinate components of curl **A** are given by cyclic permutation of (x,y,z). Thus

$$(\nabla \times \mathbf{A})_x = \frac{\partial A_z}{\partial y} - \frac{\partial A_y}{\partial z}$$

$$(\nabla \times \mathbf{A})_y = \frac{\partial A_x}{\partial z} - \frac{\partial A_z}{\partial x}$$

The complete expression for curl **A** in rectangular coordinates is then given by

$$\nabla \times \mathbf{A} = \mathbf{u}_x\left(\frac{\partial A_z}{\partial y} - \frac{\partial A_y}{\partial z}\right) + \mathbf{u}_y\left(\frac{\partial A_x}{\partial z} - \frac{\partial A_z}{\partial x}\right) + \mathbf{u}_z\left(\frac{\partial A_y}{\partial x} - \frac{\partial A_x}{\partial y}\right)$$

$$= \begin{vmatrix} \mathbf{u}_x & \mathbf{u}_y & \mathbf{u}_z \\ \dfrac{\partial}{\partial x} & \dfrac{\partial}{\partial y} & \dfrac{\partial}{\partial z} \\ A_x & A_y & A_z \end{vmatrix} \qquad (3\text{-}22)$$

Note that, in rectangular coordinates, the ∇ can be considered to be the same vector operator

$$\nabla = \mathbf{u}_x \frac{\partial}{\partial x} + \mathbf{u}_y \frac{\partial}{\partial y} + \mathbf{u}_z \frac{\partial}{\partial z}$$

as was defined for the divergence operation; the curl of **A** then has the same form as the vector (cross) product of ∇ with **A**. This occurrence of the same vector operator both for the divergence operation and for the curl operation (and, as we shall see later, for the gradient operation) is characteristic of rectangular coordinates only. It should also be pointed out that, even though we symbolize the curl operation by a cross product, the vector $\nabla \times \mathbf{A}$ is not necessarily perpendicular to **A**.

To obtain the cylindrical-coordinate components of curl **A**, we orient infinitesimal loops as shown in Fig. 3-12. For the ρ component, we go around the loop shown in Fig. 3-12*a*, starting on the side showing the arrow dl, and obtain

$$\oint \mathbf{A} \cdot d\mathbf{l} = A_\phi \rho \, d\phi + \left(A_z + \frac{\partial A_z}{\partial \phi} d\phi \right) dz - \left(A_\phi + \frac{\partial A_\phi}{\partial z} dz \right) \rho \, d\phi - A_z \, dz$$

$$= \frac{\partial A_z}{\partial \phi} d\phi \, dz - \frac{\partial A_\phi}{\partial z} \rho \, d\phi \, dz$$

Dividing through by $\Delta s = \rho \, d\phi \, dz$ according to Eq. (3-21), we have

$$(\nabla \times \mathbf{A})_\rho = \frac{1}{\rho} \frac{\partial A_z}{\partial \phi} - \frac{\partial A_\phi}{\partial z}$$

For the ϕ component of curl **A**, we go around the loop of Fig. 3-12*b*. Since this is a rectangular path and since ρ and z are "length coordinates,"

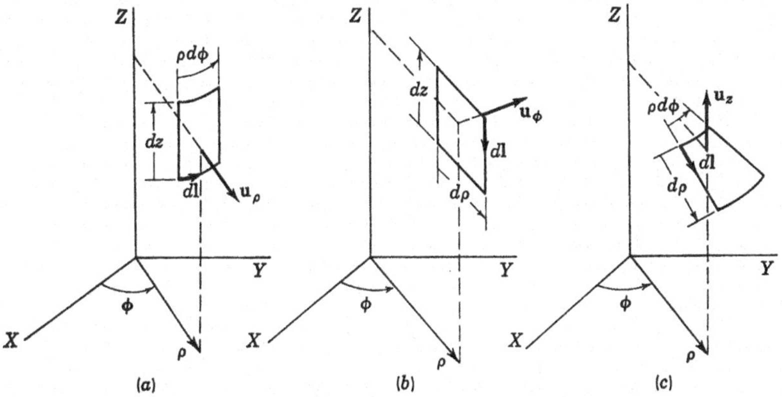

FIG. 3-12. Differential loops for determining (*a*) the ρ component, (*b*) the ϕ component, and (*c*) the z component of $\nabla \times \mathbf{A}$.

the application of Eq. (3-21) is the same as in rectangular coordinates. We therefore have

$$(\nabla \times \mathbf{A})_\phi = \frac{\partial A_\rho}{\partial z} - \frac{\partial A_z}{\partial \rho}$$

Figure 3-12*c* shows the path for determining the z component of curl **A**. Here we have the additional complication that the two sides along the ϕ-coordinate lines are of different lengths. Starting on the side showing the arrow dl, we have

$$\oint \mathbf{A} \cdot d\mathbf{l} = A_\rho \, d\rho + \left(A_\phi + \frac{\partial A_\phi}{\partial \rho} d\rho \right) (\rho + d\rho) \, d\phi$$

$$- \left(A_\rho + \frac{\partial A_\rho}{\partial \phi} d\phi \right) d\rho - A_\phi \rho \, d\phi$$

$$= A_\phi \, d\rho \, d\phi + \rho \frac{\partial A_\phi}{\partial \rho} d\rho \, d\phi - \frac{\partial A_\rho}{\partial \phi} d\phi \, d\rho$$

$$= \frac{\partial}{\partial \rho} (\rho A_\phi) \, d\rho \, d\phi - \frac{\partial A_\rho}{\partial \phi} d\phi \, d\rho$$

Dividing through by $\Delta s = \rho\, d\phi\, d\rho$ according to Eq. (3-21), we obtain

$$(\nabla \times \mathbf{A})_z = \frac{1}{\rho}\frac{\partial}{\partial \rho}(\rho A_\phi) - \frac{1}{\rho}\frac{\partial A_\rho}{\partial \phi}$$

In terms of the coordinate unit vectors, we can now write curl **A** in cylindrical coordinates as

$$\nabla \times \mathbf{A} = \mathbf{u}_\rho\left(\frac{1}{\rho}\frac{\partial A_z}{\partial \phi} - \frac{\partial A_\phi}{\partial z}\right) + \mathbf{u}_\phi\left(\frac{\partial A_\rho}{\partial z} - \frac{\partial A_z}{\partial \rho}\right) + \mathbf{u}_z\left[\frac{1}{\rho}\frac{\partial}{\partial \rho}(\rho A_\phi) - \frac{1}{\rho}\frac{\partial A_\rho}{\partial \phi}\right] \tag{3-23}$$

This can also be written in determinant form, but the resulting expression does not closely resemble the cross product of two vectors in cylindrical coordinates and is therefore little aid to our memory. We *cannot* define a single vector operator ∇ in cylindrical coordinates which would apply to both the curl and the gradient operations.

The determination of the curl operation for spherical coordinates is left as an exercise for the reader (see Prob. 3-16). The results obtained by applying Eq. (3-21) to differential coordinate loops are as follows:

$$\begin{aligned}(\nabla \times \mathbf{A})_r &= \frac{1}{r \sin \theta}\left[\frac{\partial}{\partial \theta}(\sin \theta\, A_\phi) - \frac{\partial A_\theta}{\partial \phi}\right] \\ (\nabla \times \mathbf{A})_\theta &= \frac{1}{r \sin \theta}\frac{\partial A_r}{\partial \phi} - \frac{1}{r}\frac{\partial}{\partial r}(r A_\phi) \\ (\nabla \times \mathbf{A})_\phi &= \frac{1}{r}\frac{\partial}{\partial r}(r A_\theta) - \frac{1}{r}\frac{\partial A_r}{\partial \theta}\end{aligned} \tag{3-24}$$

These components can be combined with the unit vectors $\mathbf{u}_r$, $\mathbf{u}_\theta$, $\mathbf{u}_\phi$, and a single vector equation can then be written for curl **A** in spherical components. The operation can also be written in the form of a determinant, but no ∇ operator common to the curl and divergence operations can be defined.

Circulation and curl are viewed, fundamentally, as related to the work associated with the motion of a particle about a closed path. However, for the special case in which the vector represents the force applied to the rim of a circular wheel, the circulation is proportional to the torque acting on the wheel. Such a concept is, however, restricted to circular wheels. For instance, the torque and circulation about a square "wheel" are *not* proportional to each other (see Prob. 3-26). Thus, circulation and curl should not be regarded as basically related to rotation, but rather as basically related to work. Furthermore, the curl of a vector field has no direct relationship to the field lines. For example, the lines of **A** may be straight even though curl **A** exists. Conversely, the lines of **A** may be curved even though curl **A** vanishes.

Now consider a well-behaved vector function **A** and a smooth surface S bounded by a contour C, as shown in Fig. 3-13. We divide the surface into elements, represented by the small squares in Fig. 3-13. For each element of surface, Eq. (3-21) applies. Thus, multiplying Eq. (3-21) by the elementary area $ds = \Delta s$, we have

$$\nabla \times \mathbf{A} \cdot d\mathbf{s} = \oint \mathbf{A} \cdot d\mathbf{l}$$

where the line integral is taken around the elementary contour. If we now integrate over all elements of area, the left-hand side of this equation

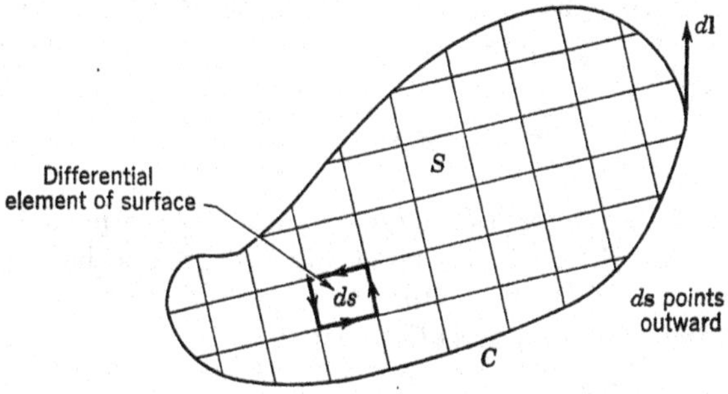

FIG. 3-13. The surface S is divided into differential elements of area.

becomes simply the surface integral of curl **A** over S. The right-hand side becomes the summation of the circulations about all the elementary contours. Since the contributions due to boundaries common to adjacent elements cancel out, we are left with only the line integral about C. Thus

$$\iint \nabla \times \mathbf{A} \cdot d\mathbf{s} = \oint \mathbf{A} \cdot d\mathbf{l} \tag{3-25}$$

where the line integral is taken about the contour bounding the surface. Equation (3-25) is called the *circulation theorem* or *Stokes' theorem*, and relates a line integral to a surface integral. In words, it states that the circulation of **A** about a contour is equal to the flow of curl **A** through any surface bounded by the contour. The application of Eq. (3-25) is restricted to functions which are well-behaved over the surface of integration.

3-7. The Maxwell-Faraday Equation. We can obtain an equation relating **E** and **B** at a point by applying Faraday's law, Eq. (3-17), to infinitesimal surface elements. Still more simply, we can apply Stokes' theorem to Faraday's law and obtain the point relationship. Stokes' theorem states that

$$\oint \mathbf{E} \cdot d\mathbf{l} = \iint \nabla \times \mathbf{E} \cdot d\mathbf{s}$$

where $d\mathbf{l}$ and $d\mathbf{s}$ are determined by the right-hand rule, Fig. 3-8. So long as the surface of integration does not vary with respect to time, we can differentiate the term on the right-hand side of Eq. (3-17) with respect to time under the integral sign. That is

$$\frac{d}{dt}\iint \mathbf{B}\cdot d\mathbf{s} = \iint \frac{\partial \mathbf{B}}{\partial t}\cdot d\mathbf{s}$$

where we use the partial derivative sign because **B** is also a function of the coordinate variables. Thus, by the above two equations, Faraday's law becomes

$$\iint \nabla \times \mathbf{E}\cdot d\mathbf{s} = -\iint \frac{\partial \mathbf{B}}{\partial t}\cdot d\mathbf{s}$$

Since we postulate that Faraday's law must hold for *every* surface, it follows that the vector integrands of the above equation must be equal. We therefore have

$$\nabla \times \mathbf{E} = -\frac{\partial \mathbf{B}}{\partial t} \tag{3-26}$$

which is called the *Maxwell-Faraday equation.* It is a partial differential equation relating **E** and **B** at a point and it expresses the same physical law as does Faraday's law. For this reason Eq. (3-26) is sometimes called Faraday's law in differential form.

An example of a *vector equation* is Eq. (3-26). It states that the two vectors, $\nabla \times \mathbf{E}$ and $-\partial\mathbf{B}/\partial t$ are everywhere equal. From the fundamental concepts of vectors, we know that if two vectors at the same point are equal, corresponding components of them are equal. Thus, Eq. (3-26) implies that

$$(\nabla \times \mathbf{E})_i = -\frac{\partial B_i}{\partial t}$$

where the subscript i denotes *any* component. For example, using rectangular components, we have

$$\begin{aligned}
(\nabla \times \mathbf{E})_x &= \frac{\partial E_z}{\partial y} - \frac{\partial E_y}{\partial z} = -\frac{\partial B_x}{\partial t} \\
(\nabla \times \mathbf{E})_y &= \frac{\partial E_x}{\partial z} - \frac{\partial E_z}{\partial x} = -\frac{\partial B_y}{\partial t} \\
(\nabla \times \mathbf{E})_z &= \frac{\partial E_y}{\partial x} - \frac{\partial E_x}{\partial y} = -\frac{\partial B_z}{\partial t}
\end{aligned} \tag{3-27}$$

where we have made use of the expressions derived in the preceding section for the components of curl **E**. Since any three noncoplanar components of a vector uniquely specify the vector, Eqs. (3-27) are equivalent to Eq. (3-26). We can write equations corresponding to Eqs. (3-27), using either cylindrical components or spherical components. The tabulation of these equations is left as an exercise for the reader.

At the beginning of this section we stated that Eq. (3-26) could be derived by applying Faraday's law to differential elements of surface. Let us see now how such a procedure would be carried out. Suppose we take the z-oriented coordinate loop shown in Fig. 3-11. In the preceding section when we derived a formula for $(\nabla \times \mathbf{E})_z$, we carried out the line integral about this loop and we obtained

$$\oint \mathbf{E} \cdot d\mathbf{l} = \left(\frac{\partial E_y}{\partial x} - \frac{\partial E_x}{\partial y}\right) dx\, dy$$

For the right-hand side of Eq. (3-26), we consider $\mathbf{B}$ to be essentially constant over the incremental area of Fig. 3-11 and we write

$$\frac{d}{dt} \iint \mathbf{B} \cdot d\mathbf{s} = \frac{\partial}{\partial t} B_z\, dx\, dy$$

Thus, Faraday's law for the loop of Fig. 3-11 reads

$$\frac{\partial E_y}{\partial x} - \frac{\partial E_x}{\partial y} = -\frac{\partial B_z}{\partial t}$$

where we have canceled the common term $dx\, dy$. This is just the last equation of Eqs. (3-27). Going through the same procedure for the other two coordinate loops, we obtain the other two of the Eqs. (3-27). Thus the set of Eqs. (3-27), which, according to basic vector concepts, are equivalent to Eq. (3-26), can be derived by three applications of Faraday's law, one for each component equation.

3-8. B and E at a Boundary. We can obtain information on the behavior of $\mathbf{B}$ and $\mathbf{E}$ at a boundary from the equations of this chapter.

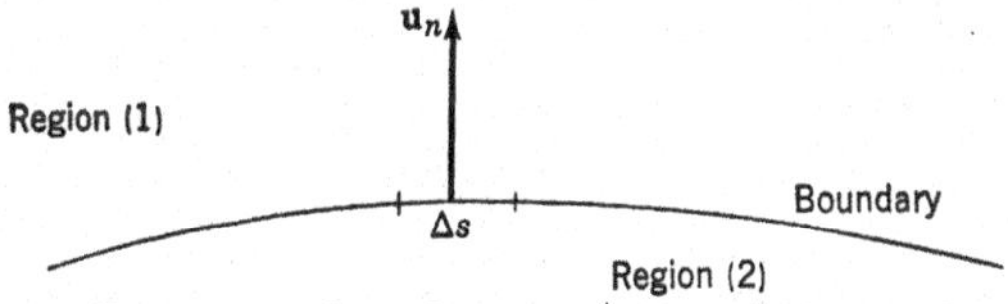

FIG. 3-14. A boundary surface in a field of B.

By the term "boundary" we mean a specified surface; it may be an interface between two different media or it may be merely a mathematical surface.

Let us first consider the magnetic flux density $\mathbf{B}$ in the vicinity of a boundary, as suggested by Fig. 3-14. Denote by region (1) the region into which $\mathbf{u}_n$ points, denote the other region by region (2), and visualize an element Δs of the surface, as shown in Fig. 3-14. Considering the element Δs to lie just inside region (1), we find that the flux intercepted is $\mathbf{B}^{(1)} \cdot \mathbf{u}_n\, \Delta s = B_n^{(1)}\, \Delta s$. This assumes that Δs is sufficiently small so that $\mathbf{B}$ is essentially constant over its extent. Similarly, if we consider the element Δs to lie just inside region (2), we find that the flux intercepted is $\mathbf{B}^{(2)} \cdot \mathbf{u}_n \Delta s = B_n^{(2)} \Delta s$. It was shown in Sec. 3-5 that the magnetic flux through a surface is equal to the magnetic flux through any other surface bounded by the same contour. Therefore, the magnetic flux must be the same regardless of whether Δs lies just inside region (1) or just inside region (2). Thus $B_n^{(1)} \Delta s = B_n^{(2)} \Delta s$, and, canceling the common Δs terms, we have

$$B_n^{(1)} = B_n^{(2)} \tag{3-28}$$

Superscripts indicate the region in which **B** is evaluated. Equation (3-28) states that *the normal component of* **B** *is continuous across any boundary.* The proof rests on the validity of Eq. (3-12) or Eq. (3-13), both of which

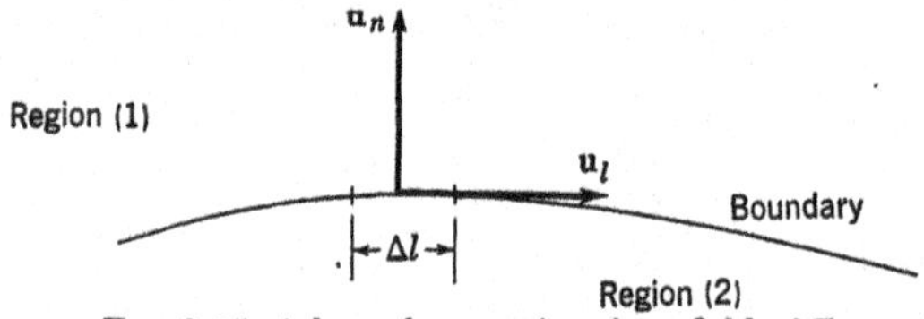

FIG. 3-15. A boundary surface in a field of **E**.

are always true in nature. When we alter Eqs. (3-12) and (3-13) for mathematical purposes in Chap. 7, we shall have to amend Eq. (3-28).

Now consider the electric intensity **E** in the vicinity of a boundary, as suggested by Fig. 3-15. Designating regions (1) and (2) related to $\mathbf{u}_n$ as before, we visualize an incremental element of length Δl along the boundary. Considering Δl to lie just inside region (1), we have the voltage along it given by $\mathbf{E}^{(1)} \cdot \mathbf{u}_l \Delta l = E_l^{(1)} \Delta l$, where $\mathbf{u}_l$ is a unit vector along Δl. This assumes that Δl is sufficiently short so that **E** is essentially constant over its length. Similarly, if we consider Δl to lie just inside region (2), the voltage along it is $\mathbf{E}^{(2)} \cdot \mathbf{u}_l \Delta l = E_l^{(2)} \Delta l$. According to Faraday's law, Eq. (3-17), the voltage about a closed path must equal the rate of decrease of the magnetic flux enclosed. Since the two paths, Δl just inside region (1) and Δl just inside region (2), together form a path enclosing virtually zero area, the magnetic flux enclosed must be negligible for **B** finite. Thus, the voltage along one path equals the voltage along the other path, and we have $E_l^{(1)} \Delta l = E_l^{(2)} \Delta l$. Canceling the common Δl terms, we obtain $E_l^{(1)} = E_l^{(2)}$. However, the direction of $\mathbf{u}_l$ is arbitrary so long as it is tangential to the given boundary. Therefore, for all components of **E** tangent to the given boundary, we have

$$E_t^{(1)} = E_t^{(2)} \tag{3-29}$$

This states that *the tangential components of* **E** *are continuous across any boundary.* The proof rests on the validity of Eq. (3-17) or Eq. (3-26), both of which are always true in nature. When we alter Eqs. (3-17) and (3-26) for mathematical purposes in Chap. 7, we shall have to amend Eq. (3-29).

QUESTIONS FOR DISCUSSION

3-1. Discuss the three common ways of writing the formula for the action of a vector, Eq. (3-1).

3-2. What difficulties would be encountered in attempting to measure electric intensity using the force relationship of Eq. (3-2)?

3-3. Discuss the difference of concept between the macroscopic **E** field and the microscopic **E** field.

3-4. In some books the formula $v = -\int \mathbf{E} \cdot d\mathbf{l}$ is found. How can this be reconciled with Eq. (3-3)?

3-5. Discuss the meaning of the terms "homogeneous," "isotropic," and "linear" as applied to conducting media.

3-6. Given a thin sheet of conducting material, what would be a logical definition of surface conductivity?

3-7. Discuss the concept of a density of power dissipation.

3-8. What is the reference direction of ψ in Eq. (3-11) if the direction of $d\mathbf{s}$ varies over the surface?

3-9. Equation (3-13) can be written concisely as $\psi = 0$. Explain this notation.

3-10. What difficulties would be encountered in attempting to measure magnetic flux density using the force relationship of Eq. (3-14)?

3-11. A common type of fluxmeter consists of an "exploring coil" and a ballistic galvanometer. What is the principle of operation of such a fluxmeter?

3-12. Discuss the difference in concept between the macroscopic **B** field and the microscopic **B** field.

3-13. Apply Eq. (3-17) to the ideal inductor of Fig. 1-2 and obtain Eq. (1-4).

3-14. Equation (3-17) can be written concisely as $v = -d\psi/dt$. Explain.

3-15. Discuss the concepts of circulation and curl.

3-16. Give an example of a vector **A** whose lines are straight although curl **A** exists.

3-17. Give an example of a vector **A** whose lines are curved although curl **A** vanishes.

3-18. Give an example for which curl **A** is perpendicular to **A**.

3-19. Give an example for which curl **A** is not perpendicular to **A**.

3-20. What might happen if Stokes' (circulation) theorem were applied to a vector function that was singular (see Prob. 3-23)?

3-21. Discuss the relationship of the Maxwell-Faraday equation, Eq. (3-26), to the Kirchhoff voltage law, Eq. (1-9).

3-22. From the analysis of Sec. 3-8, can we say anything about the behavior of the tangential components of **B** at a boundary?

3-23. From the analysis of Sec. 3-8, can we say anything about the behavior of the normal component of **E** at a boundary?

PROBLEMS

3-1. Given $\mathbf{A}(x,y,z) = \mathbf{u}_x x + \mathbf{u}_y xy + \mathbf{u}_z z^2$, determine the action of **A** along the path consisting of straight lines from the origin to (1,0,0) to (1,1,0) to (1,1,1).

3-2. For the **A** of Prob. 3-1, determine the action of **A** along the straight-line path from the origin to the point (1,1,1). (*Hint:* Express **A** in spherical components and integrate along the appropriate radius.)

3-3. Given $\mathbf{A} = \mathbf{u}_\phi$, determine the action of $\mathbf{A}$ once around a ϕ-coordinate line at arbitrary radius ρ.

3-4. Determine the voltage between any two points a and b on the same ρ-coordinate line if $\mathbf{E} = \mathbf{u}_\rho 1/\rho$.

3-5. Given $\mathbf{E} = \mathbf{u}_z \cos x \cos y \cos z \cos \omega t$, determine the voltage along a straight-line path between points $(x_0,y_0,\pi/2)$ and $(x_0,y_0,-\pi/2)$, where x_0 and y_0 are arbitrary.

3-6. What is the force on a point charge of magnitude 0.01 coulomb situated in vacuum if the field due to all other charges is 10 volts per meter?

3-7. At what rate must energy be expended if a current of 1 ampere is to be maintained along the path from a to b of Prob. 3-4?

3-8. Determine the resistance between opposite faces of a cube 10 centimeters on a side if the material is (*a*) silver, (*b*) graphite, (*c*) bakelite, (*d*) quartz (see Appendix C).

3-9. If 0.01 volt is maintained across the cube of Prob. 3-8, what is the density of power dissipation in each case?

3-10. For the field of Prob. 3-5, determine the power dissipation in the cube bounded by the six planes $x = \pi/2$, $x = -\pi/2$, $y = \pi/2$, $y = -\pi/2$, $z = \pi/2$, $z = -\pi/2$, for each of the four materials of Prob. 3-8. What is the time-average power dissipation in each case?

3-11. Given $\mathbf{B} = \mathbf{u}_r 2r^{-3} \cos\theta + \mathbf{u}_\theta r^{-3} \sin\theta$, determine the magnetic flux through the spherical cap $r = r_0$, $\theta < \theta_0$.

3-12. Determine the magnetic flux through the flat surface subtended by the rim of the cap of Prob. 3-11, (*a*) by using the divergence theorem and (*b*) by integrating over the flat surface.

3-13. Which of the following vectors are not physically realizable magnetic flux densities? (*a*) $\mathbf{B} = \mathbf{u}_z$, (*b*) $\mathbf{B} = \mathbf{u}_\rho$, (*c*) $\mathbf{B} = \mathbf{u}_\phi$, (*d*) $\mathbf{B} = \mathbf{u}_r$, (*e*) $\mathbf{B} = \mathbf{u}_\theta$.

3-14. What is the vector force on a straight section of wire 1 meter long, parallel to the x axis, carrying 10 amperes of current, if the field due to all other currents is $\mathbf{B} = \mathbf{u}_x + 2\mathbf{u}_y$?

3-15. For the $\mathbf{E}$ of Prob. 3-5, determine the rate of change of magnetic flux through the flat surface $y = 0$, $0 < x < \pi/2$, $0 < z < \pi/2$. Determine the magnetic flux, assuming that it varies sinusoidally in time.

3-16. Derive Eqs. (3-24).

3-17. Determine curl $\mathbf{A}$ for the $\mathbf{A}$ of Prob. 3-1.

3-18. Determine the circulation of $\mathbf{A}$ about the rectangular path (0,0,0) to (1,0,0) to (1,1,0) to (0,1,0) to (0,0,0) for the $\mathbf{A}$ of Prob. 3-1. Do this both ways; that is, by evaluating both sides of Eq. (3-25).

3-19. Determine curl $\mathbf{A}$ if $\mathbf{A} = \mathbf{u}_\rho \rho + \mathbf{u}_\phi \cos\phi$.

3-20. Determine curl $\mathbf{A}$ if $\mathbf{A} = \mathbf{u}_r x$.

3-21. For the $\mathbf{E}$ of Prob. 3-5, determine $\mathbf{B}$, assuming that it varies sinusoidally in time.

3-22. Write the Maxwell-Faraday equation in terms of its cylindrical-component equations and in terms of its spherical-component equations.

3-23. Given $\mathbf{A} = \mathbf{u}_\phi 1/\rho$, evaluate the circulation of $\mathbf{A}$ about the circle $\rho = 1$, $z = 0$ by evaluating the right-hand side of Eq. (3-25). Then evaluate the left-hand side of Eq. (3-25). Why are the two answers different?

3-24. If $\mathbf{E} = \mathbf{u}_x - \mathbf{u}_y$, $y < 0$, and $\mathbf{E} = K(\mathbf{u}_x + 2\mathbf{u}_y)$, $y > 0$, what must be the value of K?

3-25. If $\mathbf{B} = \mathbf{u}_x - \mathbf{u}_y$, $y < 0$, and $\mathbf{B} = K(\mathbf{u}_x + 2\mathbf{u}_y)$, $y > 0$, what must be the value of K?

3-26. Show that if $\mathbf{F}$ denotes the force per unit length acting on the rim of a circular wheel, the torque on the wheel is proportional to the circulation of $\mathbf{F}$ about the rim of the wheel. Show that this is not true for a square wheel.

CHAPTER 4

ELECTRIC FLUX DENSITY AND MAGNETIC INTENSITY

4-1. Sources of a Vector Field. The fundamental meaning of the word "source" is "cause." In science, the question of cause and effect is a difficult one; in mathematics, it is purely philosophical. What we have are equations relating two or more quantities, and we more or less arbitrarily designate one or more of the quantities to be sources. Actually, we are merely designating those quantities which we consider to be the independent parameters.

Suppose we have a well-behaved vector function **C** having nonvanishing divergence and curl. This we denote by

$$\nabla \cdot \mathbf{C} = w \qquad \nabla \times \mathbf{C} = \mathbf{W} \tag{4-1}$$

In Chap. 2, we showed how the divergence of a vector is interpreted as a source of flow. Thus, the w of Eqs. (4-1) is called the *flow source* of the vector **C**. In Chap. 3, we found that the curl of a vector is related to the circulation of the vector. If the vector represents a flow of fluid, the curl is related to the vortexes (whirls) of the fluid. Because of this, the **W** of Eqs. (4-1) is called the *vortex source* of the vector **C**. It is usually possible to calculate the vector **C** if its flow sources and vortex sources are known.*

Let us now look at the characteristics of the two types of source. Suppose in Eqs. (4-1) we have **W** zero everywhere, and w zero except within an infinitesimal sphere. This is called a *point flow source.* Since the source is spherically symmetric, the **C** vector must point in the radial direction (any other possibility would destroy the spherical symmetry). This is illustrated in Fig. 4-1. If the **C** vector points toward the source, which would be the case if w were negative, the source is often called a "sink." The point source is the simplest form of flow source. Other possibilities would be line flow sources, surface flow sources, and volume flow sources.

Now consider w to be zero everywhere in Eqs. (4-1), and **W** not zero, giving a vortex source. We first note that **W** cannot be an arbitrary

* It is more usual to call w a "source" and **W** a "vortex," but such nomenclature can be misleading.

vector, since we have the mathematical identity $\nabla \cdot \nabla \times \mathbf{C} = 0$. From the second of Eqs. (4-1) it therefore follows that $\nabla \cdot \mathbf{W} = 0$, meaning that $\mathbf{W}$ itself must be a vector with no flow sources. In terms of field lines, this states that lines of $\mathbf{W}$ can never terminate. Thus, we cannot have a point vortex source, the simplest form being an infinite line vortex source. This would be the form if $\mathbf{W}$ were zero except in a cylinder of infinitesimal cross section. We now have a cylindrically symmetric source, and the $\mathbf{C}$ field must therefore be cylindrically symmetric. Also, by applying Stokes' theorem to the second of Eqs. (4-1), we have

$$\oint \mathbf{C} \cdot d\mathbf{l} = \iint \mathbf{W} \cdot d\mathbf{s}$$

If we apply this equation to a circular path concentric with the line vortex source, we see that $\mathbf{C}$ must have a circumferential component,

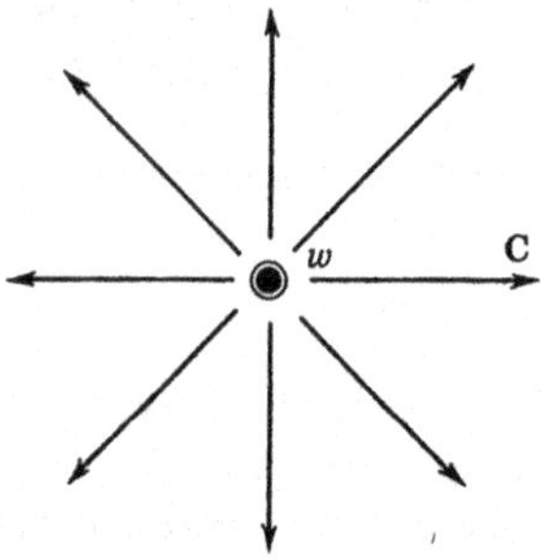

FIG. 4-1. A point flow source of **C**.

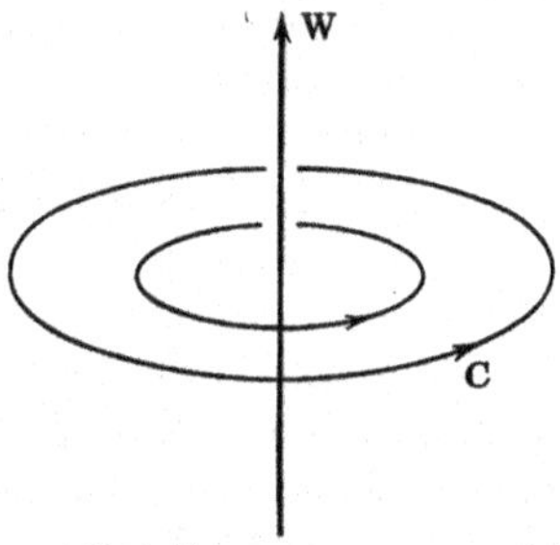

FIG. 4-2. A line vortex source of **C**.

else $\oint \mathbf{C} \cdot d\mathbf{l} = 0$. There can be no radial component of $\mathbf{C}$, for this would require flow sources and we have none. Thus, we come to the conclusion that lines of $\mathbf{C}$ encircle the vortex source as shown in Fig. 4-2. The positive reference directions of $\mathbf{C}$ and $\mathbf{W}$ must satisfy the right-hand rule of Fig. 3-8, since we have defined $\mathbf{W}$ as equal to the positive curl of $\mathbf{C}$. As mentioned earlier, the line vortex source is the simplest type, other possibilities being surface distributions and volume distributions of vortex sources.

It should be emphasized that the w and $\mathbf{W}$ of Eqs. (4-1) are mathematical sources and not necessarily actual sources. In other words, it is not important whether w and $\mathbf{W}$ cause $\mathbf{C}$ in Eqs. (4-1) or whether $\mathbf{C}$ causes w and $\mathbf{W}$. When we talk of the sources of electromagnetic fields, we do not necessarily mean the divergence and curl of the various field vectors. We usually mean ideal sources which we consider to cause the field, but which are only a few of the total mathematical sources. The rest of the mathematical sources are then viewed as being caused by the field. This is analogous to what is done in circuit theory, where a known current is viewed as a current source and is considered to cause the other currents in a network.

4-2. Electric Flux Density. We now introduce a vector function having charge density as its flow source. Thus, by definition,

$$\nabla \cdot \mathbf{D} = q_v \tag{4-2}$$

where D is called *electric flux density.** However, Eq. (4-2) does not by itself specify D uniquely and it must therefore be viewed as a partial definition of electric flux density. If q_v does not vary with respect to time and if D is spherically symmetric about each infinitesimal element of charge, then D can be found from Eq. (4-2) alone. This we show in the next section. Otherwise, we need the additional postulate given by Eq. (4-10) to complete our definition of D.

If we integrate Eq. (4-2) throughout a given region and if we apply the divergence theorem to the left-hand side, we obtain

$$\oiint \mathbf{D} \cdot d\mathbf{s} = q \tag{4-3}$$

where

$$q = \iiint q_v \, d\tau$$

is the total charge enclosed by the surface. Eq. (4-3) is called *Gauss's law;* it states that the surface integral of D over any closed surface is equal to the total charge enclosed. It is simply an alternative statement of Eq. (4-2) and is therefore true by definition. Eq. (4-2) may be called the differential form of Gauss's law, and Eq. (4-3) may be called the integral form. From Eq. (4-3) it is evident that electric flux density D has the dimensions of charge per area. The unit of D in the mksc system is the *coulomb per square meter.*

We view D fundamentally as a flow vector, and in this sense D is mathematically analogous both to electric current J and to magnetic flux density B. However, because of other equations which we shall encounter later, it is more often convenient to consider D mathematically analogous to B rather than to J. Our choice of analogous names for D and B anticipates this mathematical similarity. In view of the symmetry of equations, we define an *electric flux* ψ^e according to

$$\psi^e = \iint \mathbf{D} \cdot d\mathbf{s} \tag{4-4}$$

where the surface is not necessarily closed. Now Gauss's law can be stated as: *The electric flux through any closed surface equals the charge enclosed.* Since D has the dimensions of charge per area, it is evident from Eq. (4-4) that electric flux ψ^e has the dimensions of charge. The unit of electric flux in the mksc system is therefore the *coulomb.*

Even though D and B are analogous in the mathematical sense, there is one important difference between them. In Sec. 3-4 we stated that

* D is also called "electric displacement."

$\nabla \cdot \mathbf{B} = 0$ everywhere in the physical world. In other words, **B** has no flow sources in nature; therefore there are no magnetic charges in nature. In terms of the vector picture, this means that the lines of **B** must never end. This is in contrast with **D**, whose lines may begin and terminate on electric charges. In any region for which $q_v = 0$, the electric flux density is divergenceless and **D** is then completely analogous to the magnetic flux density **B**.

4-3. Some Properties of D. Consider a small sphere of electric charge in unbounded space, with no variation with respect to time. The problem has spherical symmetry; therefore the **D** field should be spherically symmetric. The only possible way for a vector to be spherically symmetric about a sphere is for that vector to point in the radial direction. Applying Eq. (4-3) to a spherical surface of radius r concentric with the charged sphere, we have

$$4\pi r^2 D_r = q$$

where q is the total charge on the sphere. This is valid for any radius r larger than that of the charged sphere. We therefore have the electric flux density at any point external to the charged sphere given by

$$\mathbf{D} = \frac{q}{4\pi r^2} \mathbf{u}_r \tag{4-5}$$

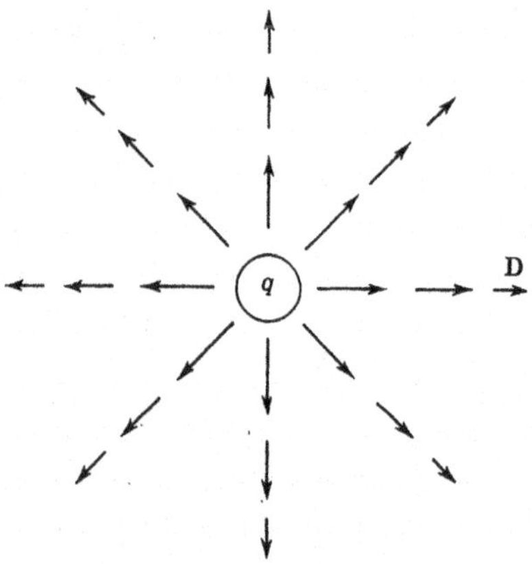

FIG. 4-3. The **D** field associated with a symmetrically charged sphere.

where $\mathbf{u}_r$ is a unit vector in the radial direction. This is illustrated by Fig. 4-3. Note that Eq. (4-5) is true regardless of the size of the charged sphere, so long as it is symmetrically charged. If we let the size of the charged sphere become vanishingly small, we have a *point charge*. Eq. (4-5) then holds everywhere, with a singularity at the point charge.

Suppose we now consider two uniformly charged spheres in unbounded space. Since Eq. (4-2) is a linear equation, it follows that the principle of superposition holds. That is, we can determine **D** for each sphere alone, and the total **D** for both spheres existing simultaneously will then be simply the sum of the individual **D**'s. This procedure is illustrated pictorially by Fig. 4-4, the **D** for each sphere being given by Eq. (4-5). Generalizing this procedure to any number of charged spheres or point charges, we have

$$\mathbf{D} = \sum_n \frac{q_n}{4\pi r_n^2} \mathbf{u}_{r,n} \tag{4-6}$$

where r_n is the distance from the nth charge to the field point, and $\mathbf{u}_{r,n}$ is the unit radial vector for the nth charge. The summation of Eq. (4-6) must, of course, be carried out vectorially.

If, instead of charged spheres and point charges, we have a continuous distribution of charge in free space, we proceed as follows. The charge distribution is broken up into infinitesimal elements of charge, and each element is then treated as a point charge. The **D** produced by the total

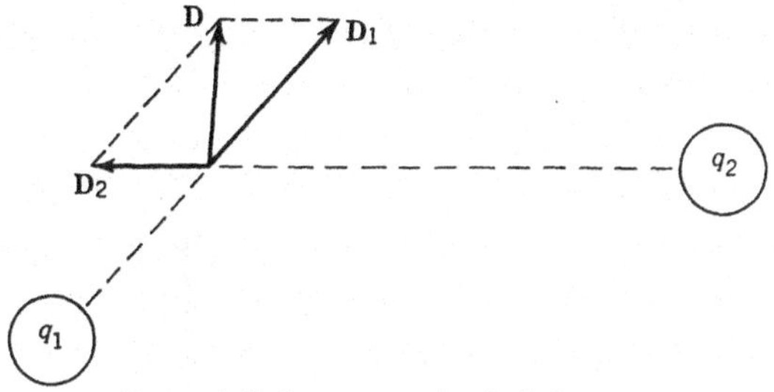

FIG. 4-4. **D** from two spherical charges.

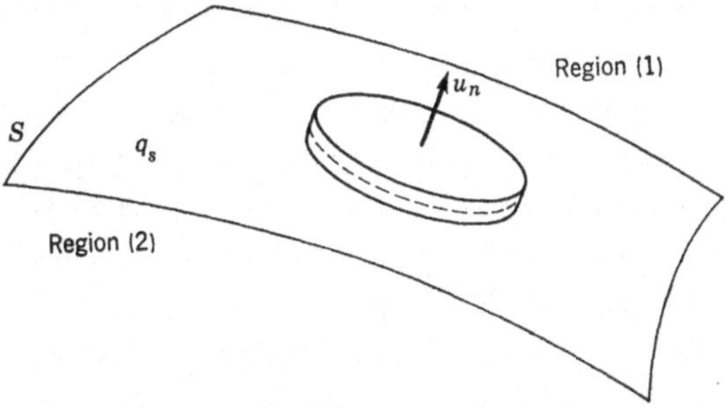

FIG. 4-5. A surface layer of charge.

charge will then be the superposition of that produced by each differential element. The resulting formula would be similar to Eq. (4-6), with the summation replaced by an integration. The same procedure holds for surface and line distributions of charge. Thus, at least in principle, we have the solution for **D** for any distribution of static electric charges in vacuum. It involves a superposition of vectors. We shall consider this solution in more detail in the next chapter and we shall also give a somewhat simpler solution in terms of a superposition of scalar functions. Solution of problems involving time-varying charges must wait until we have completed our system of equations for electromagnetic fields. The time-varying field is considered in Chap. 10.

It is important for our purposes to determine the behavior of **D** at surface distributions of charge. Consider a sheet of charge as suggested in Fig. 4-5. We choose the unit normal to the surface $\mathbf{u}_n$ to point into region (1), the region on the opposite side being designated region (2). A small coin-shaped surface is visualized as enclosing a portion of the surface charge, as shown in Fig. 4-5. We now apply Gauss's law, Eq. (4-3), to the coin-shaped surface. Assuming that **D** is essentially constant over the top of the coin-shaped surface, we have $\iint \mathbf{D} \cdot d\mathbf{s} = D_n^{(1)}A$ for the top, where $D_n^{(1)}$ denotes the normal component of **D** in region (1) and A is the cross-sectional area of the coin. Similarly, over the bottom of the coin-shaped surface, we have $\iint \mathbf{D} \cdot d\mathbf{s} = -D_n^{(2)}A$, the minus sign appearing because $\mathbf{u}_n$ points out of region (2). We assume **D** remains

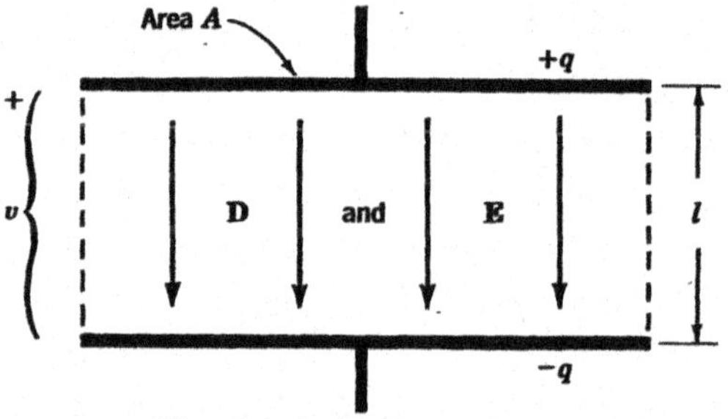

FIG. 4-6. An ideal capacitor.

finite at the surface charge, so the electric flux $\iint \mathbf{D} \cdot d\mathbf{s}$ over the side of the coin-shaped surface vanishes as we allow the height to become infinitesimal. The total charge contained by the coin-shaped surface, under the assumption that q_s is essentially constant within it, is given by q_sA. We therefore have from Gauss's law, $D_n^{(1)}A - D_n^{(2)}A = q_sA$, which, upon cancelation of the common A's, becomes

$$D_n^{(1)} - D_n^{(2)} = \mathbf{u}_n \cdot [\mathbf{D}^{(1)} - \mathbf{D}^{(2)}] = q_s \tag{4-7}$$

In other words, *at a surface layer of charge the normal component of* **D** *is discontinuous by an amount equal to the surface charge density.* This is true even if q_s is a time-varying function, for we have used only the definition of **D** in deriving Eq. (4-7).

4-4. Linear Dielectric Media. We now have two electric vector functions: **D**, which is fundamentally related to charge, and **E**, which is fundamentally related to voltage. The circuit parameter which relates charge to voltage is *capacitance.* A linear capacitor is defined according to $q = Cv$, that is, q and v are proportional to each other. We shall now show that the logical extension of this to field theory is for **D** and **E** to be proportional to each other.

Consider an ideal parallel-plate capacitor constructed of two perfectly conducting plates on opposite sides of a block of insulating material, as pictured in Fig. 4-6. We shall assume a solution for the field, and then see how it fits into our framework of equations. Suppose we take both

D and **E** inside the capacitor to be constant and perpendicular to the two conducting plates, and outside the capacitor to be zero. For Eq. (3-3) to hold, it follows that

$$E = \frac{v}{l}$$

where v is the voltage between the plates and l is the separation of the plates. For Eq. (4-7) to hold, it follows that

$$D = \frac{q}{A}$$

where the total charge q divided by the area A gives the surface density of charge on the top plate. Substituting from the above two equations into $q = Cv$, we have

$$D = \frac{Cl}{A} E$$

which states that the magnitude of **D** is proportional to that of **E**. Since **D** and **E** are assumed to be in the same direction, we can write

$$\mathbf{D} = \frac{Cl}{A} \mathbf{E} \tag{4-8}$$

The dielectric is *homogeneous* if it is the same throughout and it is *isotropic* if it behaves the same for all directions of **E** (the capacitance between opposite faces of a cube is independent of the pair chosen). For a homogeneous and isotropic block, we define the *capacitivity* or *permittivity* as

$$\epsilon = \frac{Cl}{A} \tag{4-9}$$

Thus, under the above ideal conditions, *the capacitivity ϵ of a homogeneous, isotropic dielectric material is the capacitance between opposite faces of a cube 1 meter on a side.* The dimensions of ϵ are evidently capacitance per length, and the unit in the mksc system is the *farad per meter*. The dielectric is said to be *linear* if the capacitivity is independent of the magnitude of **E**.

We now write Eq. (4-8) as

$$\mathbf{D} = \epsilon \mathbf{E} \tag{4-10}$$

This is the mathematical definition of linear, isotropic dielectric media, and a wide variety of materials are found to obey Eq. (4-10) quite closely. In the mksc system of units, the capacitivity of vacuum ϵ_0 is

$$\epsilon_0 = 8.854 \times 10^{-12} \approx \frac{1}{36\pi} \times 10^{-9} \text{ farads/meter} \tag{4-11}$$

The value of ϵ is never less than ϵ_0 for any material. In other words, a capacitor with matter between its plates will always have a higher capacitance than the same capacitor with nothing between its plates. The ratio $\epsilon_r = \epsilon/\epsilon_0$ is called the *relative capacitivity* or *dielectric constant*,* and is always greater than unity. We can interpret ϵ_r for a material to be the ratio of the capacitance of a capacitor with the material in question between its plates to that of the same capacitor with a vacuum between its plates. A table of dielectric constants for some common

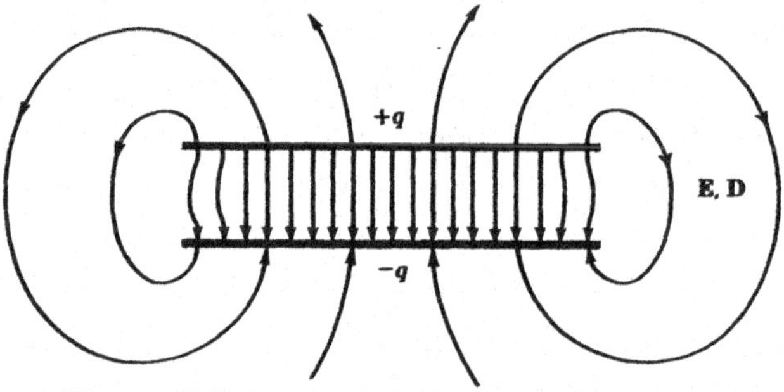

FIG. 4-7. Fringing effect for a parallel-plate capacitor.

materials is given in Appendix C. If the medium is not homogeneous, we can consider ϵ to be a function of position. In this case, we view ϵ as related to the capacitance of a cube of differential size, Δl on a side, according to

$$\epsilon = \lim_{\Delta l \to 0} \frac{\Delta C}{\Delta l} \tag{4-12}$$

Alternatively, ϵ can be viewed as the capacitance of a unit cube constructed of a material having the same capacitivity as that at the point under consideration.

Let us now inquire into the validity of our assumed solution for the ideal capacitor. If we have simply the capacitor as shown in Fig. 4-6 situated in vacuum, then we cannot have $E = v/l$ just inside the block and $E = 0$ just outside the block. This we showed in Sec. 3-8, where we found that there can be no discontinuity in a tangential component of $\mathbf{E}$ across a boundary. Therefore, our assumed solution must be somewhat incorrect. The actual field must extend into the region surrounding the capacitor, and we say that there are "fringing effects" at the edges of the capacitor. The electric field in this case is as suggested by Fig. 4-7. The smaller the separation of the plates with respect to the area of the plates, the smaller the fringing effects. The higher the capacitivity of the medium between the plates, the smaller the fringing effects. A way

* Some writers call ϵ the dielectric constant and ϵ_r the relative dielectric constant.

in which we might approximate our assumed solution quite closely would be to place something outside our capacitor to produce a field in this region. This can be done by the following fairly simple modification devised by Lord Kelvin. We extend the top and bottom plates using "guard rings" separated from the capacitor plates by very small gaps. The guard rings are maintained at the same voltage as the capacitor plates. The field within the original capacitor will then be constant as assumed, for the edge effects are removed to the edge of the guard ring. This is illustrated by Fig. 4-8.

We shall now look further into the questions what is **D**? how do we measure **D**? The two questions are, of course, intimately related. First of all, we have the fundamental concept of **D** as a vector having charge as

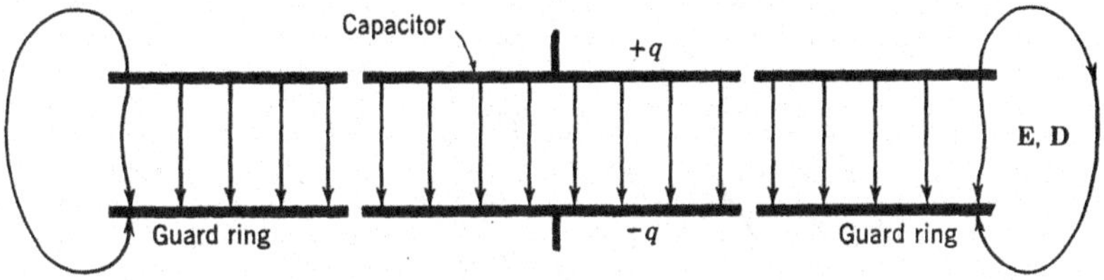

FIG. 4-8. Guard rings used to eliminate the fringing effect on a capacitor.

its source of flow. Second, for linear media, we have the relationship $\mathbf{D} = \epsilon\mathbf{E}$. However, a more meaningful interpretation of **D** can be made from Eq. (4-7). Suppose we have a surface layer of charge, with a finite value of **D** on one side and zero **D** on the other. For this to be possible in ordinarily linear media, **D** must be perpendicular to the sheet of charge, since $\mathbf{D} = \epsilon\mathbf{E}$ and we have already shown that we cannot have a discontinuity in the tangential component of **E**. Then $D = q_s$, where q_s is the charge density on the sheet, and the direction of **D** is perpendicular to the sheet. An interpretation of **D** can therefore be given as follows: *The electric flux density is a vector of magnitude equal to the surface charge density required to terminate it, and of direction perpendicular to this surface of charge.* Our ideal capacitor gives an example of a surface charge density terminating **D**. Using this capacitor, we can devise a method for measuring both **D** and **E** as follows. Suppose we construct a small "test capacitor" of two parallel plates close enough together to render fringing effects negligible. This we place in a field to be measured, as suggested by Fig. 4-9. If the plates are shorted out by a wire, charge will flow until the voltage between the two plates is zero. If the plates are rotated until a maximum charge has been transferred from one plate to the other, we must then have the plates perpendicular to **D** and **E** as shown in Fig. 4-9. (The transferred charge can be measured by means of a ballistic galvanometer in series with the shorting wire.) We can now remove the wire with no effect, since the current in it is zero. We have then effectively terminated the original field by the charge on the capacitor plates. According to the principle of superposition, we can view the total field of Fig. 4-9 as the sum of two partial fields, one produced by the original sources and the other by the test capacitor. Internal to the test capacitor these two fields must be equal and opposite, since the total field is zero. External to the capacitor the total field is due only to the original sources, for the capacitor field is negligible in this region. We now remove the capacitor from the original field, and

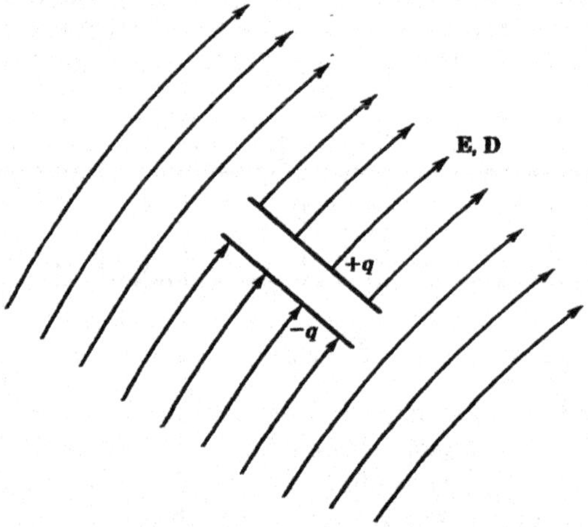

FIG. 4-9. D and E can be measured using a small test capacitor.

we measure the charge on its plates and the voltage between its plates. Internal to the capacitor we have $E = v/l$ and $D = q/A$, where A is the area of the plates and l is their separation. Since we have already concluded that the capacitor field is equal in magnitude and opposite in direction to the original field, we have effectively measured both **D** and **E** of the original field. This experiment illustrates the basic difference between **D** and **E**, that is, that **D** is related to charge and **E** to voltage. It also illustrates how, once we have the equations, our methods of measuring quantities may be deduced from the equations.

Let us now return to the problem of a pair of point charges, or charges small in extent compared to their separation. Labeling the charges q_a and q_b, the field of the a charge in vacuum is given by Eq. (4-5). Also, $\mathbf{D} = \epsilon_0\mathbf{E}$, so the electric intensity produced by the a charge is

$$\mathbf{E}^a = \frac{1}{\epsilon_0}\mathbf{D}^a = \frac{q_a}{4\pi\epsilon_0 r^2}\mathbf{u}_r$$

where the coordinate origin is at q_a. From Sec. 3-2, we have the concept that the force on the b charge is equal to the magnitude of the b charge times the electric intensity produced by the a charge. Therefore, the force exerted on the b charge is

$$\mathbf{F}^b = \frac{q_a q_b}{4\pi\epsilon_0 r^2}\mathbf{u}_r \tag{4-13}$$

where r is now the separation of the charges and $\mathbf{u}_r$ points from q_a to q_b. Eq. (4-13) is known as *Coulomb's law* and was first established experimentally. It is used as a basis for the classical exposition of electrostatic theory. If we follow analogous steps and calculate the force on the a charge due to the b charge, we get the same formula as Eq. (4-13) except that $\mathbf{u}_r$ now points from q_b to q_a. In order words, $\mathbf{F}^a = -\mathbf{F}^b$, the force on one charge being equal and opposite to the force on the other. Consideration of direction in Eq. (4-13) shows that the force is one of attraction if the two charges are of opposite sign and of repulsion if they are of the same sign.

4-5. Magnetic Intensity. If we substitute Eq. (4-2) into the equation of continuity, Eq. (2-45), we obtain

$$\nabla\cdot\left(\mathbf{J} + \frac{\partial\mathbf{D}}{\partial t}\right) = 0$$

Thus, the vector $(\mathbf{J} + \partial\mathbf{D}/\partial t)$ satisfies the condition necessary for a vortex source; that is, it has zero divergence. We now introduce a vector function having $(\mathbf{J} + \partial\mathbf{D}/\partial t)$ as its vortex source. Thus, by definition,

$$\nabla\times\mathbf{H} = \frac{\partial\mathbf{D}}{\partial t} + \mathbf{J} \tag{4-14}$$

where $\mathbf{H}$ is called the *magnetic intensity*. However, Eq. (4-14) must be viewed as a partial definition of magnetic intensity, since a further assumption is necessary to define $\mathbf{H}$ uniquely. If the quantities of Eq. (4-14) do not vary with respect to time and if $\mathbf{H}$ is rotationally symmetric about each element of current, we can then determine $\mathbf{H}$ from Eq. (4-14) alone. This will be shown for a special case in the next section. Otherwise, we need the additional postulate of Eq. (4-25) to complete our definition of $\mathbf{H}$. Eq. (4-14) is called the *Maxwell-Ampere equation.*

Suppose we now integrate Eq. (4-14) over a surface and apply Stokes' theorem to the left-hand side. If the surface of integration does not vary with time, we can remove the time derivative from under the integral sign, giving

$$\oint \mathbf{H}\cdot d\mathbf{l} = \frac{d}{dt}\iint \mathbf{D}\cdot d\mathbf{s} + \iint \mathbf{J}\cdot d\mathbf{s} \tag{4-15}$$

Making use of Eqs. (4-4) and (2-29), we can also write Eq. (4-15) as

$$\oint \mathbf{H} \cdot d\mathbf{l} = \frac{d\psi^e}{dt} + i \tag{4-16}$$

where ψ^e is the electric flux through any surface bounded by the given contour, and i is the electric current through that surface. Note the mathematical similarity between the relationships for **E** and **H**, Eqs. (3-17) and (4-15). In regions where **J** is zero, the two equations are identical except for a minus sign. It is because of this analogy that both **E** and **H** are called "intensities," and not because of any physical similarities. This symmetry of equations is made even more evident by defining a *magnetomotive force* u according to

$$u = \int \mathbf{H} \cdot d\mathbf{l} \tag{4-17}$$

which is analogous to our equation for voltage (or electromotive force as it is sometimes called). Now we can state Eq. (4-15), which is called *Ampere's law,* as: *The magnetomotive force about a closed path is equal to the sum of the electric current enclosed plus the rate of change of electric flux enclosed.*

Let us look at the dimensions of the various quantities that we have introduced. From Eq. (4-16), it is evident that magnetic intensity **H** has the dimensions of current per length. Its unit in the mksc system is therefore the *ampere per meter.* The magnetomotive force u, introduced by Eq. (4-17), evidently has the dimensions of current, its unit being the *ampere.* We have already shown that electric flux has the dimensions of charge, so $d\psi^e/dt$ has the dimensions of current. Thus, all terms of Eq. (4-16) have the dimensions of current, as they must have if our system of equations is to be consistent.

4-6. Some Properties of H. Consider the simplest type of vortex source of **H**, which would be a straight line of constant current, infinite in length. For this case, Eq. (4-16) reduces to

$$\oint \mathbf{H} \cdot d\mathbf{l} = i \tag{4-18}$$

since we have no variation with time. We assume rotational symmetry and we note that lines of **H** must encircle the current, for $\oint \mathbf{H} \cdot d\mathbf{l}$ must not be zero for paths enclosing i. It follows that **H** must exist as concentric circles enclosing the line current. Orienting a set of cylindrical axes with the z axis coincident with the line current, as shown in Fig. 4-10, we have only an H_ϕ. Applying Eq. (4-18) to any circle of constant ρ, we find

$$2\pi\rho H_\phi = i$$

where i is the magnitude of the line current. Thus, at any point external to the line source, we have

$$\mathbf{H} = \frac{i}{2\pi\rho}\mathbf{u}_\phi \tag{4-19}$$

where $\mathbf{u}_\phi$ is the unit vector in the ϕ direction. This is illustrated by Fig. 4-10. Eq. (4-19) is valid only if i does not vary with respect to time, since this is the only case for which Eq. (4-18) holds. It should also

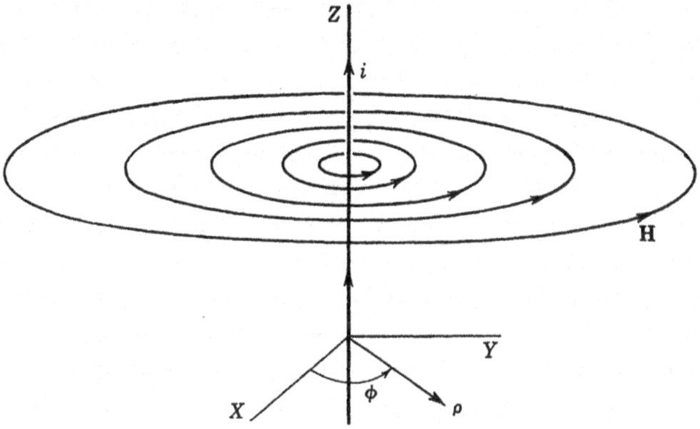

FIG. 4-10. The **H** field associated with an infinite filament of constant current.

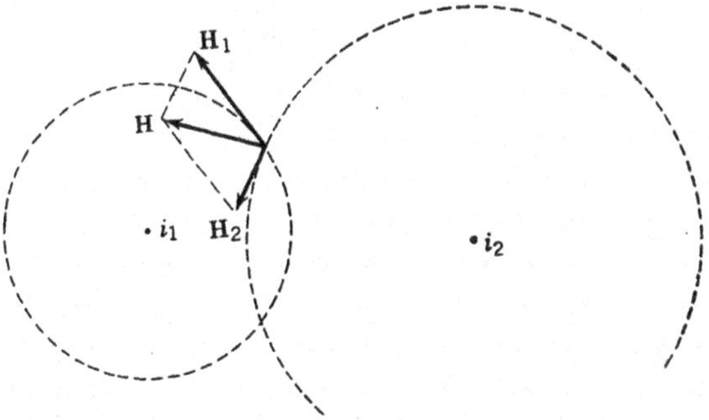

FIG. 4-11. The **H** from two filaments of constant current (i directed out of the paper).

be noted that Eq. (4-19) is valid for any cylinder of current of finite radius, so long as the current is z-directed and distributed symmetrically with respect to ϕ. In this case, i is the total current along the cylinder, and Eq. (4-19) applies only to points external to the cylinder.

If we now have two straight-line sources of current in unbounded space, we proceed as follows. Eq. (4-14) is a linear equation, so the principle of superposition applies. We find the **H** from each line source alone in unbounded space. The **H** for both together is then simply the sum of the individual **H**'s. This procedure is illustrated pictorially by Fig. 4-11 for two parallel line currents. Generalizing this procedure to any number of line currents, we have

$$\mathbf{H} = \sum_n \frac{i_n}{2\pi\rho_n} \mathbf{u}_{\phi,n} \tag{4-20}$$

where ρ_n is the radial distance from the nth line current, and $\mathbf{u}_{\phi,n}$ is the unit vector in the ϕ direction with respect to the nth line current. We could extend our formula to a continuous distribution of current filaments by replacing the summation of Eq. (4-20) by the appropriate integration. This problem is treated in more detail in Chap. 6.

One of the properties of $\mathbf{H}$ that will be important to us in our future work is its behavior at surface distributions of current. A surface distribution of electric current is suggested by Fig. 4-12. We choose to call

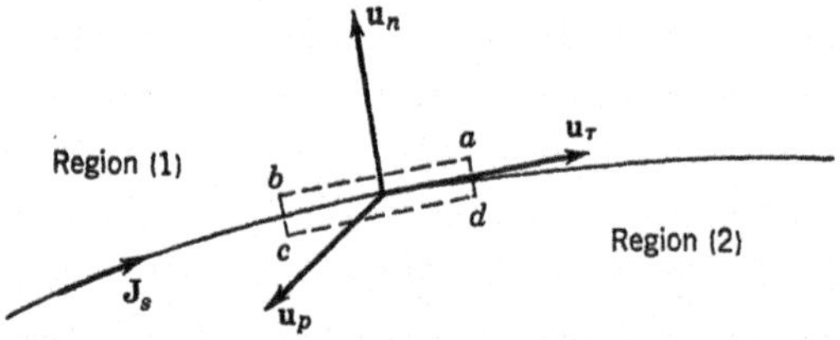

FIG. 4-12. A surface distribution of electric currents.

region (1) the region on the side of the surface into which the unit normal points, the region on the other side, region (2). A small rectangular path is visualized as enclosing a portion of the surface current, as shown in Fig. 4-12. At the midpoint of the path, which we assume to be of differential dimensions, we orient a triplet of unit vectors as follows: $\mathbf{u}_n$ is perpendicular to the surface; $\mathbf{u}_\tau$ is tangent both to the surface and to the flat surface enclosed by the rectangle; $\mathbf{u}_p$ is tangent to the surface and perpendicular to the flat surface enclosed by the rectangle such that $\mathbf{u}_p = \mathbf{u}_\tau \times \mathbf{u}_n$. This is shown in Fig. 4-12. If we apply Eq. (4-16) to the rectangular path, dividing $\oint \mathbf{H} \cdot d\mathbf{l}$ into its magnetomotive force contributions from each side of the rectangle according to Eq. (4-17), we have

$$\mathcal{U}_{ab} + \mathcal{U}_{bc} + \mathcal{U}_{cd} + \mathcal{U}_{da} = \frac{d\psi^e}{dt} + i \tag{4-21}$$

The subscripts on the $\mathcal{U}$'s indicate the segments according to the letters on Fig. 4-12, ψ^e is the electric flux enclosed by the rectangular path, and i is the electric current enclosed by the path. Assuming that $\mathbf{H}$ is essentially constant over the long sides of the rectangle, we have

$$\mathcal{U}_{ab} = -(\mathbf{H}^{(1)} \cdot \mathbf{u}_\tau)l_{ab} \qquad \mathcal{U}_{cd} = (\mathbf{H}^{(2)} \cdot \mathbf{u}_\tau)l_{cd}$$

where the superscripts (1) and (2) denote the region in which $\mathbf{H}$ is considered to be, and l_{ab} and l_{cd} are the distances a to b and c to d along the path. We let the width of the rectangle go to zero, that is, $l_{bc} \to 0$ and $l_{da} \to 0$, so that $u_{bc} \to 0$ and $u_{da} \to 0$ if **H** is finite. Also, if **D** is finite at the surface, the electric flux enclosed vanishes, so that $d\psi^e/dt \to 0$ as the width of the rectangle vanishes. For the current, however, we have a surface distribution, which means that a finite current is intercepted by the rectangle even though its width is infinitesimal. According to Eq. (2-33), the surface current intercepted by the rectangle is

$$i = (\mathbf{J}_s \cdot \mathbf{u}_p)l_{ab}$$

where we assume that the surface density of current $\mathbf{J}_s$ is essentially constant over the region enclosed by the rectangle. Also, $l_{ab} \to l_{cd}$ as the width of the rectangle vanishes, so that we have Eq. (4-21) reduced to

$$-(\mathbf{H}^{(1)} \cdot \mathbf{u}_r)l_{ab} + (\mathbf{H}^{(2)} \cdot \mathbf{u}_r)l_{ab} = (\mathbf{J}_s \cdot \mathbf{u}_p)l_{ab}$$

Canceling the common l_{ab}'s, and substituting $\mathbf{u}_r = \mathbf{u}_n \times \mathbf{u}_p$, we have

$$-(\mathbf{H}^{(1)} - \mathbf{H}^{(2)}) \cdot \mathbf{u}_n \times \mathbf{u}_p = \mathbf{J}_s \cdot \mathbf{u}_p$$

Next, we use the vector identity $\mathbf{A} \cdot \mathbf{B} \times \mathbf{C} = \mathbf{C} \cdot \mathbf{A} \times \mathbf{B}$ on the left hand term and obtain

$$-\mathbf{u}_p \cdot (\mathbf{H}^{(1)} - \mathbf{H}^{(2)}) \times \mathbf{u}_n = \mathbf{u}_p \cdot \mathbf{J}_s$$

Finally, since the orientation of the rectangle is arbitrary, the direction of $\mathbf{u}_p$ along the surface is arbitrary. The tangential vectors into which we dot $\mathbf{u}_p$ must be equal, for all their components are equal, and we have

$$\mathbf{u}_n \times (\mathbf{H}^{(1)} - \mathbf{H}^{(2)}) = \mathbf{J}_s \tag{4-22}$$

where we have eliminated the minus sign by reversing the cross product. In other words, *in crossing a surface distribution of electric currents, the component of* **H** *perpendicular to* $\mathbf{J}_s$ *and tangent to the surface suffers a discontinuity equal in magnitude to* J_s.

The interpretation of Eq. (4-22) can perhaps be made a little clearer by the following example. Suppose we have a surface z = constant on which there is a y-directed sheet of electric current. This is suggested by Fig. 4-13. Letting $\mathbf{J}_s = \mathbf{u}_y J_y$ and $\mathbf{u}_n = \mathbf{u}_z$ in Eq. (4-22), we have

$$-H_y^{(1)} + H_y^{(2)} = 0$$
$$H_x^{(1)} - H_x^{(2)} = J_y$$

Thus, the y component of **H** must be continuous, and the x component is discontinuous by the amount J_y. This is illustrated in Fig. 4-13. The plane of the paper is taken to be the plane on which $\mathbf{J}_s$ flows.

4-7. Linear Magnetic Media. In field theory, we have two magnetic vector quantities: **H**, which is fundamentally related to current, and **B**, which is fundamentally related to magnetic flux. The circuit parameter which relates magnetic flux to current is inductance. A linear inductor is defined according to $\psi = Li$, that is, ψ and i are proportional to each other. We shall now illustrate that the logical extension of this to field theory is for **B** and **H** to be proportional to each other.

Consider an ideal inductor constructed of a block of matter wound with a layer of wire, as shown in Fig. 4-14. The turns of the winding are

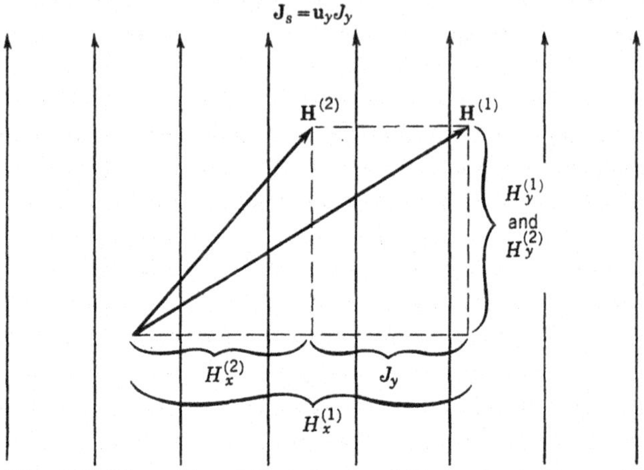

FIG. 4-13. Illustration of the behavior of **H** at a sheet of current.

close enough together so that we may view them as producing a continuous sheet of current encircling the inductor. Let us assume a reasonable solution, and see later how we may approximate this in practice. We consider **B** and **H** to be essentially constant and directed along the axis of the coil inside the coil; we consider them to be negligible outside the coil. The ψ used in the equation $\psi = Li$ is the total flux cutting a surface bounded by the wire, often called the "flux linkage." We have assumed the flux density constant in the inductor, so the flux linkage ψ is the number of turns N times the flux for a simple cross section, which is the flux density B times the cross-sectional area A. Therefore, we have

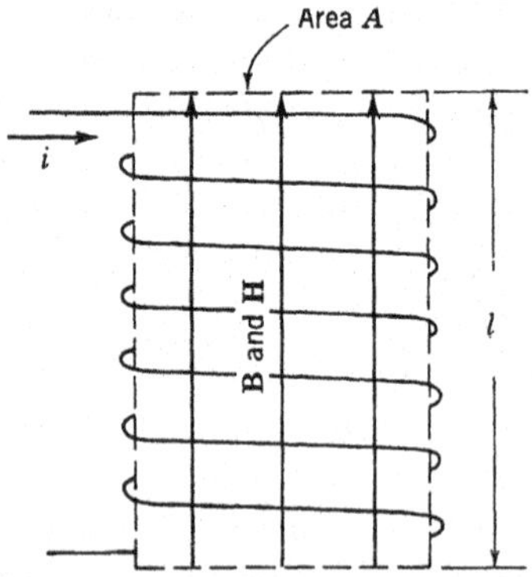

FIG. 4-14. An ideal inductor.

$$B = \frac{\psi}{NA}$$

internal to the ideal inductor. The coil is effectively a sheet of current equal in strength to the current i times the number of turns per unit length N/l, where l is the length of the inductor. According to Eq. (4-22), the strength of the current sheet must equal the magnitude of the discontinuity in tangential **H**, which in this case is simply H internal, since **H** is assumed to be zero external. We therefore have

$$H = \frac{iN}{l}$$

internal to the ideal inductor. Substitution of the above two equations into $\psi = Li$ gives

$$B = \frac{Ll}{N^2A} H$$

which states that the magnitude of **B** is proportional to that of **H**. Since we have **B** and **H** in the same direction, we can include this information by writing

$$\mathbf{B} = \frac{Ll}{N^2A} \mathbf{H} \tag{4-23}$$

The block is *homogeneous* if it is of the same material throughout, and *isotropic* if the material has the same magnetic properties regardless of the direction of the magnetic field. For a homogeneous and isotropic block, we define the *inductivity* or *permeability* as

$$\mu = \frac{Ll}{N^2A} \tag{4-24}$$

Thus, *the inductivity μ of a homogeneous isotropic material can be viewed as the inductance of an ideal inductor according to Eq.* (4-24). If we visualize this inductor as being a single-turn block ($N = 1$), then μ can be thought of as the inductance of a unit cube. The dimensions of μ are inductance per length, as shown by Eq. (4-24). The unit of μ in the mksc system is the *henry per meter*. The inductivity is said to be *linear* if μ is independent of the magnitude of **B**.

We now write Eq. (4-23) as

$$\mathbf{B} = \mu\mathbf{H} \tag{4-25}$$

which we view as the mathematical definition of linear isotropic media (in the magnetic sense). In the absence of matter, that is, in vacuum, we have μ in the mksc system given by

$$\mu_0 = 4\pi \times 10^{-7} \text{ henrys/meter} \tag{4-26}$$

where the subscript zero indicates vacuum. The mksc system of electromagnetic units is determined by the arbitrary choice of one electrical unit in addition to the three mechanical units (mass, length, and time). This choice is usually that of Eq. (4.26), the size of other units being a consequence of this. Most materials have a value of inductivity very close to that of vacuum. A class of materials, called *diamagnetic*, has value of μ slightly less than μ_0 (of the order of 0.01 per cent less). Another class of materials, called *paramagnetic*, has values of μ slightly greater than μ_0 (of the order of 0.01 per cent greater). A third class of materials, called *ferromagnetic*, has values of μ much greater than μ_0. However, most ferromagnetic materials are actually nonlinear, so representing them by a constant μ must be considered only a rough approximation. The ratio $\mu_r = \mu/\mu_0$ is called *relative inductivity* or *relative permeability*. We can interpret μ_r for a material to be the ratio of the inductance of an inductor filled with that material to that of the same inductor in vacuum. A table of relative permeabilities for some common materials is given in

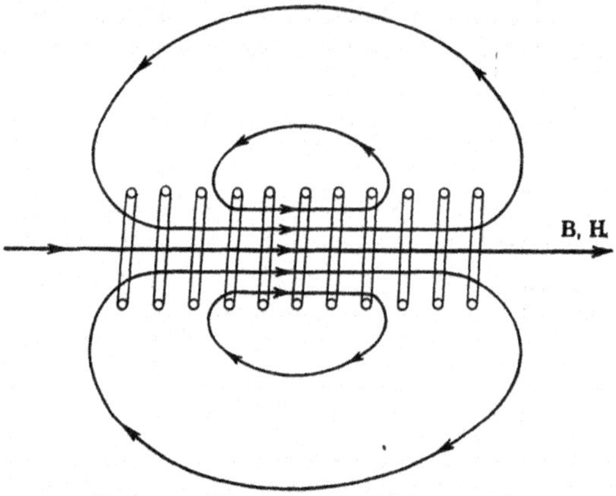

FIG. 4-15. The magnetic field of a coil.

Appendix C. For most engineering applications, we can consider all materials except the ferromagnetic ones to have $\mu = \mu_0$, and we can call them nonmagnetic. If a material is not homogeneous, we consider μ to be a function of position. In this case, we visualize μ as being related to the inductance of an infinitesimal inductor, analogous to the infinitesimal capacitor used for capacitivity.

We shall now inquire into the validity of our assumed solution for the ideal inductor. If the inductor is situated alone in vacuum, we cannot have **B** and **H** constant inside and zero outside, for the lines of **B** must be continuous. This we showed in Sec. 3-8. There must be a field external to the inductor, the lines closing around from one end to the other. This is suggested by Fig. 4-15. Thus, even though **B** and **H** may be small (because the lines are spread out over so large a region) external to the inductor, they do not vanish completely. Also, **B** and **H** differ internally from the assumed constant solution near the ends of the coil. If the coil is long compared to its diameter, the field is close to the ideal solution except near the ends of the coil. We can remove this "end effect" by placing "guard coils" on each end of the inductor, keeping the ampere-turns per meter the same. For example, the field in the center part of the coil of Fig. 4-15 approximates the assumed solution, and the center part could be viewed as an ideal inductor with the rest of the coil acting as the guard coils. Better yet, we could consider the ideal inductor to be a small section of a toroidal coil, in which case the external field is identically zero (see Prob. 4-17).

Let us now look more closely at the interpretation of **H**. Our fundamental concept of **H** is that it is a vector having $(\mathbf{J} + \partial\mathbf{D}/\partial t)$ as its vortex

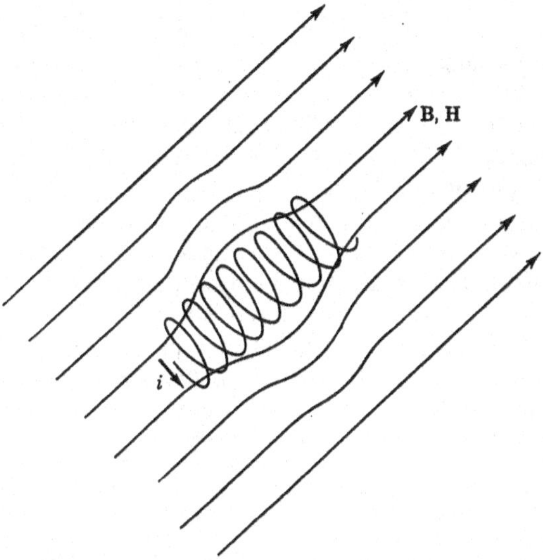

FIG. 4-16. B and H can be measured using a small test coil.

source. When there is no variation with respect to time, it has simply **J** as its vortex source. Also, in linear media, $\mathbf{B} = \mu\mathbf{H}$. Using Eq. (4-22), we can obtain a further understanding of what **H** is. Suppose we have a surface distribution of electric currents, with a value of **H** on one side and zero **H** on the other. For this to be possible, **H** must be tangent to the sheet of current. According to Eq. (4-22), $\mathbf{u}_n \times \mathbf{H} = \mathbf{J}_s$, where $\mathbf{J}_s$ is the surface current and where $\mathbf{u}_n$ points into the region where **H** exists. We can therefore interpret **H** as follows: *The magnetic intensity is a vector of magnitude equal to the surface current required to terminate it, of direction both parallel to this surface and perpendicular to the direction of current.*

Using the solenoid (coil) as a test instrument, we can devise a method for measuring **H** and **B**. Suppose we have a coil long enough so that **H** at its center is equal to the surface current density (ampere-turns per meter) of the coil. This we place in a magnetic field to be measured. We then adjust the current in the solenoid and vary its orientation until the magnetic field at its center is zero. This is suggested by Fig. 4-16. (We can think of the current as being fed to the coil via twisted leads so that their effect is nil. The existence of zero field internal to the coil could be detected by rotating a small coil connected to an a-c voltmeter. This voltage, according to Faraday's law, vanishes when the **B** field vanishes.) Note that in Fig. 4-16 the field lines go around the test coil, for they cannot terminate. The field pictured is the vector sum of the original field to be measured and the field of the solenoid. At the midpoint of the coil, the total field is zero, so the field due to the solenoid must be equal and opposite to the original field. If we now remove the coil, keeping the current unchanged, we can determine **H** and **B** according to our previous results. Thus, we have $H = iN/l$ (N is the number of turns, l the length of the solenoid). The total flux linkage of the coil could be found by integrating its terminal voltage as the current is reduced to zero. The magnetic flux density is then $B = \psi/NA$ (ψ is the flux linkage, A the cross-sectional area of the solenoid). Since the solenoid field is equal in magnitude to the original field, we have effectively measured the original field. This experiment illustrates the basic difference between **H** and **B**, that is, that **H** is related to current and **B** to flux.

Let us now return to the problem of a pair of parallel filaments of current. Labeling the currents i_a and i_b (with reference directions the same), we have the field of the a current in vacuum given by Eq. (4-19). Also, $\mathbf{B} = \mu_0\mathbf{H}$, so the magnetic flux density produced by the a current is

$$\mathbf{B}^a = \mu_0\mathbf{H}^a = \frac{\mu_0 i_a}{2\pi\rho}\mathbf{u}_\phi$$

where the z axis lies along i_a. According to Eq. (3-14), the force on a length l of the b current is given by

$$\mathbf{F}^b = i_b\mathbf{l} \times \mathbf{B}^a$$

where $\mathbf{l} = \mathbf{u}_z l$. Substituting for $\mathbf{B}^a$ and performing the vector product, we have

$$\mathbf{F}^b = -\frac{\mu_0 i_a i_b l}{2\pi\rho}\mathbf{u}_\rho \tag{4-27}$$

where ρ is now the separation of the current filaments and $\mathbf{u}_\rho$ points from i_a to i_b. Eq. (4-27) is known as *Ampere's force law*, and was first established experimentally. It can be used as the starting point for a development of magnetostatics which emphasizes the force aspect of fields. If we follow analogous steps and calculate the force on a length l of the a current due to the b current, we obtain the same formula as Eq. (4-27) except that $\mathbf{u}_\rho$ now points from i_b to i_a. In other words, $\mathbf{F}^a = -\mathbf{F}^b$, the force on a length l of one current being equal and opposite to the force on a length l of the other current. Consideration of direction in Eq.

(4-27) shows that the force is one of attraction if the two currents are in the same direction, and of repulsion if they are in opposite directions.

4-8. Energy. The circuit element which stores electric energy is the capacitor, and that which stores magnetic energy is the inductor. Starting from the concept of ideal capacitors and inductors, we can arrive at expressions for electric and magnetic energy in terms of the field vectors.

First, consider the capacitor and electric energy. From circuit concepts, the energy associated with a capacitor is given by

$$W_e = \tfrac{1}{2} qv$$

If we have an ideal capacitor (**E** and **D** constant in a block, zero external), we showed in Sec. 4-4 that

$$E = \frac{v}{l} \qquad D = \frac{q}{A}$$

Substituting these expressions into the energy relationship, we have

$$W_e = \tfrac{1}{2}DEAl$$

But Al is simply the volume τ of the capacitor, so that the energy is given by

$$W_e = \tfrac{1}{2}DE\tau = \tfrac{1}{2}\epsilon E^2\tau \tag{4-28}$$

where the last form makes use of $\mathbf{D} = \epsilon\mathbf{E}$. Since **D** and **E** are constant within the ideal capacitor, we can consider the energy to be distributed throughout the capacitor with a *density of electric energy* w_e given by

$$w_e = \tfrac{1}{2}\mathbf{D}\cdot\mathbf{E} = \tfrac{1}{2}\epsilon\mathbf{E}\cdot\mathbf{E} \tag{4-29}$$

The total electric energy associated with the ideal capacitor is then the energy density times the volume of the capacitor, which is Eq. (4-28).

Suppose we now have an electric field existing throughout a region. If we break up the region into incremental elements of volume, the field throughout each volume element is essentially constant. A typical volume element $\Delta\tau$ is illustrated by Fig. 4-17. Within $\Delta\tau$ we have **D** and **E** the same as assumed for the ideal capacitor, so we view each volume element as an ideal capacitor of incremental size. The electric energy associated with each volume element is then given by Eq. (4-28) with τ replaced by $\Delta\tau$. Summing over all volume elements, we have the electric energy associated with the entire field. In the limit as we let the volume elements tend to zero, the summation is replaced by an integration, giving

$$W_e = \iiint w_e\, d\tau \tag{4-30}$$

where w_e is as given by Eq. (4-29). The energy formulas of this section are valid for linear isotropic media.

Let us now consider the inductor and magnetic energy. From circuit concepts, the energy associated with an inductor is given by

$$W_m = \tfrac{1}{2}\psi i$$

Our ideal inductor has an assumed solution of **B** and **H** constant inside, and zero outside. In Sec. 4-7, it was shown that in this case

$$B = \frac{\psi}{NA} \qquad H = \frac{iN}{l}$$

Substituting these expressions into the energy relationship, we have

$$W_m = \tfrac{1}{2}BHAl$$

But Al is simply the volume τ of the block, so we have the energy given by

$$W_m = \tfrac{1}{2}BH\tau = \tfrac{1}{2}\mu H^2\tau \tag{4-31}$$

where the last form uses the relationship $\mathbf{B} = \mu\mathbf{H}$. Since **B** and **H** are taken to be constant within the inductor, we can consider the energy to be distributed throughout the inductor with a *density of magnetic energy* w_m given by

$$w_m = \tfrac{1}{2}\mathbf{B}\cdot\mathbf{H} = \tfrac{1}{2}\mu\mathbf{H}\cdot\mathbf{H} \tag{4-32}$$

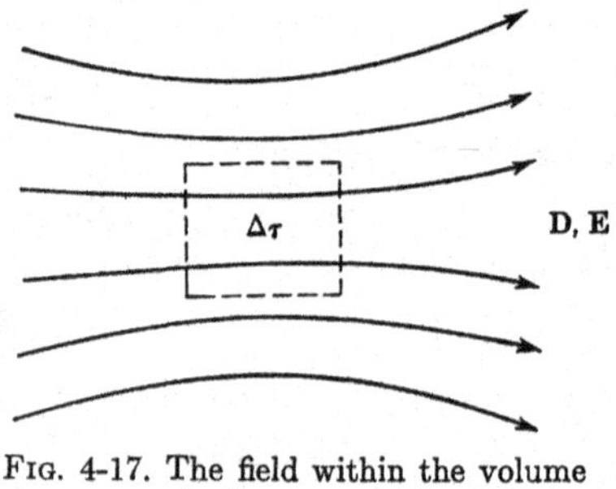

FIG. 4-17. The field within the volume element $\Delta\tau$ is the same as the field for the ideal capacitor.

The total magnetic energy associated with the ideal block inductor is then the energy density times the volume, which is Eq. (4-31).

If we have a magnetic field existing throughout a region, we can break up the region into incremental elements of volume, and treat each volume element as an ideal inductor. Figure 4-17 with the arrows relabeled **B, H** would illustrate a typical volume element $\Delta\tau$. The magnetic energy associated with each such volume element would be given by Eq. (4-31) with τ replaced by $\Delta\tau$. Summing over all volume elements, we obtain the magnetic energy associated with the entire field. Taking the volume elements to be of differential size, the summation becomes an integration, and we have

$$W_m = \iiint w_m \, d\tau \tag{4-33}$$

where w_m is given by Eq. (4-32). These formulas are valid for linear isotropic media.

In defining energy density functions according to Eqs. (4-29) and (4-32), we are considering the energy to be distributed throughout the field. Since we have no method of directly measuring energy, we have no way of determining the location of energy. It can therefore be neither proved nor disproved that the energy is distributed according to Eqs. (4-29) and (4-32). We postulate that the energy is so distributed because to do so is useful. So long as this postulate leads to no contradictions in the theory (and none have ever been found), we can regard it as being true.

4-9. Maxwell's Equations. We have now postulated all of the fundamental equations of macroscopic (large with respect to atomic dimensions) electromagnetic field theory. The group of equations

$$\begin{aligned} \nabla \times \mathbf{E} &= -\frac{\partial \mathbf{B}}{\partial t} \qquad & \nabla \cdot \mathbf{B} &= 0 \\ \nabla \times \mathbf{H} &= \frac{\partial \mathbf{D}}{\partial t} + \mathbf{J} \qquad & \nabla \cdot \mathbf{D} &= q_v \end{aligned} \tag{4-34}$$

each of which we have discussed in the preceding sections, are known as *Maxwell's equations*. They are not all independent relations, for the divergence equations are implied by the curl equations. Taking the divergence of both sides of the first of Eqs. (4-34), and using the vector identity $\nabla \cdot \nabla \times \mathbf{A} = 0$, we have

$$\nabla \cdot \frac{\partial \mathbf{B}}{\partial t} = \frac{\partial}{\partial t} (\nabla \cdot \mathbf{B}) = 0$$

This states that $\nabla \cdot \mathbf{B}$ must at least be constant with respect to time. Since a nonzero value of the divergence of $\mathbf{B}$ has never been found in nature, this constant is postulated to be zero, which gives us the first divergence equation of Maxwell. Taking the divergence of both sides of the second curl equation of Maxwell, we have

$$\nabla \cdot \left(\frac{\partial \mathbf{D}}{\partial t} + \mathbf{J} \right) = 0$$

Integrating this with respect to time, we obtain

$$\nabla \cdot \mathbf{D} = - \int_{-\infty}^{t} \nabla \cdot \mathbf{J} \, dt = q_v$$

which is the last of Eqs. (4-34). Note that the time derivative of the above equation is the equation of continuity, so this is also implied by Maxwell's equations.

Maxwell's equations alone do not contain sufficient information for obtaining solutions (except in special cases), because of the many quantities involved. It is usually necessary to obtain additional relationships among $\mathbf{E}$, $\mathbf{D}$, $\mathbf{B}$, $\mathbf{H}$, and $\mathbf{J}$ before we have a determinate set of equations. For linear media, we have

$$\mathbf{D} = \epsilon \mathbf{E} \qquad \mathbf{B} = \mu \mathbf{H} \qquad \mathbf{J} = \sigma \mathbf{E} \tag{4-35}$$

which serve to complete the system of equations. These are known as *constitutive relationships*, because they specify the properties (or constitution) of the media being considered. Substituting Eqs. (4-35) back into Maxwell's equations, we obtain two independent relations (the two curl equations) in two unknowns (**E** and **H**).

TABLE 4-1. CORRESPONDENCES BETWEEN CIRCUIT CONCEPTS AND FIELD CONCEPTS

Circuit concepts	*Field concepts*
Voltage, v	Electric intensity, **E**
Current, i	Electric current density, **J** Magnetic intensity, **H**
Magnetic flux, ψ	Magnetic flux density, **B**
Charge, q	Electric charge density, q_v Electric flux density, **D**
$\sum_n v_n = -\frac{d\psi}{dt}$	$\nabla \times \mathbf{E} = -\frac{\partial \mathbf{B}}{\partial t}$
$\sum_n i_n = -\frac{dq}{dt}$	$\nabla \cdot \mathbf{J} = -\frac{\partial q_v}{\partial t}$ $\nabla \times \mathbf{H} = \frac{\partial \mathbf{D}}{\partial t} + \mathbf{J}$
Linear elements: Resistors, $i = (1/R)v$ Capacitors, $q = Cv$ Inductors, $\psi = Li$	Linear media: $\mathbf{J} = \sigma\mathbf{E}$ $\mathbf{D} = \epsilon\mathbf{E}$ $\mathbf{B} = \mu\mathbf{H}$
Power dissipation in resistors: $(1/R)v^2$	Power dissipation in conductors: σE^2
Energy: Capacitors, $\frac{1}{2}qv = \frac{1}{2}Cv^2$ Inductors, $\frac{1}{2}\psi i = \frac{1}{2}Li^2$	Energy density: Electric, $\frac{1}{2}\mathbf{D} \cdot \mathbf{E} = \frac{1}{2}\epsilon E^2$ Magnetic, $\frac{1}{2}\mathbf{B} \cdot \mathbf{H} = \frac{1}{2}\mu H^2$

We have shown that Maxwell's equations are a logical extension of Kirchhoff's equations for circuits. Indeed, our whole development of electromagnetic field theory was based upon the extension of circuit concepts. We should not claim to have derived field theory from circuit theory, for some bold postulates were necessary to effect the transition. In a sense, field theory is more complete than circuit theory, because it gives us information about the electrical state of all points in space. Circuit theory is concerned only with voltages between terminals and currents in wires. This does not mean, however, that field theory is more exact than circuit theory. In fact, circuit theory can be rigorously derived from field theory, and therefore must be equally exact. There are, of course, limits to the use of linear circuit elements to approximate actual circuit elements. However, there are also limits to the use of linear media to approximate actual media in field theory.

If we keep in mind the various correspondences between circuit concepts and field concepts, we have an effective aid for remembering the latter. Table 4-1 summarizes some of the correspondences. Note that for some circuit quantities there are two analogous field quantities. This arises from the extension of the circuit concepts necessary in order to pass to the field concepts. In the second column of Table 4-1 we have placed the more obvious correspondences first.

QUESTIONS FOR DISCUSSION

4-1. What is the meaning of the following statement? A source is a quantity which we consider to be an independent parameter.

4-2. Why is it impossible to have a point vortex source?

4-3. Can both flow sources and vortex sources exist at the same place?

4-4. Gauss's law, Eq. (4-3), can be written concisely as $\psi^e = q$. Explain this notation.

4-5. What are some physical limitations to the concept of a point charge?

4-6. Can **D** have vortex sources as well as flow sources in linear media?

4-7. Would Eq. (4-5) be valid if the medium were not homogeneous?

4-8. How does the vector addition of Eq. (4-6) differ from scalar addition?

4-9. What would be the behavior of the normal component of **D** at a surface on which there is no electric charge?

4-10. Explain the following statement: Eq. (3-28) is a direct consequence of the absence of flow sources of **B** in nature.

4-11. Discuss the concept of capacitivity, or permittivity.

4-12. Would you expect an exact analysis of a capacitor (including fringing effects) to give a higher or a lower value of capacitance than the value given by the idealized analysis of Sec. 4-4?

4-13. What are some practical limitations to the use of a test capacitor to measure **D** and **E**?

4-14. The name "apparent surface-charge density" has been suggested for **D**. Discuss the merits of this suggestion.

4-15. Ampere's law, Eq. (4-16), can be written concisely as $u = i + d\psi^e/dt$. Explain this notation.

4-16. In some books the unit of u is called the "ampere-turn." How does this compare to our unit of u, the "ampere"?

4-17. What are some physical limitations to the concept of a line current?

4-18. Would Eq. (4-19) be valid if the medium were not homogeneous?

4-19. What would be the behavior of the tangential components of **H** at a boundary on which there are no electric currents?

4-20. A possible name for **H** might be "apparent surface-current density." Discuss the merits of this suggestion.

4-21. Would you expect an exact analysis of an inductor to give a higher or a lower value of inductance than the value given by the idealized analysis of Sec. 4-7?

4-22. What are some practical limitations to the use of a test coil to measure **H** and **B**?

4-23. Discuss the concept of inductivity, or permeability.

4-24. Why can we neither prove nor disprove the energy density postulates, Eqs. (4-29) and (4-32)?

4-25. Which of Eqs. (4-34) are postulated as a result of experimental evidence, and which are true by definition?

4-26. Discuss the correspondences between circuit quantities and field quantities, represented by Table 4-1.

PROBLEMS

4-1. Find the electric charge distribution if (*a*) $\mathbf{D} = \mathbf{u}_x x$; (*b*) $\mathbf{D} = \mathbf{u}_r$; (*c*) $\mathbf{D} = \mathbf{u}_\rho/\rho$; (*d*) $\mathbf{D} = \mathbf{u}_r/r^2$.

4-2. Find the charge contained within the sphere $r = 1$ for each case of Prob. 4-1.

4-3. What is the electric flux through the flat surface subtended by the circle $\rho = 1$, $z = 1$, for each case of Prob. 4-1?

4-4. Calculate the electric flux density and the electric intensity 5 meters from a point charge of 0.001 coulomb in vacuum.

4-5. Calculate the electric flux density and the electric intensity midway between two point charges of magnitude 0.01 and 0.001 coulomb in vacuum, separated 1 meter.

4-6. Calculate the force between the two charges of Prob. 4-5.

4-7. Given $\mathbf{D} = \mathbf{u}_x 2 + \mathbf{u}_y x$, $x < 0$, and $\mathbf{D} = \mathbf{u}_x y + \mathbf{u}_y x$, $x > 0$, find the surface-charge density over the plane $x = 0$.

4-8. Determine the approximate capacitance of a parallel-plate capacitor having plate area of 0.1 square meter and plate separation of 0.001 meter in air. Repeat for the case in which mica fills the region between the plates.

4-9. Given that $\mathbf{D} = \mathbf{u}_r K/r^2$, with K a constant, is the field between a pair of concentric spherical conductors of radii a and b, $b > a$; determine K in terms of the charge q on the inner sphere. Determine the voltage v between the two spheres, assuming vacuum between them. Show that the **D** field external to both spheres is zero if the outer sphere carries a charge equal in magnitude but opposite in sign to that on the inner sphere. Evaluate the capacitance between the spheres according to $C = q/v$.

4-10. If the field is static (no time-variation), determine the current distribution if (*a*) $\mathbf{H} = \mathbf{u}_x y$, (*b*) $\mathbf{H} = \mathbf{u}_\phi$, (*c*) $\mathbf{H} = \mathbf{u}_\phi/\rho$.

4-11. Calculate the magnetic intensity and magnetic flux density 0.5 meter from a long straight wire carrying 5 amperes in vacuum.

4-12. Calculate the magnetic intensity and magnetic flux density midway between two long parallel wires in vacuum, separated by 1 meter, carrying currents of 5 and 10 amperes in the same direction.

4-13. Calculate the force per unit length between the two currents of Prob. 4-12.

4-14. Given $\mathbf{H} = \mathbf{u}_x(y + 1)$, calculate the magnetomotive force about the circle $\rho = 1$, $z = 0$.

4-15. Given $\mathbf{H} = \mathbf{u}_x x + \mathbf{u}_y y$, $x < 0$, and $\mathbf{H} = \mathbf{u}_x x - \mathbf{u}_y y$, $x > 0$, determine the surface-current distribution over the plane $x = 0$.

4-16. Calculate the approximate inductance of a 100-turn coil 0.5 meter long having 0.01 square meter cross section, situated in air. Repeat for the case in which the coil is immersed in iron having a relative permeability of 1000.

4-17. The magnetic field of a closely wound toroidal coil is confined to the interior of the toroid (see Fig. 4-18). Treating the winding as a current sheet, show that the magnetic field $\mathbf{H} = \mathbf{u}_\phi Ni/2\pi\rho$ internal to the toroid satisfies Eq. (4-22) at the winding (N is the number of turns, i is the current in the wire, ρ is the radial distance from the axis of the toroid). Taking a square cross section, with dimensions as shown in Fig. 4-18, show that the magnetic flux through a simple cross section of the toroid is

$$\psi = \frac{Ni}{2\pi} \mu c \log \frac{b}{a}$$

Remembering that the flux linkage is N times the flux through a simple cross section, determine the inductance of the toroid according to Eq. (1-5). Show that when

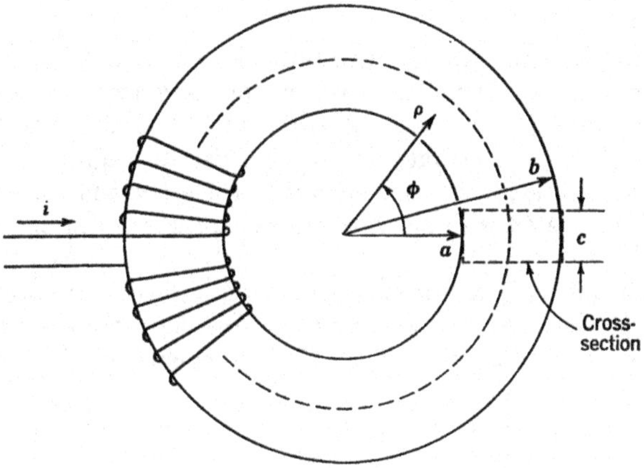

FIG. 4-18. A closely wound toroidal coil (winding covers the entire toroid).

$a \approx b$, the inductance is given by

$$L \approx \frac{\mu N^2(b - a)c}{\pi(b + a)}$$

Compare this answer with Eq. (4-24).

4-18. Given $\mathbf{E} = \mathbf{u}_x + 2\mathbf{u}_y$, determine the electric energy density. What is the total energy in a unit cube centered at the origin?

4-19. Given that $\mathbf{E} = \mathbf{u}_x y \cos z \cos \omega t$ in vacuum, and knowing that all field vectors vary sinusoidally in time, determine $\mathbf{D}$, $\mathbf{B}$, $\mathbf{H}$, w_e, and w_m.

4-20. Write Maxwell's equations explicitly in terms of rectangular components and coordinates. Repeat for cylindrical and spherical components and coordinates.

4-21. Suppose that the sheet of current $\mathbf{J}_s = \mathbf{u}_y J_0$, where J_0 is a constant over the plane $x = 0$, is the only source of $\mathbf{H}$ in vacuum. (*a*) By symmetry, justify that $\mathbf{H} = \mathbf{H}(x)$. (*b*) From $\nabla \cdot \mathbf{H} = 0$ and $\nabla \times \mathbf{H} = 0$ (except at $x = 0$), show that $\mathbf{H}$ = constant vector. (*c*) Using Eq. (4-22) evaluate H_z in the two regions $x > 0$ and $x < 0$. (*d*) By symmetry, justify that $H_x = H_y = 0$.

CHAPTER 5

THE FIELD OF STATIC CHARGES

5-1. Introduction. Charge distributions that do not vary with time are called *static charges*. The electric intensity and the electric flux density associated with static electric charges are also independent of time. Static electric charges do not give rise to magnetic fields, there being no interaction between time-invariant electric and magnetic fields. The field associated with static electric charges is called an *electrostatic field*.

The fundamental equations of electrostatic fields are obtained from the general relationships by setting all time derivatives equal to zero. Since there is only an electric field, the field equations become

$$\nabla \times \mathbf{E} = 0 \qquad \nabla \cdot \mathbf{D} = q_v \qquad \mathbf{D} = \epsilon \mathbf{E} \tag{5-1}$$

In this chapter, we shall investigate solutions to Eqs. (5-1) only for linear, homogeneous, and isotropic media of infinite extent. A vector is said to be *irrotational* in regions of vanishing curl. Therefore, the static electric intensity is everywhere irrotational. The integral form of the first of Eqs. (5-1) is obtained by setting the time derivative of Eq. (3-17) equal to zero. Thus

$$\oint \mathbf{E} \cdot d\mathbf{l} = 0 \tag{5-2}$$

which is valid for every closed contour, is mathematically equivalent to $\nabla \times \mathbf{E} = 0$. The integral equivalent to the second of Eqs. (5-1) is Eq. (4-3), repeated here for the sake of completeness.

$$\oiint \mathbf{D} \cdot d\mathbf{s} = q \tag{5-3}$$

where

$$q = \iiint q_v \, d\tau$$

The charge q is, of course, the charge contained within the surface to which Eq. (5-3) is applied.

5-2. The Field Integral. In Sec. 4-3, we solved for the field associated with a point charge and we indicated that the solution could be extended to include an arbitrary charge distribution. We shall now formalize this extension mathematically.

We have the field of a point charge q given by

$$\mathbf{D} = \frac{q}{4\pi r^2} \mathbf{u}_r \tag{5-4}$$

where r is the distance from q to the point at which $\mathbf{D}$ is evaluated and where $\mathbf{u}_r$ is a unit vector pointing radially outward from q. If q is situated at the coordinate origin, then r is the usual spherical coordinate and $\mathbf{u}_r$ is the usual radial unit coordinate vector. If q is not at the coordinate origin, then the r of Eq. (5-4) is not that of the spherical coordinate system. Suppose that the charge is located at the point (x',y',z') in space and that we wish to evaluate the field at the point (x,y,z) in space. We shall henceforth always use primed coordinates to designate the coordinates of the source, calling them *source coordinates*, and we shall use unprimed coordinates to designate those of the field point, calling them *field coordinates*. We now define *radius vectors*

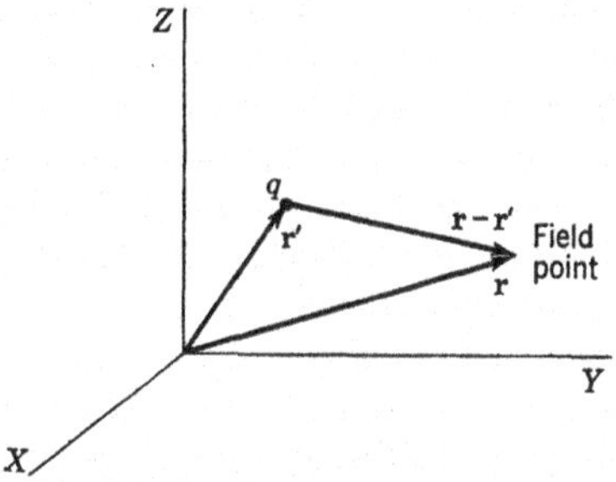

FIG. 5-1. A point charge at (x',y',z').

$$\begin{aligned} \mathbf{r} &= \mathbf{u}_x x + \mathbf{u}_y y + \mathbf{u}_z z \\ \mathbf{r}' &= \mathbf{u}_x x' + \mathbf{u}_y y' + \mathbf{u}_z z' \end{aligned} \tag{5-5}$$

which are vectors extending from the coordinate origin to the field point and source point, respectively. This is shown in Fig. 5-1. The radius vector from the source to the field point is $\mathbf{r} - \mathbf{r}'$, as we can see from Fig. 5-1. From our elementary vector concepts, we can now determine the distance from the source to the field point, which is the magnitude of $\mathbf{r} - \mathbf{r}'$. Thus,

$$|\mathbf{r} - \mathbf{r}'| = \sqrt{(x - x')^2 + (y - y')^2 + (z - z')^2} \tag{5-6}$$

replaces the factor r of Eq. (5-4) if the charge is not at the coordinate origin. The unit vector pointing from q to the field point is simply the vector $\mathbf{r} - \mathbf{r}'$ divided by its magnitude. Therefore, we can now generalize Eq. (5-4) to read

$$\mathbf{D} = \frac{q}{4\pi|\mathbf{r} - \mathbf{r}'|^2} \frac{\mathbf{r} - \mathbf{r}'}{|\mathbf{r} - \mathbf{r}'|} \tag{5-7}$$

Substituting for $\mathbf{r}$ and $\mathbf{r}'$ from Eqs. (5-5) into Eq. (5-7), we obtain the functional form for the field at any point (x,y,z) due to a charge at any point (x',y',z'). To emphasize that $\mathbf{D}$ is evaluated at any unprimed-coordinate point and that q is situated at the primed-coordinate point, we shall use the notation

$$\mathbf{D} = \mathbf{D}(\mathbf{r}) \qquad q = q(\mathbf{r}') \tag{5-8}$$

This means merely that **D** is a function of (x,y,z), and, rather than put these three variables in parentheses, we use the single symbol $(\mathbf{r})$. A similar interpretation holds for the notation $q(\mathbf{r}')$. Thus, our final equation for **D** from a point charge becomes

$$\mathbf{D}(\mathbf{r}) = \frac{q(\mathbf{r}')}{4\pi|\mathbf{r} - \mathbf{r}'|^3}(\mathbf{r} - \mathbf{r}') \tag{5-9}$$

where we have combined the magnitude factors in the denominator of Eq. (5-7) into a single term.

We have built up Eq. (5-9) using fundamental vector concepts. Perhaps it would be well to look at this equation in terms of the explicit coordinate variables so that we may be sure of its meaning. In terms of rectangular components, Eq. (5-9) is equivalent to the three equations

$$\begin{aligned} D_x(x,y,z) &= \frac{q(x',y',z')(x - x')}{4\pi[(x - x')^2 + (y - y')^2 + (z - z')^2]^{3/2}} \\ D_y(x,y,z) &= \frac{q(x',y',z')(y - y')}{4\pi[(x - x')^2 + (y - y')^2 + (z - z')^2]^{3/2}} \\ D_z(x,y,z) &= \frac{q(x',y',z')(z - z')}{4\pi[(x - x')^2 + (y - y')^2 + (z - z')^2]^{3/2}} \end{aligned} \tag{5-10}$$

Suppose we illustrate the interpretation of Eq. (5-9), or of the above equivalent equations, by a numerical example. Let us assume that a charge of 10 coulombs is situated at $(x',y',z') = (1,3,2)$ and let us evaluate the field at $(x,y,z) = (2,5,4)$. Substituting these values directly into Eqs. (5-10), we find that the field is given by

$$D_x = \frac{1}{10.8\pi} \qquad D_y = \frac{2}{10.8\pi} \qquad D_z = \frac{2}{10.8\pi}$$

Writing this in vector form, we have

$$\mathbf{D} = \frac{1}{10.8\pi}(\mathbf{u}_x + \mathbf{u}_y 2 + \mathbf{u}_z 2)$$

The magnitude of **D** is given by the square root of the sum of the squares of its components, being

$$D = \frac{3}{10.8\pi}$$

which, of course, could be obtained more simply from Eq. (5-4).

If we had several point charges, $q_1, q_2, \ldots, q_N$, located at the points $\mathbf{r}_1', \mathbf{r}_2', \ldots, \mathbf{r}_N'$, we could obtain the field at any point **r** by superimposing the fields due to each charge. The total **D** is thus a summation over terms of the form of Eq. (5-9), one for each charge, giving

$$\mathbf{D}(\mathbf{r}) = \frac{1}{4\pi} \sum_{n=1}^{N} \frac{q_n(\mathbf{r}'_n)}{|\mathbf{r} - \mathbf{r}'_n|^3} (\mathbf{r} - \mathbf{r}'_n) \tag{5-11}$$

This can be reduced to three equations for the rectangular components of $\mathbf{D}$, similar to Eqs. (5-10). Thus, Eq. (5-11) means that we can add the rectangular components of $\mathbf{D}$ from all charges, the sums being the rectangular components of the total $\mathbf{D}$.

Now consider a volume distribution of charge q_v, as suggested by Fig. 5-2. Since superposition holds, we can subdivide the charge distribution into differential volume elements, obtain the field from each element, and then sum over all these partial fields to obtain the total field. We continue to use primed coordinates for the source and unprimed for the field point. A representative volume element we denote by $d\tau'$, signifying $dx'\,dy'\,dz'$. The charge contained within such an element is evidently $q_v\,d\tau'$. At the field point, the volume element of charge looks like a point charge, for it is of differential size. Therefore, the field at $\mathbf{r}$ due to the differential element of charge at $\mathbf{r}'$ is given by Eq. (5-9) with $q(\mathbf{r}')$ replaced by $q_v(\mathbf{r}')\,d\tau'$. Summing over the entire charge distribution, we have the total $\mathbf{D}$ at the field point given by

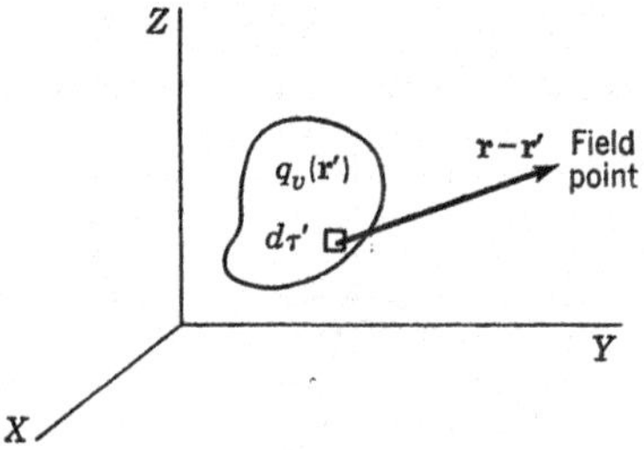

FIG. 5-2. A volume distribution of charge.

$$\mathbf{D}(\mathbf{r}) = \frac{1}{4\pi} \iiint \frac{q_v(\mathbf{r}')(\mathbf{r} - \mathbf{r}')}{|\mathbf{r} - \mathbf{r}'|^3}\, d\tau' \tag{5-12}$$

This formula we call the *field integral*, since it is essentially an integration over partial fields. We may be able to interpret Eq. (5-12) more easily if we look at the component equations equivalent to the single vector equation. These are obtained in the same way as Eqs. (5-10). For example, the x component of $\mathbf{D}$ is given by

$$D_x(x,y,z) = \frac{1}{4\pi} \iiint \frac{q_v(x',y',z')(x - x')\, dx'\, dy'\, dz'}{[(x - x')^2 + (y - y')^2 + (z - z')^2]^{3/2}}$$

and corresponding equations hold for the y and z components. These are just scalar integrals, the notation having the same meaning as in ordinary calculus.

We might wish to treat problems for which the charge is distributed in surface layers q_s or in line distributions q_l. The construction of the field integral in such cases follows the same logic as that used for volume distribution, except that now the differential elements of charge are $q_s\,ds'$ and $q_l\,dl'$. The summation over all elements now results in

$$\mathbf{D}(\mathbf{r}) = \frac{1}{4\pi}\iint \frac{q_s(\mathbf{r}')(\mathbf{r}-\mathbf{r}')}{|\mathbf{r}-\mathbf{r}'|^3}\,ds' \tag{5-13}$$

for surface charges, and in

$$\mathbf{D}(\mathbf{r}) = \frac{1}{4\pi}\int \frac{q_l(\mathbf{r}')(\mathbf{r}-\mathbf{r}')}{|\mathbf{r}-\mathbf{r}'|^3}\,dl' \tag{5-14}$$

for line charges. If in a single problem we have a combination of volume, surface, line, and point charges, the total field is just the superposition of the contributions from each type.

Perhaps an example would serve to illustrate the field integrals. Consider a sheet of charge, of uniform density q_s, in the shape of a circular disk, as shown in Fig. 5-3. We shall use the field integral to evaluate the field along the axis of the disk (the z axis). From the symmetry of the problem, we can conclude that $\mathbf{D}$ along the axis has only a z component, for there is no preferred direction for any component perpendicular to z. (It could also be shown by mathematical means, of course, that the x and y components are zero.) The coordinates of the field point are now specialized to $(0,0,z)$, and those of the source to $(x',y',0)$. Because the source has cylindrical symmetry, it is convenient to express the source in terms of cylindrical coordinates. As is evident from Fig. 5-3, the transformation is

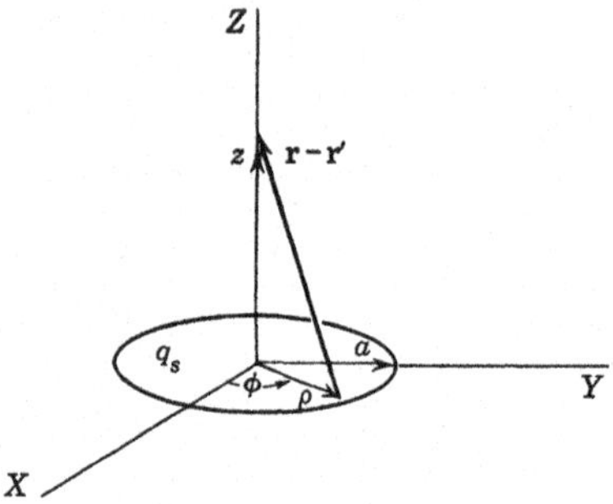

FIG. 5-3. A disk of uniform charge density.

$$x' = \rho\cos\phi \qquad y' = \rho\sin\phi$$

The factor $|\mathbf{r}-\mathbf{r}'|$ can be evaluated either by direct substitution into Eq. (5-6) as

$$|\mathbf{r}-\mathbf{r}'| = \sqrt{\rho^2\cos^2\phi + \rho^2\sin^2\phi + z^2} = \sqrt{\rho^2+z^2}$$

or by consideration of the geometry of Fig. 5-3. The differential element of surface on the source is

$$ds' = \rho\,d\phi\,d\rho$$

and the integration must extend from zero to 2π on ϕ and from zero to a on ρ. Thus, for the z component of Eq. (5-13), we have

$$D_z(0,0,z) = \frac{1}{4\pi}\int_0^{2\pi}\int_0^a \frac{q_s z}{(\rho^2+z^2)^{3/2}}\,\rho\,d\phi\,d\rho$$

We can remove q_s from under the integral signs because it is a constant. We can remove z from under the integral signs because it is a field coordinate, not a source coordinate (or integration variable). The ϕ integration gives 2π, so the above equation reduces to

$$D_z(0,0,z) = \frac{q_s z}{2} \int_0^a \frac{\rho \, d\rho}{(\rho^2 + z^2)^{3/2}}$$

This remaining integral can now be evaluated, giving

$$D_z(0,0,z) = \frac{q_s z}{2} \left(\frac{1}{|z|} - \frac{1}{\sqrt{a^2 + z^2}} \right) \tag{5-15}$$

Note that in going from a point just below the disk to a point just above the disk, D_z is discontinuous by the amount q_s, a result consistent with Eq. (4-7).

5-3. Gradient. Consider a well-behaved, single-valued, scalar function of position, which we shall denote by w. Suppose we take a small displacement in the w field and denote it by $\Delta \mathbf{l}$. The *directional derivative in the direction of* $\Delta \mathbf{l}$ is then defined as

$$\frac{dw}{dl} = \lim_{\Delta l \to 0} \frac{\Delta w}{\Delta l} \tag{5-16}$$

This will, in general, be different for different directions of displacement. Eq. (5-16) gives the rate of increase of w in the chosen direction. We define a vector, called the *gradient* of w, as a vector with magnitude equal to the maximum directional derivative and with the same direction as the maximum directional derivative. Our defining equation for the gradient, symbolized by grad w or ∇w, is thus

$$\text{grad } w = \nabla w = \left(\frac{dw}{dl} \mathbf{u}_l \right)_{\text{max}} \tag{5-17}$$

where $\mathbf{u}_l$ is a unit vector in the direction of $d\mathbf{l}$. The subscript "max" indicates that both dw/dl and $\mathbf{u}_l$ are associated with the maximum directional derivative.

For purposes of discussion, let us refer our quantities to a rectangular coordinate system. An incremental displacement $d\mathbf{l}$ can be written in rectangular components as

$$d\mathbf{l} = \mathbf{u}_x \, dx + \mathbf{u}_y \, dy + \mathbf{u}_z \, dz \tag{5-18}$$

The unit vector $\mathbf{u}_l$ is, of course, $d\mathbf{l}$ divided by its magnitude dl, so

$$\mathbf{u}_l = \mathbf{u}_x \frac{dx}{dl} + \mathbf{u}_y \frac{dy}{dl} + \mathbf{u}_z \frac{dz}{dl} \tag{5-19}$$

For a given $d\mathbf{l}$, the incremental change in w is given by

$$dw = \frac{\partial w}{\partial x}\,dx + \frac{\partial w}{\partial y}\,dy + \frac{\partial w}{\partial z}\,dz \tag{5-20}$$

Finally, the directional derivative in the direction of $d\mathbf{l}$ is Eq. (5-20) divided by dl, so we have

$$\frac{dw}{dl} = \frac{\partial w}{\partial x}\frac{dx}{dl} + \frac{\partial w}{\partial y}\frac{dy}{dl} + \frac{\partial w}{\partial z}\frac{dz}{dl} \tag{5-21}$$

We now note that if we define a vector

$$\mathbf{u}_x\frac{\partial w}{\partial x} + \mathbf{u}_y\frac{\partial w}{\partial y} + \mathbf{u}_z\frac{\partial w}{\partial z}$$

Eq. (5-21) is the scalar product of this vector with $\mathbf{u}_l$. Thus, we see that the directional derivatives of w are the components of the above vector. From our elementary vector concepts, it follows that the magnitude of the above vector must be the maximum directional derivative. Therefore, this vector must be the expression for the gradient of w in rectangular coordinates, that is

$$\text{grad } w = \nabla w = \mathbf{u}_x\frac{\partial w}{\partial x} + \mathbf{u}_y\frac{\partial w}{\partial y} + \mathbf{u}_z\frac{\partial w}{\partial z} \tag{5-22}$$

Note that, in rectangular coordinates, we can view the gradient as the same "del" operator that we found in the divergence and curl operations,

$$\nabla = \mathbf{u}_x\frac{\partial}{\partial x} + \mathbf{u}_y\frac{\partial}{\partial y} + \mathbf{u}_z\frac{\partial}{\partial z}$$

operating on a scalar function according to Eq. (5-22). In general, the notation ∇w symbolizes grad w, for a single differential operator common to the gradient, divergence, and curl operations can be found only in rectangular coordinates. From Eq. (5-22) it follows that the maximum rate of increase of w at a point is

$$|\nabla w| = \left(\frac{dw}{dl}\right)_{\max} = \sqrt{\left(\frac{\partial w}{\partial x}\right)^2 + \left(\frac{\partial w}{\partial y}\right)^2 + \left(\frac{\partial w}{\partial z}\right)^2} \tag{5-23}$$

We have already noted that

$$\frac{dw}{dl} = \nabla w \cdot \mathbf{u}_l \tag{5-24}$$

that is, that the components of grad w are the directional derivatives. Also the incremental change in w for an arbitrary $d\mathbf{l}$ is given by

$$dw = \nabla w \cdot d\mathbf{l} \tag{5-25}$$

which can be seen from Eqs. (5-18), (5-20), and (5-22). We have used rectangular coordinates to prove Eqs. (5-24) and (5-25). These equations must, however, be valid in all coordinate systems. To illustrate the geometric interpretation of the equations in this paragraph, we show ∇w and $d\mathbf{l}$ and their component vectors for a possible situation in Fig. 5-4.

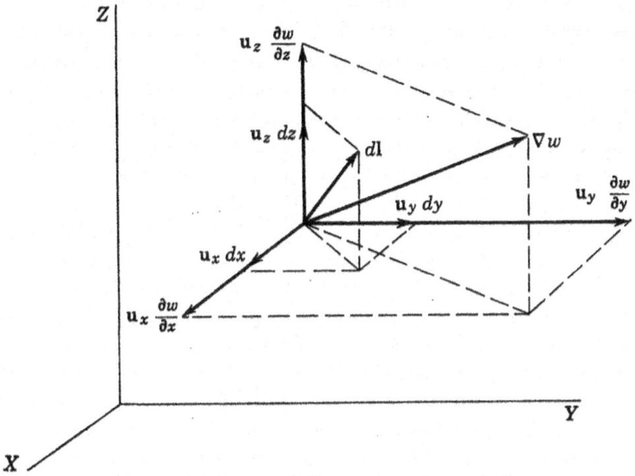

FIG. 5-4. Picture of the vectors ∇w and $d\mathbf{l}$.

Eq. (5-25) provides us with a convenient method of determining grad w in other coordinate systems. In cylindrical coordinates, the differential element of length is given by

$$d\mathbf{l} = \mathbf{u}_\rho\, d\rho + \mathbf{u}_\phi\, \rho\, d\phi + \mathbf{u}_z\, dz \tag{5-26}$$

and the differential change in w is

$$dw = \frac{\partial w}{\partial \rho}\, d\rho + \frac{\partial w}{\partial \phi}\, d\phi + \frac{\partial w}{\partial z}\, dz \tag{5-27}$$

If Eq. (5-25) is valid for all possible $d\mathbf{l}$'s, it follows that

$$\nabla w = \mathbf{u}_\rho \frac{\partial w}{\partial \rho} + \mathbf{u}_\phi \frac{1}{\rho} \frac{\partial w}{\partial \phi} + \mathbf{u}_z \frac{\partial w}{\partial z} \tag{5-28}$$

in cylindrical coordinates. Similarly, in spherical coordinates,

$$d\mathbf{l} = \mathbf{u}_r\, dr + \mathbf{u}_\theta\, r\, d\theta + \mathbf{u}_\phi\, r \sin\theta\, d\phi \tag{5-29}$$

and

$$dw = \frac{\partial w}{\partial r}\, dr + \frac{\partial w}{\partial \theta}\, d\theta + \frac{\partial w}{\partial \phi}\, d\phi \tag{5-30}$$

Therefore, the expression for grad w in spherical coordinates is

$$\nabla w = \mathbf{u}_r \frac{\partial w}{\partial r} + \mathbf{u}_\theta \frac{1}{r} \frac{\partial w}{\partial \theta} + \mathbf{u}_\phi \frac{1}{r \sin\theta} \frac{\partial w}{\partial \phi} \tag{5-31}$$

The important formulas derived in this section are summarized in Appendix B.

By the use of the gradient operation, we can derive a vector field grad w from any scalar w. A picture of the vector field gives us information about the scalar field, for lines of grad w are in the direction of the maximum rate of increase of w. Since the directional derivatives of w are the components of grad w, it follows that there is no rate of change of w in a direction perpendicular to grad w. Thus, starting at a point for which $w = w_o$ (a constant), we can go in any direction perpendicular to grad w and w is unchanged. This gives us a *surface of constant* w, which must everywhere be perpendicular to grad w. A picture of the surfaces

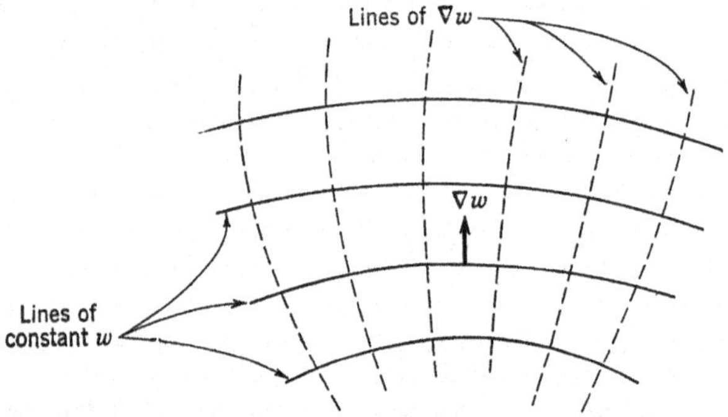

FIG. 5-5. Lines of grad w are everywhere perpendicular to lines (or surfaces) of constant w.

of constant w gives us information about the vector grad w, for it is everywhere perpendicular to these surfaces. For a two-dimensional case, the surfaces of constant w become lines of constant w, and the situation is easier to visualize. Fig. 5-5 shows lines of constant w and lines of grad w for a possible two-dimensional case. A familiar example of such a two-dimensional plot is the so-called "topographic map." In this case, the scalar function is height above sea level, $h = h(x,y)$. The lines of constant h are the contour lines, and the lines of ∇h would be the paths of steepest ascent. The latter are usually not shown, but can be readily sketched since they are everywhere perpendicular to the contour lines.

The vector grad w has some special properties, which we shall now consider. Suppose we form the usual line integral of the vector grad w along some path between two points in space, denoted by a and b. This is written

$$\int_a^b \nabla w \cdot d\mathbf{l}$$

where a specific path is implied. Using Eq. (5-25), we are able to reduce this integral to

$$\int_{w(a)}^{w(b)} dw$$

where $w(a)$ and $w(b)$ are the values of w at the points a and b. Integrating this, we find that

$$\int_a^b \nabla w \cdot d\mathbf{l} = w(b) - w(a) \tag{5-32}$$

showing that the value of the line integral depends only on the values of w at the end points. Thus, *the line integral of grad w between two points is independent of the path chosen.* From this it follows that the line integral of grad w about any closed path is zero, for then $a = b$ and $w(a) = w(b)$ in Eq. (5-32). Thus

$$\oint \nabla w \cdot d\mathbf{l} = 0 \tag{5-33}$$

is a mathematical identity, holding for all closed paths. Recalling the fundamental definition of the curl of a vector, it follows from Eq. (5-33) that

$$\nabla \times \nabla w = 0 \tag{5-34}$$

is also a mathematical identity. Eq. (5-34) can also be readily shown by expanding it out in rectangular coordinates. Thus, *the gradient of a scalar is an irrotational vector.* We shall show in the next section that the converse is also true. That is, *if the curl of a vector is zero, then that vector is the gradient of some scalar.* It follows from Eq. (5-22) that two scalars which differ only by a constant have the same gradient. Thus, an irrotational vector is the gradient of an infinity of scalars, differing from one another by constants.

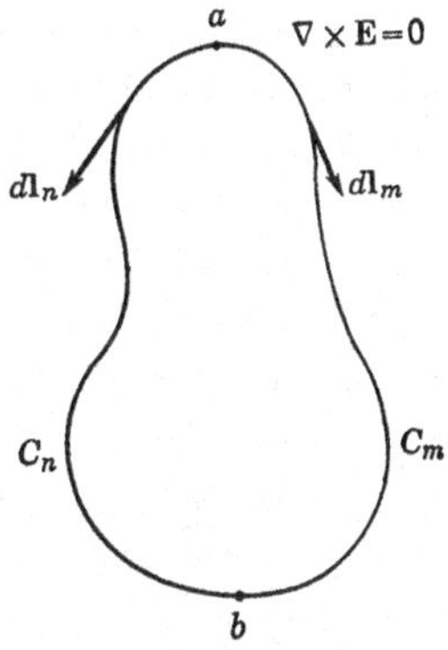

FIG. 5-6. Two paths connecting a and b in an irrotational field.

5-4. Electric Scalar Potential. The electric intensity due to static charges has no curl; that is, $\nabla \times \mathbf{E} = 0$ everywhere. According to Stokes' theorem, this is equivalent to Eq. (5-2) for all closed contours. Consider two arbitrary points in space, denoted by a and b. Along every path joining a and b we have a voltage according to Eq. (3-3). Two representative paths are shown in Fig. 5-6. This pair of paths form a closed loop, to which we can apply Eq. (5-2). Thus

$$\oint_{C_n+C_m} \mathbf{E} \cdot d\mathbf{l} = \int_{C_n} \mathbf{E} \cdot d\mathbf{l}_n - \int_{C_m} \mathbf{E} \cdot d\mathbf{l}_m = 0$$

Denoting the voltage between a and b along the path C_n by $v_n(a,b)$, this becomes

$$v_n(a,b) = v_m(a,b)$$

for all m and n; it states that *the voltage between a and b is independent of the path.* Thus, there exists a unique voltage between every two points, and we may drop the subscripts and denote this voltage by

$$v(a,b) = \int_a^b \mathbf{E} \cdot d\mathbf{l} \tag{5-35}$$

Eq. (5-35) is valid for all paths joining a and b. If we have three points, a, b, and c, and apply Eq. (5-2) to a loop joining these three points, we have

$$v(a,b) + v(b,c) + v(c,a) = 0$$

This is, of course, the usual statement of Kirchhoff's voltage law for d-c circuits. In the usual parlance of circuit theory, $v(a,b)$ is called "the voltage drop from a to b."

Once we pick a fixed point b, the quantity $v(a,b)$ is a function of the point a. This we emphasize by writing

$$v(a,b) = V_b(a)$$

FIG. 5-7. A differential displacement $d\mathbf{l}$.

In other words, V_b is simply a scalar function of position. Suppose we now take a differential displacement in space, $d\mathbf{l}$, and denote its end points by a_1 and a_2, as shown in Fig. 5-7. The differential voltage *rise* along $d\mathbf{l}$ is given by

$$\begin{aligned} dV_b &= V_b(a_2) - V_b(a_1) = v(a_2,b) - v(a_1,b) \\ &= v(a_2,a_1) = -\mathbf{E} \cdot d\mathbf{l} \end{aligned}$$

But in the preceding section, we found that, for any scalar field,

$$dV_b = \nabla V_b \cdot d\mathbf{l}$$

Thus, all components of $-\mathbf{E}$ are equal to those of ∇V_b, and from our elementary concept of the equality of vectors, it follows that

$$\mathbf{E} = -\nabla V_b$$

In words, this states that the electric intensity is the negative gradient of the scalar function V_b. But the point b can be picked at random; so $\mathbf{E}$ is the negative gradient of an infinity of scalar functions. However, any two of these scalars differ only by a constant, for

$$V_b(a) - V_c(a) = v(a,b) - v(a,c) = v(c,b)$$

and $v(c,b)$ is constant since b and c are fixed points. This checks with our observation of the preceding section that an infinity of scalars, all differing by a constant, may have the same gradient. Incidentally, the above analysis constitutes a mathematical proof that if a vector has zero curl it is the gradient of some scalar function.

We have found that if **E** is irrotational, which is the case in electrostatics, then

$$\mathbf{E} = -\nabla V \tag{5-36}$$

where V is called an *electric scalar potential.* There are, in general, an infinity of V functions differing from each other only by constants, so we say that V is "indeterminate" by a constant. The voltage between any two points in space is unique, given by Eq. (5-35). Substituting from Eq. (5-36), we have

$$\begin{aligned} v(a,b) &= \int_a^b \mathbf{E} \cdot d\mathbf{l} = -\int_a^b \nabla V \cdot d\mathbf{l} \\ &= -\int_a^b \frac{dV}{dl}\, dl = -\int_{V(a)}^{V(b)} dV \end{aligned}$$

which upon integration becomes

$$v(a,b) = V(a) - V(b) \tag{5-37}$$

This is the same for all V's, since the indeterminate constant subtracts out. In this case the quantity $v(a,b)$ is called the *potential difference.* If we arbitrarily specify the potential at one point in space, we are effectively specifying the constant term in V, and the scalar potential is then a unique function. When V is taken to be zero at some point, we call V the potential with respect to that point. A common choice is to specify V to be zero at "infinity," that is, at great distance from the charge, and V is then called *the potential with respect to infinity.*

In linear media, that is, when $\mathbf{D} = \epsilon\mathbf{E}$ with ϵ independent of the field vectors, the Maxwell divergence relationship becomes

$$\nabla \cdot (\epsilon\mathbf{E}) = q_v$$

Substituting for **E** from Eq. (5-36), we can obtain a partial differential equation for V. This is

$$\nabla \cdot (\epsilon \nabla V) = -q_v \tag{5-38}$$

which is valid even if ϵ is a function of position. We shall be interested primarily in solutions to Eq. (5-38) in homogeneous space, that is, when ϵ is a constant. In this case, ϵ can be removed from under the divergence operation, giving

$$\nabla^2 V = -\frac{q_v}{\epsilon} \tag{5-39}$$

where

$$\nabla^2 V = \nabla \cdot \nabla V \tag{5-40}$$

Eq. (5-39) is called *Poisson's equation,* and the differential operator of Eq. (5-40) is called the *Laplacian operator.* In regions of vanishing charge, Eq. (5-39) reduces to

$$\nabla^2 V = 0 \tag{5-41}$$

which is called *Laplace's equation.*

It should be emphasized that the notation $\nabla \cdot \nabla V$ is shorthand for div grad V, that is, we have to take the divergence of the gradient of V. In rectangular coordinates, the divergence and gradient operators take the same form, and we have the Laplacian of V given by

$$\begin{aligned}\nabla^2 V &= \left(\mathbf{u}_x \frac{\partial}{\partial x} + \mathbf{u}_y \frac{\partial}{\partial y} + \mathbf{u}_z \frac{\partial}{\partial z}\right) \cdot \left(\mathbf{u}_x \frac{\partial V}{\partial x} + \mathbf{u}_y \frac{\partial V}{\partial y} + \mathbf{u}_z \frac{\partial V}{\partial z}\right) \\ &= \frac{\partial^2 V}{\partial x^2} + \frac{\partial^2 V}{\partial y^2} + \frac{\partial^2 V}{\partial z^2}\end{aligned} \tag{5-42}$$

In cylindrical coordinates, the gradient and divergence operators are different in form; for convenience, they are repeated below.

$$\nabla V = \mathbf{u}_\rho \frac{\partial V}{\partial \rho} + \mathbf{u}_\phi \frac{1}{\rho} \frac{\partial V}{\partial \phi} + \mathbf{u}_z \frac{\partial V}{\partial z}$$

$$\nabla \cdot \mathbf{A} = \frac{1}{\rho} \frac{\partial}{\partial \rho} (\rho A_\rho) + \frac{1}{\rho} \frac{\partial A_\phi}{\partial \phi} + \frac{\partial A_z}{\partial z}$$

Substituting the vector ∇V into the divergence formula, we obtain for the Laplacian of V in cylindrical coordinates,

$$\nabla^2 V = \frac{1}{\rho} \frac{\partial}{\partial \rho} \left(\rho \frac{\partial V}{\partial \rho}\right) + \frac{1}{\rho^2} \frac{\partial^2 V}{\partial \phi^2} + \frac{\partial^2 V}{\partial z^2} \tag{5-43}$$

Similarly, the Laplacian of V in spherical coordinates is found to be

$$\nabla^2 V = \frac{1}{r^2} \frac{\partial}{\partial r}\left(r^2 \frac{\partial V}{\partial r}\right) + \frac{1}{r^2 \sin \theta} \frac{\partial}{\partial \theta}\left(\sin \theta \frac{\partial V}{\partial \theta}\right) + \frac{1}{r^2 \sin^2 \theta} \frac{\partial^2 V}{\partial \phi^2} \tag{5-44}$$

These formulas for $\nabla^2 V$ are summarized in Appendix B.

5-5. Point Charges. Charge contained within a region whose dimensions are small compared with the distance from the charge to the field point can be considered to be a point charge. We have already given consideration to the field due to point charges, but we wish to reconsider the problem bringing the scalar potential into the picture.

Consider a point charge of strength q situated at the coordinate origin in an unbounded homogeneous region. The problem has complete spherical symmetry, so we assume that the scalar potential V depends only upon the radial distance r from the charge. Letting $V = V(r)$, for Eq. (5-39) we have

$$\nabla^2 V = \frac{1}{r^2} \frac{d}{dr}\left(r^2 \frac{dV}{dr}\right) = 0$$

which holds everywhere except at the origin, where the charge exists. This is simply $\nabla^2 V$ in spherical coordinates, with θ and ϕ derivatives set equal to zero and with partial derivatives replaced by total derivatives since r is the only variable. The derivative of the term in the parentheses is zero, so the term itself must be a constant. Thus

$$r^2 \frac{dV}{dr} = K_1$$

where K_1 is a constant. Dividing through by r^2 and integrating, we have

$$V = \int \frac{K_1}{r^2}\, dr = -\frac{K_1}{r} + K_2$$

where K_2 is also a constant. This is the general functional form for the potential associated with the point charge.

The electric intensity is given by

$$\mathbf{E} = -\nabla V = -\mathbf{u}_r \frac{dV}{dr} = -\mathbf{u}_r \frac{K_1}{r^2}$$

where we have used the formula for ∇V in spherical coordinates. The constant K_1 can now be evaluated by applying Gauss's law to a sphere concentric with the point charge. This gives

$$q = \oint\!\!\oint \mathbf{D} \cdot d\mathbf{s} = \oint\!\!\oint \epsilon \mathbf{E} \cdot \mathbf{u}_r\, ds$$
$$= -\int_0^{2\pi} \int_0^{\pi} \epsilon K_1 \sin\theta\, d\theta\, d\phi = -4\pi\epsilon K_1$$

Substituting this value of K_1 into the equation for $\mathbf{E}$, we have

$$\mathbf{E} = \mathbf{u}_r \frac{q}{4\pi\epsilon r^2} \tag{5-45}$$

and substituting into the equation for V, we have

$$V = \frac{q}{4\pi\epsilon r} + K_2 \tag{5-46}$$

Eq. (5-45) is, of course, the same solution as Eq. (4-5). The constant K_2 does not enter into the expression for the field, and any choice of K_2 is permissible. Thus, the scalar potential is indeterminate by a constant, which is the same conclusion we reached in the preceding section.

If we arbitrarily specify V at some point to be zero, we then have the potential with respect to that point, and K_2 is determined. The most useful and reasonable choice is to specify that the potential vanishes as $r \to \infty$. In this case $K_2 = 0$, and we have the potential with respect to infinity given by

$$V = \frac{q}{4\pi\epsilon r} \tag{5-47}$$

for a point charge at the origin. Note that spheres concentric with the point charge are *equipotential surfaces*, that is, surfaces of constant V. The electric intensity is, of course, given by $\mathbf{E} = -\nabla V$.

If the charge is not situated at the coordinate origin, the scalar potential is still given by Eq. (5-47) if r is interpreted as the distance from the charge to the field point. This we can emphasize by recasting the formula in terms of the notation introduced in Sec. 5-2. The picture for a point charge at arbitrary coordinates is Fig. 5-1. Again we use the notation $q(\mathbf{r}')$ to signify that the charge is located at the point (x',y',z') and the notation $V(\mathbf{r})$ to signify that the potential is evaluated at the point (x,y,z). The distance from the source to the field point is $|\mathbf{r} - \mathbf{r}'|$, so the formula for V due to a point charge q becomes

$$V(\mathbf{r}) = \frac{q(\mathbf{r}')}{4\pi\epsilon|\mathbf{r} - \mathbf{r}'|} \tag{5-48}$$

Eq. (5-47) can be viewed as the special case for which $\mathbf{r}' = 0$. In rectangular coordinates, the distance $|\mathbf{r} - \mathbf{r}'|$ is the familiar

$$|\mathbf{r} - \mathbf{r}'| = \sqrt{(x - x')^2 + (y - y')^2 + (z - z')^2}$$

For the source point, or the field point, or both, expressed in other than rectangular coordinates, we can obtain corresponding formulas for $|\mathbf{r} - \mathbf{r}'|$ by substituting the appropriate coordinate transformations into the above equation. The several forms for $|\mathbf{r} - \mathbf{r}'|$ corresponding to the possible permutations of rectangular, cylindrical, and spherical coordinates are summarized in Appendix B.

If we have several point charges, $q_1, q_2, \ldots, q_N$, located at the various points $\mathbf{r}'_1, \mathbf{r}'_2, \ldots, \mathbf{r}'_N$, we can apply superposition to the individual potentials and obtain the potential for the entire system. The potential for each charge is given by Eq. (5-48), so the potential for a system of point charges is given by

$$V(\mathbf{r}) = \frac{1}{4\pi\epsilon} \sum_{n=1}^{N} \frac{q_n(\mathbf{r}'_n)}{|\mathbf{r} - \mathbf{r}'_n|} \tag{5-49}$$

The electric intensity from the system is then given by $\mathbf{E} = -\nabla V$. Note that by using the scalar potential we get only a single scalar superposition, whereas the superposition of fields is a vector operation. This is one of the advantages of using the scalar potential.

To illustrate the application of Eq. (5-49), let us consider two point charges, q_1 and q_2, separated by a distance d in an infinite homogeneous medium. Evidently, the field from these charges is rotationally symmetric about the line joining the charges. We therefore need to consider only the field in a plane containing the charges. In one such plane, we orient our axes so that the charges lie on the x axis and so that the origin is midway between the charges. This is shown in Fig. 5-8.

Using Eq. (5-49), we have the potential with respect to infinity given by

$$V = \frac{1}{4\pi\epsilon}\left(\frac{q_1}{r_1} + \frac{q_2}{r_2}\right) \tag{5-50}$$

where r_1 and r_2 are the distances from q_1 and q_2, respectively, to the field point. An equipotential surface is by definition a surface for which V has everywhere the same value. Setting V in Eq. (5-50) equal to specific values, say 0, 1, 2, etc., defines the various equipotential surfaces of the field. The mathematical equations for these surfaces may be found by substituting into Eq. (5-50) for r_1 and r_2 expressed in terms of the coordinates, but the resulting equation is rather complex. To plot the equipotential surfaces, which in the x-y plane become equipotential lines, it is simpler to use a graphical construction. Rearranging Eq. (5-50), we have

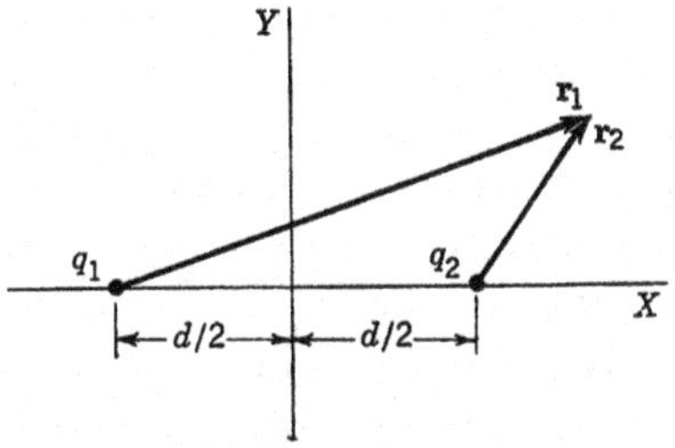

FIG. 5-8. Two point charges.

$$r_2 = \frac{q_2}{4\pi\epsilon V - q_1/r_1} \tag{5-51}$$

For a given value of V, we can choose an r_1 and determine r_2 by Eq. (5-51). The intersections of r_1 and r_2 then give us two points on the equipotential line in the x-y plane, one above the x axis and one below it. By choosing various values of r_1, we can determine a number of points on the equipotential lines and we can obtain the lines by connecting these points. A rotation of the x-y plane picture about the x axis gives the equipotential surfaces. The **E** lines are everywhere perpendicular to the equipotential surfaces and can be sketched if sufficient equipotential surfaces are known. Fig. 5-9 shows the equipotential lines in the x-y plane for the cases $q_2 = q_1$, $q_2 = 4q_1$, $q_2 = -4q_1$, and $q_2 = -q_1$.

If the charges are of opposite sign, the surface for which $V = 0$ has the special property that it is a sphere if $q_2 \neq -q_1$ and a plane if $q_2 = -q_1$. This can be seen as follows: Choose $|q_1| \geq |q_2|$, and let

$$q_2 = -\alpha q_1$$

where $0 < \alpha \leq 1$ (we can always label the charges so that $|q_1| \geq |q_2|$). Setting $V = 0$ in Eq. (5-50), we have

$$-\frac{q_2}{q_1} = \alpha = \frac{r_2}{r_1} \tag{5-52}$$

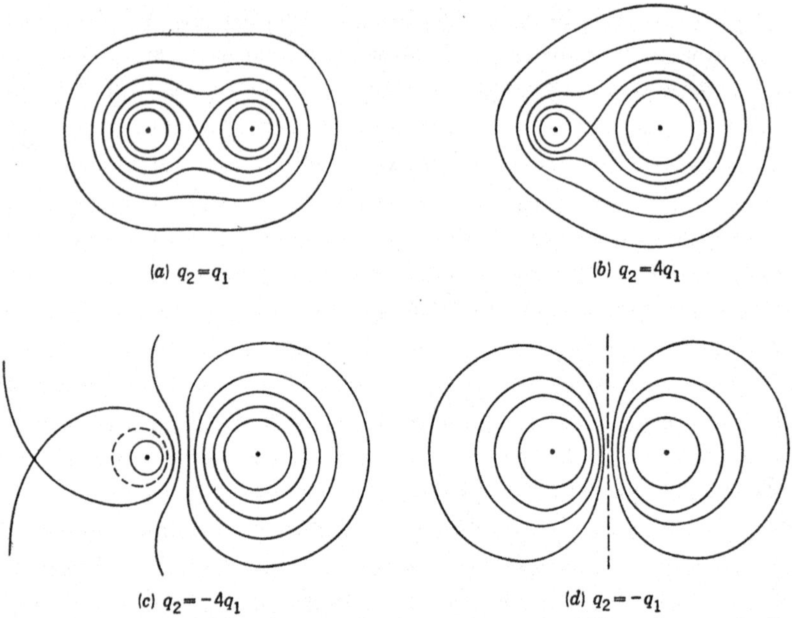

FIG. 5-9. Equipotential lines for two point charges (zero-potential lines shown dashed).

If $\alpha = 1$, that is, if $q_2 = -q_1$, this defines the plane $x = 0$ of Fig. 5-8, since $r_1 = r_2$ everywhere on this plane. To show that Eq. (5-52) defines a circle if $\alpha < 1$, we express r_1 and r_2 in terms of the coordinates. Referring to Fig. 5-8, we have

$$r_1^2 = \left(x + \frac{d}{2}\right)^2 + y^2 \qquad r_2^2 = \left(x - \frac{d}{2}\right)^2 + y^2$$

Substituting these into the square of Eq. (5-52), we obtain

$$\alpha^2 = \frac{(x - d/2)^2 + y^2}{(x + d/2)^2 + y^2}$$

This can be rearranged in the form

$$\left[x - \frac{d}{2}\left(\frac{1 + \alpha^2}{1 - \alpha^2}\right)\right]^2 + y^2 = \left(\frac{\alpha d}{1 - \alpha^2}\right)^2 \tag{5-53}$$

which is the equation of a circle with its center at

$$x = \frac{d(1 + \alpha^2)}{2(1 - \alpha^2)} \qquad y = 0$$

and with radius $\alpha d/(1 - \alpha^2)$. This zero potential surface is a sphere surrounding the smaller of the two charges, and is shown dashed in Fig. 5-9c.

5-6. The Scalar Potential Integral. We now consider a volume distribution of charge q_v which is in an unbounded homogeneous region, as illustrated by Fig. 5-2. By the same procedure we employed for the field integral, we subdivide the charge distribution into differential elements. Each element we view as a point charge. The potential from each element of charge is then given by Eq. (5-48) with $q(\mathbf{r}')$ replaced by $q_v(\mathbf{r}')\,d\tau'$. Summing over the entire charge distribution, we have the total potential at the point $\mathbf{r}$ given by

$$V(\mathbf{r}) = \frac{1}{4\pi\epsilon} \iiint \frac{q_v(\mathbf{r}')}{|\mathbf{r} - \mathbf{r}'|}\,d\tau' \tag{5-54}$$

This is called the *electric scalar potential integral.* The electric intensity is obtained from $\mathbf{E} = -\nabla V$, giving us a formal solution for the electrostatic field.

We shall view Eq. (5-54) as the general solution for the potential due to charge in unbounded homogeneous regions. The cases of point charges, line charges, and surface charges we consider to be special cases of Eq. (5-54) obtainable by suitable limiting processes. For point charges, q_v is zero except at discrete points and the volume integral reduces to the summation of Eq. (5-49). For line charges, the volume integral reduces to a line integral over the filament of charge. Thus, the electric potential from a filament of charge q_l is given by

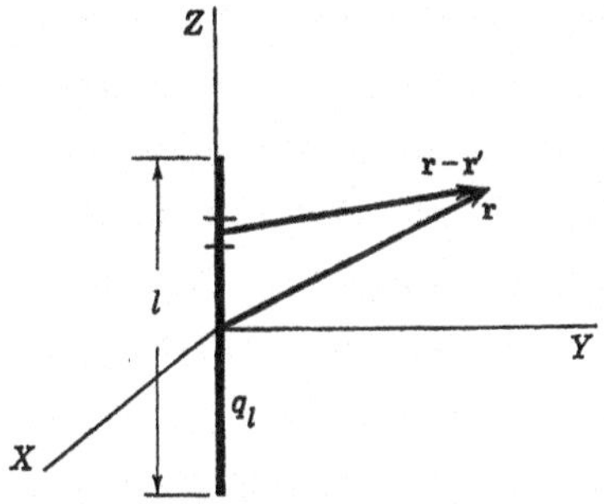

FIG. 5-10. A filament of charge.

$$V(\mathbf{r}) = \frac{1}{4\pi\epsilon} \int \frac{q_l(\mathbf{r}')}{|\mathbf{r} - \mathbf{r}'|}\,dl' \tag{5-55}$$

When the charge exists in surface distributions, the volume integral reduces to a surface integral. Thus, the electric potential from a surface charge q_s is given by

$$V(\mathbf{r}) = \frac{1}{4\pi\epsilon} \iint \frac{q_s(\mathbf{r}')}{|\mathbf{r} - \mathbf{r}'|}\,ds' \tag{5-56}$$

If we have a combination of point charges, line charges, surface charges, and volume charges, the total potential is just the superposition of the potentials for each type.

To illustrate the application of the scalar potential integral, we shall treat the problem of a line distribution of charge of finite length. We have a uniform linear density of charge q_l distributed along the z axis in an infinite homogeneous medium, as shown in Fig. 5-10. A representative differential element of charge and the radius vector from this element to the field point are also shown.

For this problem we use Eq. (5-55), which is a summation over all differential elements of the line charge. From considerations of symmetry, we infer that the potential should be cylindrically symmetric about the line charge. The source coordinates are $(0,0,z')$, and q_l is

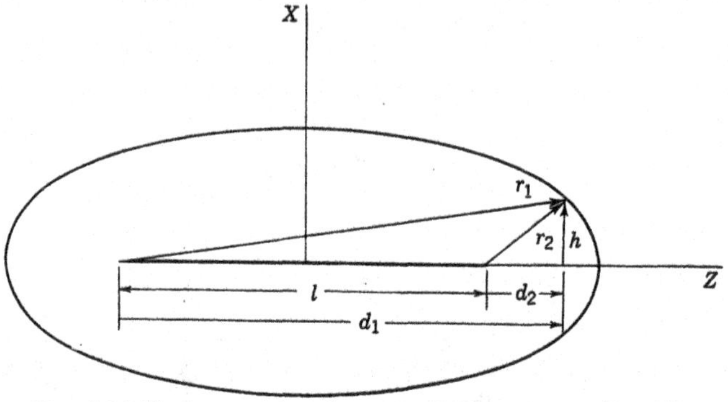

FIG. 5-11. Equipotential surfaces of a line charge are ellipsoids.

constant over the filament of charge, so Eq. (5-55) reduces to

$$V = \frac{q_l}{4\pi\epsilon} \int_{-\frac{l}{2}}^{\frac{l}{2}} \frac{1}{\sqrt{x^2 + y^2 + (z - z')^2}}\, dz'$$

This is readily integrable, and we obtain

$$V = \frac{q_l}{4\pi\epsilon} \log \left[\frac{z + l/2 + \sqrt{x^2 + y^2 + (z + l/2)^2}}{z - l/2 + \sqrt{x^2 + y^2 + (z - l/2)^2}} \right] \tag{5-57}$$

We shall now show that the equipotential surfaces form a set of prolate ellipsoids with foci at the ends of the line charge.

We define the parameters

$$r_1 = \sqrt{x^2 + y^2 + (z + l/2)^2} \qquad d_1 = z + \frac{l}{2}$$

$$r_2 = \sqrt{x^2 + y^2 + (z - l/2)^2} \qquad d_2 = z - \frac{l}{2}$$

and rewrite Eq. (5-57) as

$$V = \frac{q_l}{4\pi\epsilon} \log \left(\frac{d_1 + r_1}{d_2 + r_2} \right)$$

(note that d_1 and d_2 are algebraic quantities). The x-z plane is taken to be a representative plane containing the line charge, and the various parameters are shown in Fig. 5-11. An equipotential surface is therefore defined by

$$\frac{d_1 + r_1}{d_2 + r_2} = k \tag{5-58}$$

where

$$\log k = \frac{4\pi\epsilon V}{q_l}$$

Using the relationships (see Fig. 5-11)

$$r_1^2 - d_1^2 = r_2^2 - d_2^2 = h^2$$

we can reduce Eq. (5-58) to the form

$$r_1 + r_2 = l\frac{k+1}{k-1} \tag{5-59}$$

Thus, the sum of the distances from the ends of the line charge to a point on the equipotential surface is constant. This is a property of the ellipse, so the cross sections of the equipotential surfaces must be elliptical. The complete surfaces are obtained by rotating these cross sections about the z axis, generating prolate ellipsoids.

5-7. Two-dimensional Charge Distributions. When the charge distribution is unvarying with respect to one direction in space, we say that it is a two-dimensional source. In other words, letting the z direction be the direction of no variation, we have the charge distribution a function of x and y, or else of ρ and ϕ. It is of infinite extent in the z direction. Such source distributions cannot be realized in actuality, but can be used to approximate sources very long in one direction.

The simplest two-dimensional source is the infinitely long straight filament of constant charge. To analyze this type of source, we shall use cylindrical coordinates with the z axis taken along the line charge. The problem has complete cylindrical symmetry, so we may assume that the scalar potential V depends only upon the radial distance from the line charge. Letting $V = V(\rho)$, and using $\nabla^2 V$ in cylindrical coordinates, we obtain from Eq. (5-39)

$$\nabla^2 V = \frac{1}{\rho}\frac{d}{d\rho}\left(\rho\frac{dV}{d\rho}\right) = 0$$

everywhere except at the z axis, where the charge exists. Since the derivative of the quantity in parentheses is zero, that quantity must be a constant, so

$$\rho\frac{dV}{d\rho} = K_1$$

Dividing through by ρ and integrating, we have

$$V = \int \frac{K_1}{\rho}\,d\rho = K_1 \log \rho + K_2$$

where K_1 and K_2 are constants. (All logarithms are to the base e unless otherwise specified.) This is the general functional form for the potential associated with a uniform line charge.

The electric intensity is given by

$$\mathbf{E} = -\nabla V = -\mathbf{u}_\rho \frac{\partial V}{\partial \rho} = -\mathbf{u}_\rho \frac{K_1}{\rho}$$

We can now evaluate K_1 by Gauss's law applied to the surface shown in Fig. 5-12. This surface is a cylinder concentric with the line charge, of height h, bounded by planes over the top and bottom. We have $\mathbf{D} = \epsilon\mathbf{E}$ parallel to the top and bottom surfaces, so the surface integral over this part is zero. The only contribution comes from the cylindrical wall, so we have

$$q = q_l h = \oint\!\!\oint \mathbf{D} \cdot d\mathbf{s} = h \int_0^{2\pi} \epsilon E_\rho \, \rho \, d\phi$$
$$= -h\epsilon K_1 \int_0^{2\pi} d\phi = -h2\pi\epsilon K_1$$

where q_l is the linear charge density along the filament. The h's cancel, giving

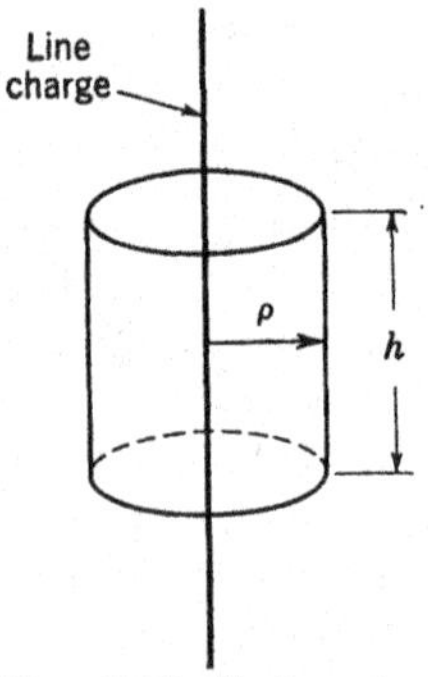

FIG. 5-12. Surface to which Gauss's law is applied.

$$K_1 = \frac{-q_l}{2\pi\epsilon}$$

Substituting this into the equation for $\mathbf{E}$, we have

$$\mathbf{E} = \mathbf{u}_\rho \frac{q_l}{2\pi\epsilon\rho} \tag{5-60}$$

Thus, the electric field points radially outward from the filament and varies inversely with the distance from the cylinder. A substitution for K_1 into the equation for V gives for the scalar potential

$$V = \frac{-q_l}{2\pi\epsilon} \log \rho + K_2 \tag{5-61}$$

The constant K_2 does not enter into the field expression, and any choice of K_2 is permissible.

If we arbitrarily specify V to be zero at some point, we have the potential with respect to that point, and K_2 is determined. Note that the first term of Eq. (5-61) becomes negatively infinite as $\rho \to \infty$. Thus, no finite value of K_2 can give us zero potential at infinity. For many purposes, it is convenient to let $K_2 = 0$, in which case we have the potential with respect to a unit cylinder given by

$$V = \frac{-q_l}{2\pi\epsilon} \log \rho \tag{5-62}$$

This occurrence of an infinite potential at an infinitely remote distance is characteristic of sources of infinite extent. Note that the equipotential surfaces for the uniform line charge are cylinders concentric with the source.

If the line charge is not located along the z axis but is displaced parallel to it, Eq. (5-62) would still be valid if ρ were interpreted as the radial distance from the line source to the field point. The formula for V in this case can be put into more descriptive form by making use of the two-dimensional radius vector. Suppose the line source is located at the coordinates (x',y'), and the field point at (x,y), as suggested by Fig. 5-13. The two-dimensional radius vectors to these points are

$$\begin{aligned} \boldsymbol{\varrho} &= \mathbf{u}_x x + \mathbf{u}_y y \\ \boldsymbol{\varrho}' &= \mathbf{u}_x x' + \mathbf{u}_y y' \end{aligned} \tag{5-63}$$

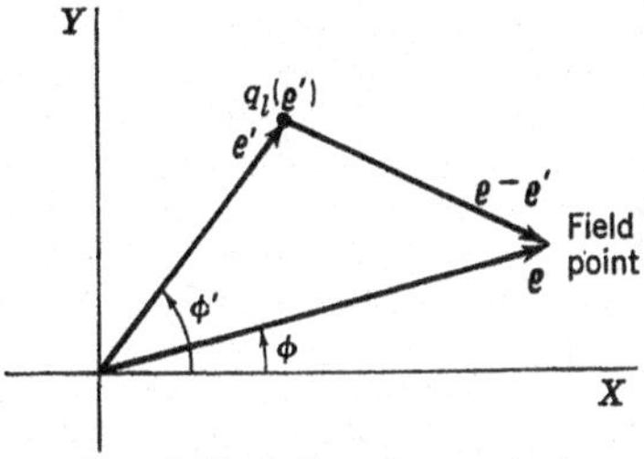

FIG. 5-13. A line charge at ρ'.

as shown in Fig. 5-13. The radius vector from the line source to the field point is now evidently $\boldsymbol{\varrho} - \boldsymbol{\varrho}'$, the magnitude of which replaces ρ in Eq. (5-62). To emphasize that q_l is at the primed-coordinate point and that V is evaluated at the unprimed-coordinate point, we use the notation $q_l(\boldsymbol{\varrho}')$ and $V(\boldsymbol{\varrho})$. Thus, the general formula for the potential due to an infinite line charge is

$$V(\boldsymbol{\varrho}) = \frac{-q_l(\boldsymbol{\varrho}')}{2\pi\epsilon} \log |\boldsymbol{\varrho} - \boldsymbol{\varrho}'| \tag{5-64}$$

Eq. (5-62) is now the special case for which $\boldsymbol{\varrho}' = 0$. If rectangular coordinates are used for both the source and the field points, then $|\boldsymbol{\varrho} - \boldsymbol{\varrho}'|$ assumes the familiar form

$$|\boldsymbol{\varrho} - \boldsymbol{\varrho}'| = \sqrt{(x - x')^2 + (y - y')^2} \tag{5-65}$$

It is sometimes convenient to express the source coordinates, or the field coordinates, or both, in polar coordinates (which are the usual cylindrical coordinates in a plane transverse to the z axis). The several forms for $|\boldsymbol{\varrho} - \boldsymbol{\varrho}'|$ corresponding to the permutations of rectangular and polar coordinates may be obtained from the corresponding formulas for $|\mathbf{r} - \mathbf{r}'|$ by letting $z = z' = 0$ (see Appendix B).

If we have several parallel line charges, $q_{l,1}, q_{l,2}, \ldots, q_{l,N}$, located at the various positions $\boldsymbol{\varrho}_1', \boldsymbol{\varrho}_2', \ldots, \boldsymbol{\varrho}_N'$, we can apply superposition and obtain the potential for the entire system. Eq. (5-64) applies for each line charge, so for N line charges we have

$$V(\boldsymbol{\varrho}) = \frac{-1}{2\pi\epsilon} \sum_{n=1}^{N} q_{l,n}(\boldsymbol{\varrho}_n') \log |\boldsymbol{\varrho} - \boldsymbol{\varrho}_n'| \tag{5-66}$$

The electric intensity from the system of line charges is then given by $\mathbf{E} = -\nabla V$.

If we have a two-dimensional volume distribution of charge, that is, q_v independent of z, we can divide the source into line elements of differential cross section. A representative differential element of cross-sectional area of the source we denote by ds', the prime emphasizing that source coordinates are involved. Such a differential area specifies a line charge of strength $q_v(\varrho')\,ds'$. The potential associated with the differential line charge is given by Eq. (5-64) with $q_l(\varrho')$ replaced by $q_v(\varrho')\,ds'$. Summing over all differential elements, we have the total potential from the source given by

$$V(\varrho) = \frac{-1}{2\pi\epsilon}\iint q_v(\varrho')\log|\varrho - \varrho'|\,ds' \tag{5-67}$$

This is the two-dimensional electric scalar potential integral. The electric intensity is, of course, given by $\mathbf{E} = -\nabla V$.

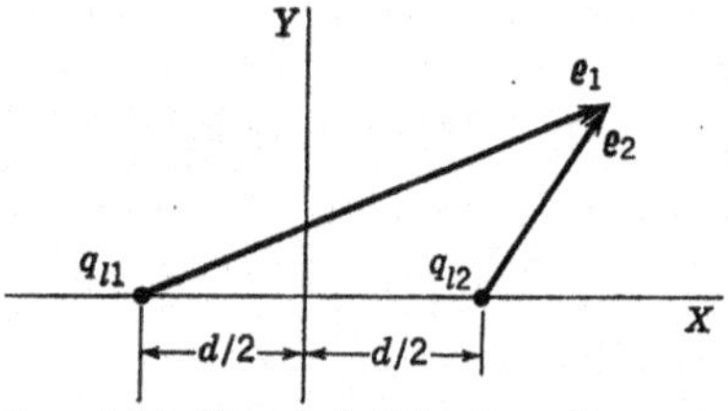

FIG. 5-14. Two infinitely long filaments of charge.

We view Eq. (5-67) as the general formula for two-dimensional charge distributions. The formula for discrete line charges, Eq. (5-66), is the special case obtained when the volume charge density is zero except at discrete points in the cross-sectional plane. We might also have a two-dimensional surface distribution of charge, in which case Eq. (5-67) reduces to a line integral. In other words, if we have $q_s(\varrho')$ independent of z, the differential elements of source are $q_s(\varrho')\,dl'$, and the total potential is given by

$$V(\varrho) = \frac{-1}{2\pi\epsilon}\int q_s(\varrho')\log|\varrho - \varrho'|\,dl' \tag{5-68}$$

If we have a combination of two-dimensional line charges, surface charges, and volume charges, the total potential is just the superposition of the contributions from each type of source.

Let us now illustrate the application of our two-dimensional formulas by considering two parallel line charges, of infinite length, separated by a distance d, in an infinite homogeneous medium. The field in this case is evidently the same over all planes perpendicular to the line charges. In such plane, we orient axes as shown in Fig. 5-14. The z axis lies midway between the two line charges $q_{l,1}$ and $q_{l,2}$.

According to Eq. (5-66), we have a potential associated with the line charges given by

$$V = \frac{-1}{2\pi\epsilon} [q_{l,1} \log (\rho_1) + q_{l,2} \log (\rho_2)] \tag{5-69}$$

where ρ_1 and ρ_2 are the radial distances from $q_{l,1}$ and $q_{l,2}$, respectively, to the field point. The equipotential surfaces are defined by setting V equal to various values in Eq. (5-69). In general, the equipotential surfaces, which are lines in the cross-sectional planes, can be geometrically constructed by a method similar to the method demonstrated for two point charges in Sec. 5-5.

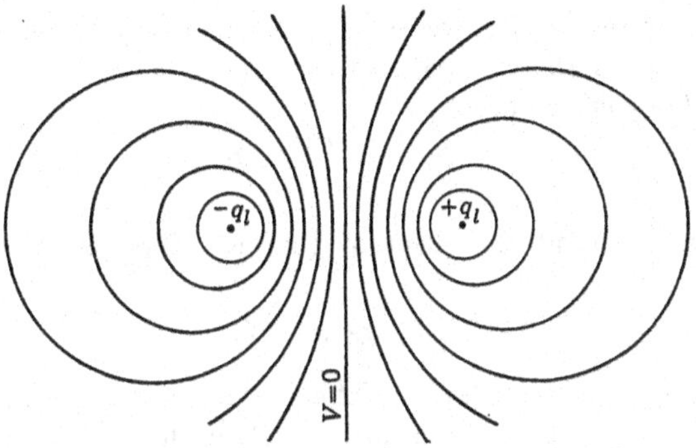

FIG. 5-15. Cross section of the equipotential surfaces of two line charges $+q_l$ and $-q_l$

An important special case occurs when $q_{l,1} = -q_{l,2} = q_l$. Now Eq. (5-69) reduces to

$$V = \frac{q_l}{2\pi\epsilon} \log \left(\frac{\rho_2}{\rho_1}\right) \tag{5-70}$$

and the equipotential surfaces are defined by

$$\frac{\rho_2}{\rho_1} = k \tag{5-71}$$

where

$$\log k = \frac{2\pi\epsilon V}{q_l}$$

At the end of Sec. 5-5, we showed that Eq. (5-71) defines a circle with its center at $x = d(1 + k^2)/2(1 - k^2)$, $y = 0$, and of radius $k\,d/(1 - k^2)$, if $k < 1$. If $k > 1$, we merely take the reciprocal of Eq. (5-71), and the same analysis applies. If $k = 1$, we have the line (or plane) $x = 0$ defined. Thus, *all* equipotential lines in the x-y plane are circles, and the complete equipotential surfaces are circular cylinders. (The plane corresponding to $V = 0$ is viewed as a cylinder of infinite radius.) A plot of the equipotential lines from two equal but opposite line charges is shown in Fig. 5-15.

5-8. One-dimensional Charge Distributions. If the charge distribution varies only with respect to one direction in space, we call it a one-dimensional source. We shall take the y and z directions to be the directions of no variation, so the charge distribution is a function only of x. It is, of course, infinite in extent in the y and z directions. One-dimensional distributions can be used to approximate sources which extend for large but finite distances along a plane.

The elementary one-dimensional source is the infinite flat plane of uniform surface charge density q_s. Orienting rectangular axes so that the x axis is perpendicular to the plane, with origin at the plane, we have symmetry and we see that the scalar potential should depend only upon the distance from the charged sheet. Thus, we set $V = V(x)$ and we obtain for Poisson's equation,

$$\nabla^2 V = \frac{d^2 V}{dx^2} = 0$$

everywhere except at the sheet of charge. This result requires that dV/dx be a constant, so

$$V = \int K_1\, dx = K_1 x + K_2$$

where K_1 and K_2 are constants. This is the general functional form for the potential associated with a uniform plane sheet of charge of infinite extent.

The problem we are now considering differs from the previous ones in that the sheet of charge divides space into two regions, namely, the region for which $x < 0$ and the region for which $x > 0$. In each region the potential must be of the above form, but the constants K_1 and K_2 may be different. We have postulated that, because of symmetry, V must depend only upon the distance from the charged sheet. Therefore K_1 in the region $x < 0$ must be equal to $-K_1$ in the region $x > 0$, and K_2 must be the same in both regions. This gives us

$$V = \begin{cases} K_1 x + K_2 & x > 0 \\ -K_1 x + K_2 & x < 0 \end{cases}$$

The electric intensity is now given by

$$\mathbf{E} = -\nabla V = \begin{cases} -\mathbf{u}_x K_1 & x > 0 \\ \mathbf{u}_x K_1 & x < 0 \end{cases}$$

We can evaluate K_1 by satisfying the boundary condition

$$D_n^{(1)} - D_n^{(2)} = q_s$$

at the charged sheet. Taking region (1) to be $x > 0$ and region (2) to be $x < 0$, we have

$$q_s = \epsilon E_x^{(1)} - \epsilon E_x^{(2)} = -2\epsilon K_1$$

Substituting for K_1 into the equation for **E**, we have for the electric intensity

$$\mathbf{E} = \begin{cases} \mathbf{u}_x \dfrac{q_s}{2\epsilon} & x > 0 \\ -\mathbf{u}_x \dfrac{q_s}{2\epsilon} & x < 0 \end{cases} \tag{5-72}$$

Thus, the **E** field points perpendicularly outward from the charged sheet and is independent of the distance from the sheet. Substituting for

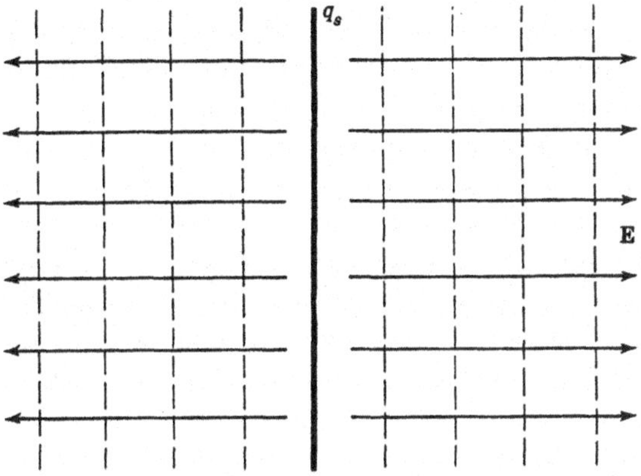

FIG. 5-16. The field lines and equipotential surfaces (dashed) for a sheet of uniform charge.

K_1 in the equation for V, we have for the potential

$$V = \begin{cases} \dfrac{-q_s}{2\epsilon} x + K_2 & x > 0 \\ \dfrac{q_s}{2\epsilon} x + K_2 & x < 0 \end{cases} \tag{5-73}$$

The constant K_2 does not enter into the field expression and is arbitrary.

The uniform sheet of charge is a source of infinite extent, and the potential, understandably, becomes infinite at infinite distance from the sheet. A convenient choice for K_2 is to set $K_2 = 0$, in which case we have the potential with respect to $x = 0$ given by

$$V = \frac{-q_s}{2\epsilon} |x| \tag{5-74}$$

The magnitude signs on x make Eq. (5-74) valid everywhere. The equipotential surfaces in this case are planes parallel to the charged sheet. Fig. 5-16 shows the field lines and equipotential surfaces (dashed lines) associated with the infinite sheet of uniform charge.

If the sheet of charge is not at the coördinate origin but is displaced along the x axis, we must modify Eq. (5-74) by replacing $|x|$ by the distance from the sheet to the field point. Denoting the coordinate of the sheet by x', and that of the field point by x, the general formula for the potential associated with a sheet of uniform charge density is

$$V(x) = \frac{-q_s(x')}{2\epsilon}|x - x'| \tag{5-75}$$

If we have several parallel sheets of charge, $q_{s,1}, q_{s,2}, \ldots, q_{s,N}$, situated at the positions $x'_1, x'_2, \ldots, x'_N$, we can use superposition and obtain an expression for the potential due to the entire system. Each sheet of charge has a potential of the form of Eq. (5-75), so the total potential due to several sheets of charge is given by

$$V(x) = \frac{-1}{2\epsilon}\sum_{n=1}^{N} q_{s,n}(x'_n)|x - x'_n| \tag{5-76}$$

If we have a continuous distribution of charge q_v which is independent of the y and z directions, we can superimpose the potentials from charge sheets of differential thickness. The surface charge contained within a representative increment dx' is $q_v(x')\,dx'$. The potential associated with each differential sheet of charge is given by Eq. (5-75) with q_s replaced by $q_v(x')\,dx'$. Summing over all differential sheets, we have the total potential given by

$$V(x) = \frac{-1}{2\epsilon}\int q_v(x')|x - x'|\,dx' \tag{5-77}$$

This is the one-dimensional scalar potential integral. We view Eq. (5-77) as the general form, with Eq. (5-76) the special case where q_v is zero except in discrete planes. The electric intensity is still given by $\mathbf{E} = -\nabla V$.

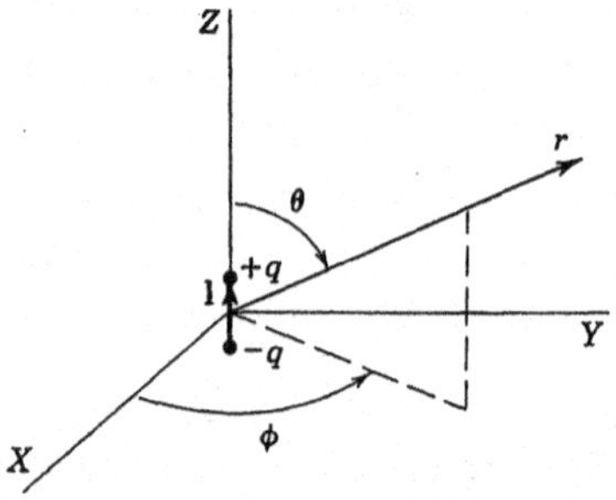

FIG. 5-17. A charge dipole $q\mathbf{l}$.

5-9. Charge Dipoles. Two point charges, of equal strength but of opposite sign, separated by a small distance, form an elementary *charge dipole*. Such a dipole is shown in Fig. 5-17, where the two charges $+q$ and $-q$ are separated by the vector increment $\mathbf{l}$ directed from $-q$ to $+q$. The coordinate system is chosen so that the dipole is located at the origin with $\mathbf{l}$ pointing in the $+z$ direction. The *electric dipole moment* is defined as the vector $q\mathbf{l}$. A *point dipole* is formed according to a limiting process $l \to 0$, $q \to \infty$, such that ql remains finite. Note that a dipole source is a vector source, for it has both magnitude and direction.

The electric potential associated with the dipole of Fig. 5-17 is determined according to Eq. (5-49). Considering the expanded view of the dipole given in Fig. 5-18, we see that the radius vector to the positive charge is $\mathbf{u}_z l/2$ and that the radius vector to the negative charge is $-\mathbf{u}_z l/2$. Therefore, the potential is given by

$$V(r,\theta,\phi) = \frac{1}{4\pi\epsilon}\left(\frac{q}{|\mathbf{r} - \mathbf{u}_z l/2|} - \frac{q}{|\mathbf{r} + \mathbf{u}_z l/2|}\right) \qquad (5\text{-}78)$$

If the field point (r,θ,ϕ) is far from the dipole, the radius vectors from the positive charge, the origin, and the negative charge, to the field point

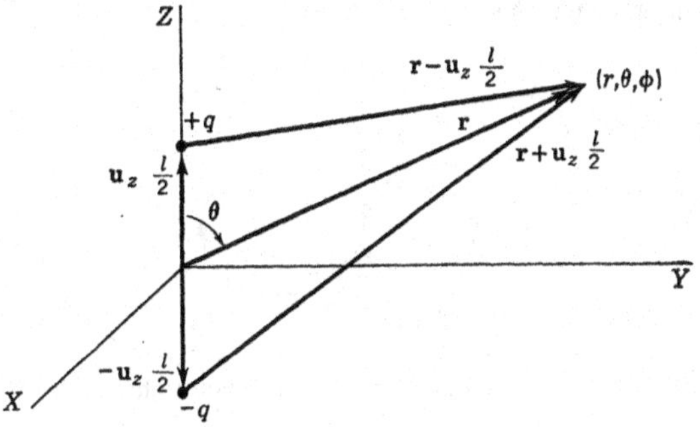

FIG. 5-18. Expanded view of the charge dipole.

become approximately parallel. Thus, if $r \gg l$, we have

$$\left|\mathbf{r} - \mathbf{u}_z\frac{l}{2}\right| \approx r - \frac{l}{2}\cos\theta$$

$$\left|\mathbf{r} + \mathbf{u}_z\frac{l}{2}\right| \approx r + \frac{l}{2}\cos\theta$$

In this case

$$V \approx \frac{q}{4\pi\epsilon}\left(\frac{1}{r - l/2\cos\theta} - \frac{1}{r + l/2\cos\theta}\right)$$

$$= \frac{q}{4\pi\epsilon}\left[\frac{l\cos\theta}{r^2 - (l/2\cos\theta)^2}\right]$$

If we proceed to the limit of a point dipole, that is, let $l \to 0$, the above equations become exact and the second term in the denominator vanishes.* This gives for the point dipole

$$V(r,\theta,\phi) = \frac{ql\cos\theta}{4\pi\epsilon r^2} \qquad (5\text{-}79)$$

which is also valid at large distances from a finite-size dipole. Note that V varies inversely as the square of the distance from the dipole, in con-

* The limiting process may also be accomplished by expanding Eq. (5-78) in a Maclaurin series in l and then letting $l \to 0$.

trast with the reciprocal-distance law of the point charge. Also, V is no longer spherically symmetric.

The electric intensity is now given by $\mathbf{E} = -\nabla V$. Performing the gradient operation on Eq. (5-79) in spherical coordinates, we have

$$\mathbf{E} = \frac{ql}{4\pi\epsilon r^3} (\mathbf{u}_r\, 2 \cos \theta + \mathbf{u}_\theta \sin \theta) \tag{5-80}$$

Thus, the electric intensity varies inversely with the cube of the distance from the dipole, in contrast with the inverse-square law of the point charge. A sketch of the lines of electric intensity associated with the point dipole is given in Fig. 5-19.

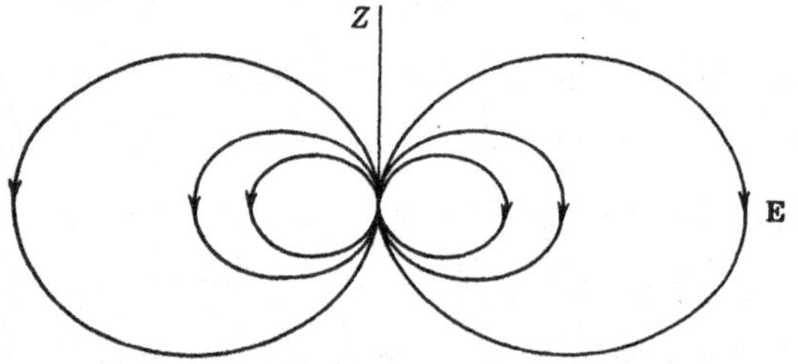

FIG. 5-19. The field associated with a point-charge dipole.

We can generalize our formulas to apply where $\mathbf{l}$ does not point in the z direction and where the dipole is not at the coordinate origin, as follows. Suppose we have at the point $\mathbf{r}'$ a dipole $q\mathbf{l}$ of arbitrary orientation. Eq. (5-79) still applies if θ is considered to be the angle between the direction of the dipole moment and the direction of the radius vector $\mathbf{r} - \mathbf{r}'$ from the dipole to the field point $\mathbf{r}$. From our elementary vector concepts, we know that $q\mathbf{l} \cdot (\mathbf{r} - \mathbf{r}') = ql|\mathbf{r} - \mathbf{r}'| \cos \eta$, where η is the angle between the direction of $\mathbf{l}$ and that of $\mathbf{r} - \mathbf{r}'$. Also, the r in the denominator of Eq. (5-79) must be considered as the distance from the dipole to the field point, $|\mathbf{r} - \mathbf{r}'|$. Thus, the expression for the potential associated with a dipole of arbitrary location and orientation is given by

$$V(\mathbf{r}) = \frac{q\mathbf{l}(\mathbf{r}') \cdot (\mathbf{r} - \mathbf{r}')}{4\pi\epsilon|\mathbf{r} - \mathbf{r}'|^3} \tag{5-81}$$

If we have several dipoles, we can obtain the total potential by a superposition of the potentials from each one. The dipole concept can be generalized to include volume distributions, surface distributions, and line distribution of dipole moments.

A charge quadrupole is formed when two dipoles are nearly coincident and their moments are antiparallel, as illustrated in Fig. 5-20. In the limit, as the separation of the charges goes to zero and $q \to \infty$, we obtain the point quadrupole. The dipole fields then cancel, giving rise to a quadrupole field. The potential in this case vanishes as r^{-3} as $r \to \infty$, in contrast with the r^{-2} law for dipoles, and the r^{-1} law for point charges. A detailed analysis of quadrupoles requires the use of tensor analysis. If two nearly coincident quadrupoles are oriented so that their quadrupole fields cancel, we obtain the elementary charge octupole. This procedure can be carried on indefinitely, obtaining higher-order charge multipoles. The potential vanishes inversely as r to one-higher-integer power for each successive multipole.

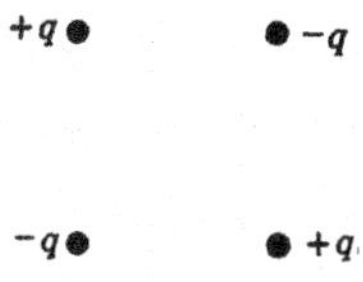

FIG. 5-20. A charge quadrupole.

Suppose we have a distribution of charge of finite extent in a homogeneous medium of infinite extent. If the distribution has a net charge, at great enough distance from the charge the field reduces to that of a point charge. If the distribution has zero net charge, but a net dipole moment, then at great enough distance from the charge the field reduces to that of a charge dipole. If both the net charge and the net dipole moment are zero, the quadrupole field becomes dominant at large distances, and so on. In fact, given an arbitrary charge distribution contained within a sphere of radius R, the same field external to $r = R$ can be produced by a distribution of multipoles at the coordinate origin. An excellent treatment of the general theory of multipoles can be found in Stratton.*

5-10. Reciprocity. It is evident from Eq. (5-48) that the potential (with respect to infinity) at $\mathbf{r}$ due to a point charge at $\mathbf{r}'$ is equal to the potential at $\mathbf{r}'$ due to a point charge of the same magnitude at $\mathbf{r}$. This is the concept of reciprocity as applied to two point charges. We shall now consider a more general statement of reciprocity for electrostatic fields.

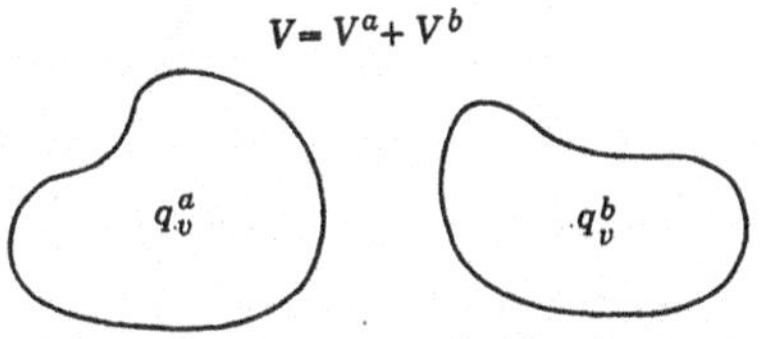

FIG. 5-21. Two charge distributions acting together.

Suppose we consider two sets of charges q_v^a and q_v^b, as represented by Fig. 5-21. If they are immersed in a homogeneous medium of infinite extent, the "partial potentials" V^a and V^b are given by Eq. (5-54) as applied to each set, that is,

$$V^a(\mathbf{r}) = \frac{1}{4\pi\epsilon} \iiint \frac{q_v^a(\mathbf{r}')}{|\mathbf{r} - \mathbf{r}'|} \, d\tau'$$

$$V^b(\mathbf{r}) = \frac{1}{4\pi\epsilon} \iiint \frac{q_v^b(\mathbf{r}')}{|\mathbf{r} - \mathbf{r}'|} \, d\tau'$$

The total potential of the two sets of charges acting simultaneously is, of course, the sum of the partial potentials. Taking the expression for

* J. A. Stratton, "Electromagnetic Theory," McGraw-Hill Book Company, Inc., New York, 1941, pp. 172–183.

V^b, multiplying it by q_v^a, and integrating over all space, we have

$$\iiint q_v^a V^b \, d\tau = \frac{1}{4\pi\epsilon} \iiint d\tau \iiint d\tau' \frac{q_v^a(\mathbf{r}) q_v^b(\mathbf{r}')}{|\mathbf{r} - \mathbf{r}'|}$$

Note that the right-hand side is symmetrical in $\mathbf{r}$ and $\mathbf{r}'$ since

$$|\mathbf{r} - \mathbf{r}'| = |\mathbf{r}' - \mathbf{r}|$$

Therefore, its value is the same if we interchange a and b, so

$$\iiint q_v^a V^b \, d\tau = \iiint q_v^b V^a \, d\tau \tag{5-82}$$

This is the statement of reciprocity for electrostatics. We have proved it for the charges in unbounded homogeneous media and their potentials with respect to infinity. It can, however, also be proved for nonhomogeneous media (see Prob. 5-24).

The reciprocity theorem can be used as a mathematical tool to "measure" the potential at a point. Suppose the source of the b field is a point charge of unit magnitude at the point $\mathbf{r}'$. We then have

$$\iiint V^a q_v^b \, d\tau = V^a(\mathbf{r}') \tag{5-83}$$

and

$$V^b(\mathbf{r}) = \frac{1}{4\pi\epsilon|\mathbf{r} - \mathbf{r}'|} \tag{5-84}$$

Now Eq. (5-82) reduces to

$$\iiint \frac{q_v^a(\mathbf{r}) \, d\tau}{4\pi\epsilon|\mathbf{r} - \mathbf{r}'|} = V^a(\mathbf{r}') \tag{5-85}$$

which is identical with the potential integral solution except that the source coordinates are now unprimed and the field coordinates primed. However, $|\mathbf{r} - \mathbf{r}'|$ is symmetrical with respect to primed and unprimed coordinates, so Eq. (5-85) is indeed identical with the potential integral. Thus, there is no mathematical difference in the two solutions, but there is a conceptual difference. In using the potential integral, we are summing over the partial potentials from the individual elements of the source. In using reciprocity, we place a unit point charge at the point where we want to evaluate the field, and evaluate Eq. (5-82). The reciprocity theorem is very closely related to the mathematical theorem called Green's theorem.*

An example should serve to illustrate the use of reciprocity as outlined above. Consider the charge dipole of the previous section to be the source of field a, and place a unit charge at an arbitrary field point to serve as the source of field b. This procedure is illustrated in Fig. 5-22. The integral of Eq. (5-85) now reduces to

$$V^a(\mathbf{r}') = \frac{q}{4\pi\epsilon r^+} - \frac{q}{4\pi\epsilon r^-}$$

* *Ibid.*, p. 165.

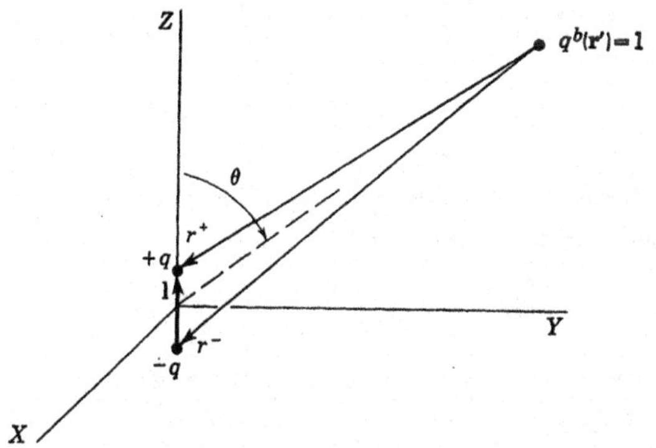

FIG. 5-22. Reciprocity used to evaluate the field of a dipole.

But if $r' \gg l$, then

$$r^+ \approx r' - \frac{l}{2}\cos\theta \qquad r^- \approx r' + \frac{l}{2}\cos\theta$$

and the solution reduces to that of Sec. 5-9 except for r replaced by r'. It makes no difference whether we denote the field point by primed or unprimed coordinates; so the two solutions are identical.

We can also devise a mathematical test source to measure the components of $\mathbf{E}$ directly, using the reciprocity integral. Suppose the source of the b field is a point dipole of unit dipole moment at $\mathbf{r}'$. An expanded view of this test dipole at a point in the a field is shown in Fig. 5-23. The integral on the right of Eq. (5-82) now becomes

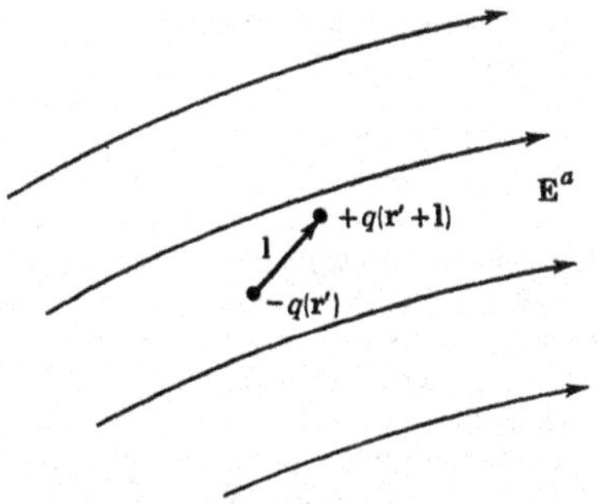

FIG. 5-23. A unit test dipole, $ql = 1$.

$$\begin{aligned}\iiint V^a q_v{}^b\, d\tau &= -qV^a(\mathbf{r}') + qV^a(\mathbf{r}' + \mathbf{l}) \\ &= -qV^a(\mathbf{r}') + q\left(1 + l\frac{\partial}{\partial l}\right)V^a(\mathbf{r}') \\ &= ql\frac{\partial}{\partial l}V^a(\mathbf{r}') = \frac{\partial}{\partial l}V^a(\mathbf{r}') \\ &= -E_l{}^a(\mathbf{r}')\end{aligned}$$

The potential of this test dipole is given by Eq. (5-81) with $ql = 1$, that is

$$V^b(\mathbf{r}) = \frac{\mathbf{u}_l \cdot (\mathbf{r} - \mathbf{r}')}{4\pi\epsilon|\mathbf{r} - \mathbf{r}'|^3}$$

Using these results in Eq. (5-82), we have

$$E_l^a(\mathbf{r}') = -\mathbf{u}_l \cdot \iiint \frac{q_v^a(\mathbf{r})(\mathbf{r} - \mathbf{r}')}{4\pi\epsilon|\mathbf{r} - \mathbf{r}'|^3}\, d\tau \tag{5-86}$$

Since the direction of $\mathbf{u}_l$ is arbitrary, Eq. (5-86) is valid for all components, so the vectors themselves must be equal. Replacing $(\mathbf{r} - \mathbf{r}')$ by $(\mathbf{r}' - \mathbf{r})$ and combining components, we have

$$\mathbf{E}^a(\mathbf{r}') = \iiint \frac{q_v^a(\mathbf{r})(\mathbf{r}' - \mathbf{r})}{4\pi\epsilon|\mathbf{r}' - \mathbf{r}|^3}\, d\tau \tag{5-87}$$

Thus, we arrive at the same formula that we called the field integral, Eq. (5-12), except that $\mathbf{r}$ and $\mathbf{r}'$ are interchanged.

QUESTIONS FOR DISCUSSION

5-1. A force field that is irrotational is also called "conservative." What is the basis for this latter term?

5-2. Is the radius vector a "point function," that is, does it have a definite value associated with each point in space?

5-3. Is $\mathbf{r} - \mathbf{r}'$ a point function?

5-4. Does the field integral, Eq. (5-12), hold for points within q_v, or does it apply only external to q_v?

5-5. Discuss the concepts of directional derivative and gradient.

5-6. Eq. (5-23) gives the maximum rate of increase of the scalar function at a point. What is the maximum rate of decrease?

5-7. Given an arbitrary well-behaved vector function, can we always find a scalar function such that its gradient is the original vector function?

5-8. Under what conditions is the voltage between a pair of terminals independent of the path between the terminals?

5-9. Is the potential difference of Eq. (5-37) reference-positive at terminal a or at terminal b?

5-10. Poisson's equation is a partial differential equation of what order? Is it linear? Is it homogeneous?

5-11. What are the restrictions on the medium surrounding the point charge if Eq. (5-47) is to be valid?

5-12. Is the scalar potential defined by Eq. (5-54) the only one that correctly gives the electric intensity according to $\mathbf{E} = -\nabla V$?

5-13. Is the following statement true? Equipotential surfaces are always either closed surfaces or surfaces that extend to infinity.

5-14. Discuss the concept of a two-dimensional charge distribution.

5-15. Is an infinitely long filament of charge on which the charge varies along its length a two-dimensional charge distribution?

5-16. If a long line charge were approximated by an infinite line charge, what restrictions must be placed on the use of the solution?

5-17. Given two parallel line charges of unequal strength, are the equipotential surfaces cylinders?

5-18. Discuss the concept of a one-dimensional charge distribution.

5-19. Of what shape are the equipotential surfaces of two parallel planes of uniform surface-charge density?

5-20. In Eq. (5-77), the charge distribution is a volume density, yet the integral is a line integral. Explain this.

5-21. Give some physical limitations to the concept of a point charge dipole.

5-22. Interpret Eq. (5-82) in terms of work.

5-23. What is the difference in philosophy between Eqs. (5-54) and (5-85)?

5-24. What is the difference in philosophy between Eqs. (5-12) and (5-87)?

PROBLEMS

5-1. Determine the field of the finite filament of charge, Fig. 5-10, using the field integral, Eq. (5-14). Check the answer in the plane $z = 0$ with that obtained from Eq. (5-57) according to $\mathbf{E} = -\nabla V$.

5-2. Determine the field due to point charges $q_1 = 0.01$ coulomb at (0,1,0) and $q_2 = 0.05$ coulomb at (1,0,0). Evaluate $\mathbf{D}$ at (0,0,1).

5-3. Given the scalar function $w = xe^y \sin z$, determine the gradient of w. Find the directional derivatives of w in the x, y, and z directions. Evaluate these directional derivatives at (1,1,1) and find the maximum directional derivative at this point.

5-4. Evaluate the gradient of the following scalar functions: (*a*) $w = x$, (*b*) $w = \phi$, (*c*) $w = r$, (*d*) $w = r\theta$.

5-5. Evaluate the action of the gradient of $w = x^2ye^z$ between the points (1,1,1) and (2,1,1).

5-6. Show that $\nabla \cdot \mathbf{r} = 3$ and $\nabla \times \mathbf{r} = 0$.

5-7. Show that $\nabla \cdot (\mathbf{r} - \mathbf{r}') = -\nabla' \cdot (\mathbf{r} - \mathbf{r}')$, where ∇' indicates differentiation with respect to the primed coordinates.

5-8. Show that $\nabla(1/|\mathbf{r} - \mathbf{r}'|) = -\nabla'(1/|\mathbf{r} - \mathbf{r}'|)$.

5-9. Prove Eq. (5-34) by expanding it in rectangular components.

5-10. Given $V = e^x \sin y$, determine the electric intensity. Evaluate the voltage between (1,1,0) and (1,2,0) by actual integration and by taking the potential difference.

5-11. Given that the electric intensity is

$$\mathbf{E} = \mathbf{u}_x 4 \sin (4x) \sin (3y)e^{-5z} - \mathbf{u}_y 3 \cos (4x) \cos (3y)e^{-5z} + \mathbf{u}_z 5 \cos (4x) \sin (3y)e^{-5z}$$

find a possible scalar potential.

5-12. Derive Eq. (5-44), starting from the equations for divergence and gradient in spherical coordinates.

5-13. Solve Prob. (5-2) by first obtaining the scalar potential and then differentiating it.

5-14. Consider the point charges $q_1 = 0.05$ coulomb and $q_2 = -0.01$ coulomb situated 0.1 meter apart, as shown in Fig. 5-8. Graphically construct several equipotential lines in the $z = 0$ plane, and determine the potential of each line.

5-15. Sketch the $\mathbf{E}$ lines for the field of Prob. 5-14.

5-16. Determine the scalar potential along the z axis for the disk of Fig. 5-3. Differentiate the potential to obtain the field, and compare the answer with Eq. (5-15).

5-17. Obtain the potential and the field of the "line dipole" formed by two infinitely-long parallel line charges separated by an incremental distance l. The two filaments are charged oppositely with magnitude q_l. Consider the dipole moment $q_l l$ to point along the x axis, and take the limit $q_l l \to$ constant as $l \to 0$.

5-18. If all charge lies between $x = -a$ and $x = a$ in a one-dimensional problem, show that

$$v(a,-a) = V(a) - V(-a) = \frac{1}{\epsilon} \int_{-a}^{a} x q_v \, dx$$

5-19. Determine the potential in the one-dimensional problem where $q_v = x$, $-a < x < a$, and zero elsewhere. Determine the potential difference between $x = a$ and $x = -a$.

5-20. Determine the potential of the charge quadrupole shown in Fig. 5-24. Take the limit $ql^2 \rightarrow$ constant as $l \rightarrow 0$. (All charges are in the $x = 0$ plane.)

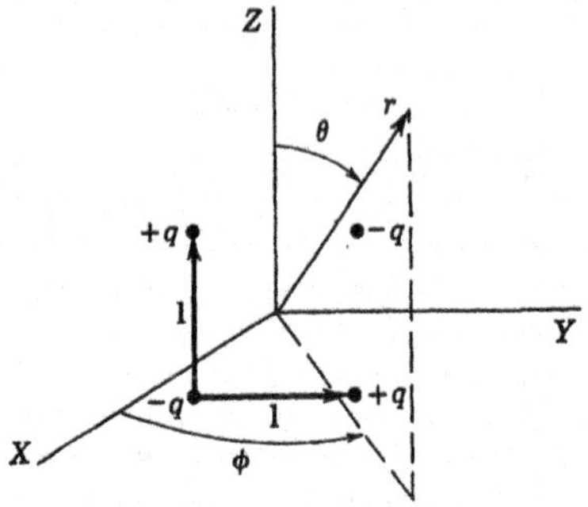

FIG. 5-24. Geometry for Prob. 5-20.

5-21. Show that the potential of the point dipole of Fig. 5-17 is given by $V = -l \frac{\partial}{\partial z} (V_p)$, where V_p is the potential of a single point charge. Give a physical interpretation of this result.

5-22. Evaluate V for the filament of charge, Fig. 5-10, using reciprocity.

5-23. Evaluate $\mathbf{E}$ along the z axis for the disk of charge, Fig. 5-3, using reciprocity.

5-24. Prove that the reciprocity relationship, Eq. (5-82), is valid for nonhomogeneous media providing

$$\oint\!\!\oint (V^a\mathbf{D}^b - V^b\mathbf{D}^a) \cdot d\mathbf{s} = 0$$

over the surface surrounding the region of integration. (*Hint:* Start directly from the field equations as they apply to the a and b fields.)

5-25. Given a sphere of radius a, with surface-charge density $q_s = q_0 \cos \theta$, where q_0 is a constant, determine the scalar potential in integral form. Evaluate this potential for the case $r \gg a$, where r is the radius to the field point (use spherical coordinates).

CHAPTER 6

THE FIELD OF STEADY CURRENTS

6-1. Introduction. Currents that do not vary with respect to time are called *steady currents*. Charge and current are related by the equation of continuity, Eq. (2-45), which states that charge must accumulate unless the current is divergenceless. We are not yet prepared to deal with charges that vary with respect to time, so we shall restrict consideration to the case for which

$$\nabla \cdot \mathbf{J} = 0 \tag{6-1}$$

In other words, there is no accumulation of charge due to the action of a divergenceless steady current. The magnetic intensity and the magnetic flux density associated with divergenceless steady currents are independent of time. The field associated with a divergenceless steady electric current is called a *magnetostatic field*.

The fundamental equations of magnetostatic fields are obtained from the general relationships by setting all time derivatives equal to zero. The equations for the magnetostatic field are therefore

$$\nabla \times \mathbf{H} = \mathbf{J} \qquad \nabla \cdot \mathbf{B} = 0 \qquad \mathbf{B} = \mu \mathbf{H} \tag{6-2}$$

Note that the first equation implies Eq. (6-1), for $\nabla \cdot \nabla \times \mathbf{H} = 0$ is a mathematical identity. In this chapter, we shall investigate solutions to Eqs. (6-2) only for linear, homogeneous, and isotropic media of infinite extent. The integral form of the first of Eqs. (6-2) is obtained from Eq. (4-16) by setting the time derivative equal to zero. Thus

$$\oint \mathbf{H} \cdot d\mathbf{l} = i \tag{6-3}$$

where

$$i = \iint \mathbf{J} \cdot d\mathbf{s}$$

is valid for every closed contour and is equivalent to $\nabla \times \mathbf{H} = \mathbf{J}$. The current i is the current through any surface bounded by the contour of Eq. (6-3), for the current is divergenceless. This divergenceless character of the current is equivalent to Kirchhoff's current law as applied to d-c circuits. The integral equivalent to the second of Eqs. (6-2) is just Eq. (3-13), which, repeated here for convenience, is

$$\oint\!\!\!\oint \mathbf{B} \cdot d\mathbf{s} = 0 \tag{6-4}$$

This applies to all closed surfaces.

6-2. Magnetic Vector Potential. The magnetic flux density in the magnetostatic field has no divergence, that is, $\nabla \cdot \mathbf{B} = 0$ everywhere. We recall the vector identity $\nabla \cdot \nabla \times \mathbf{A} = 0$, that is, the vector $\nabla \times \mathbf{A}$ is divergenceless. The converse of this is also true:* *If a vector has zero divergence, it is the curl of some other vector function.* Since $\nabla \cdot \mathbf{B} = 0$, we may let

$$\mathbf{B} = \nabla \times \mathbf{A} \tag{6-5}$$

where **A** is called a *magnetic vector potential.* However, since $\nabla \times \nabla w = 0$ is a mathematical identity, we also have $\mathbf{B} = \nabla \times (\mathbf{A} + \nabla w)$, where w is *any* scalar. We can therefore form any number of vector potentials, differing from each other by the gradient of a scalar, all of which give the same **B** according to Eq. (6-5). Thus, **A** is not unique, and we say that it is indeterminate by a gradient. A quantity that is of interest to us is the magnetic flux, given by

$$\psi = \iint \mathbf{B} \cdot d\mathbf{s} \tag{6-6}$$

It was shown in Sec. 3-5 that ψ was the same for all surfaces bounded by a given contour. In other words, the magnetic flux linking a given closed contour is unique. Substituting Eq. (6-5) into Eq. (6-6), and applying Stokes' theorem, we obtain

$$\psi = \oint \mathbf{A} \cdot d\mathbf{l} \tag{6-7}$$

Thus, the magnetic flux linking a closed contour is the line integral of *any* magnetic vector potential about that contour.

If we substitute from $\mathbf{B} = \mu\mathbf{H}$ into the equation $\nabla \times \mathbf{H} = \mathbf{J}$, we have

$$\nabla \times (\mu^{-1}\mathbf{B}) = \mathbf{J}$$

Expressing **B** in terms of a vector potential, as in Eq. (6-5), we obtain

$$\nabla \times (\mu^{-1}\nabla \times \mathbf{A}) = \mathbf{J} \tag{6-8}$$

This is the partial differential equation for determining magnetic vector potentials. If we now specialize Eq. (6-8) to homogeneous isotropic media, μ becomes a constant and can be removed from under the curl operation. This gives

$$\nabla \times \nabla \times \mathbf{A} = \mu\mathbf{J} \tag{6-9}$$

which is the basic equation for **A** in homogeneous isotropic media.

Let us now look at the operation $\nabla \times \nabla \times \mathbf{A}$. Expanding it out in

* L. Brand, "Vector and Tensor Analysis," John Wiley & Sons, Inc., New York, 1947, p. 201.

rectangular coordinates, we have for the x component

$$\begin{aligned}(\nabla \times \nabla \times \mathbf{A})_x &= \frac{\partial^2 A_y}{\partial x \partial y} - \frac{\partial^2 A_x}{\partial y^2} - \frac{\partial^2 A_x}{\partial z^2} + \frac{\partial^2 A_z}{\partial z \partial x} \\ &= -\left(\frac{\partial^2 A_x}{\partial x^2} + \frac{\partial^2 A_x}{\partial y^2} + \frac{\partial^2 A_x}{\partial z^2}\right) + \frac{\partial}{\partial x}\left(\frac{\partial A_x}{\partial x} + \frac{\partial A_y}{\partial y} + \frac{\partial A_z}{\partial z}\right) \\ &= -\nabla^2 A_x + (\nabla \nabla \cdot \mathbf{A})_x\end{aligned}$$

where ∇^2 is the usual Laplacian operator. The above equation with x replaced by y gives the y component of $\nabla \times \nabla \times \mathbf{A}$, and, similarly, x replaced by z gives the z component. If we define *in rectangular components* the operation

$$\nabla^2 \mathbf{A} = \mathbf{u}_x \nabla^2 A_x + \mathbf{u}_y \nabla^2 A_y + \mathbf{u}_z \nabla^2 A_z \tag{6-10}$$

then we can write concisely

$$\nabla \times \nabla \times \mathbf{A} = \nabla(\nabla \cdot \mathbf{A}) - \nabla^2 \mathbf{A} \tag{6-11}$$

For **A** expressed in other than rectangular components, Eq. (6-11) can be considered as the *defining equation* for the operation $\nabla^2 \mathbf{A}$. In cylindrical coordinates, the ρ and ϕ components of $\nabla^2 \mathbf{A}$ are *not* equal to $\nabla^2 A_\rho$ and $\nabla^2 A_\phi$; similarly for spherical components. This can be readily shown by expanding Eq. (6-11) in cylindrical components and by comparing the result with the ∇^2 operator given in Sec. 5-4. We can, however, express the ∇^2 operators of Eq. (6-10) in cylindrical or spherical coordinates, without affecting the validity of Eq. (6-10). In other words, it is not the coordinate system that is important; it is the manner in which **A** is divided into components that matters. Another way of putting it is that the unit vectors associated with the component vectors must be constant vectors for $(\nabla^2 \mathbf{A})_i = \nabla^2 A_i$ to be true.

Returning now to the equation for **A**, Eq. (6-9), we can use the relationship of Eq. (6-11) and write

$$\nabla(\nabla \cdot \mathbf{A}) - \nabla^2 \mathbf{A} = \mu \mathbf{J} \tag{6-12}$$

It is recalled that we have interpreted only the curl of **A**, according to Eq. (6-5). Since the divergence of a vector involves different derivatives than does the curl operation, we might suspect that one can arbitrarily specify the divergence of **A**. A convenient choice would be for $\nabla \cdot \mathbf{A}$ to vanish, for then the first term of Eq. (6-12) would vanish. Let us see if this can be done. Recall that **A** is indeterminate by the gradient of an arbitrary scalar. Suppose we have a vector **A** which correctly gives the field according to Eq. (6-5), but it has a finite divergence. We can then form a new vector potential **A′** according to

$$\mathbf{A}' = \mathbf{A} + \nabla w$$

where we desire $\nabla \cdot \mathbf{A}' = 0$. Taking the divergence of the above equation, we want

$$\nabla \cdot \mathbf{A}' = 0 = \nabla \cdot \mathbf{A} + \nabla \cdot \nabla w$$

The operation on w is recognized as the Laplacian, so we have

$$\nabla^2 w = -\nabla \cdot \mathbf{A}$$

as an equation for determining w. The gradient of a solution to the above Poisson's equation added to the original vector potential gives a new vector potential which is divergenceless. We therefore lose no generality if we specify that $\mathbf{A}$ be divergenceless. It is usually, but not always, desirable to do so. *If* we choose

$$\nabla \cdot \mathbf{A} = 0 \tag{6-13}$$

then Eq. (6-12) reduces to

$$\nabla^2 \mathbf{A} = -\mu \mathbf{J} \tag{6-14}$$

This is known as the *vector Poisson equation.* Written in rectangular components, the component equations are

$$\begin{aligned} \nabla^2 A_x &= -\mu J_x \\ \nabla^2 A_y &= -\mu J_y \\ \nabla^2 A_z &= -\mu J_z \end{aligned} \tag{6-15}$$

Thus, the choice of Eq. (6-13) reduces Eq. (6-12) to a set of three scalar Poisson equations involving the *rectangular* components of $\mathbf{A}$. This is the principal advantage of choosing $\mathbf{A}$ to be divergenceless. In regions of vanishing current, Eq. (6-14) reduces to

$$\nabla^2 \mathbf{A} = 0 \tag{6-16}$$

which is called the *vector Laplace equation.* The rectangular components of $\mathbf{A}$ in a region of vanishing current satisfy the scalar Laplace equation, obtained by setting the right-hand terms of Eqs. (6-15) equal to zero.

6-3. The Vector Potential Integral. We now have a magnetic vector potential specified by the vector Poisson equation, each rectangular component of which is a scalar Poisson equation. In the preceding chapter, we had the scalar electric potential specified by the scalar Poisson equation, and we constructed various solutions to it. Since the equations relating rectangular components of $\mathbf{A}$ to corresponding rectangular components of $\mu\mathbf{J}$ have the same mathematical form as the equation relating V to q_v/ϵ, the solutions take the same mathematical form. We view the three-dimensional scalar potential integral, Eq. (5-54), as giving the general solution to the scalar Poisson equation. We need only replace V

by A_x, A_y, or A_z, and q_v/ϵ by μJ_x, μJ_y, or μJ_z; and we obtain

$$\begin{aligned} A_x(\mathbf{r}) &= \frac{\mu}{4\pi} \iiint \frac{J_x(\mathbf{r}')}{|\mathbf{r} - \mathbf{r}'|} \, d\tau' \\ A_y(\mathbf{r}) &= \frac{\mu}{4\pi} \iiint \frac{J_y(\mathbf{r}')}{|\mathbf{r} - \mathbf{r}'|} \, d\tau' \\ A_z(\mathbf{r}) &= \frac{\mu}{4\pi} \iiint \frac{J_z(\mathbf{r}')}{|\mathbf{r} - \mathbf{r}'|} \, d\tau' \end{aligned} \tag{6-17}$$

We can now combine Eqs. (6-17) into a single vector equation

$$\mathbf{A}(\mathbf{r}) = \frac{\mu}{4\pi} \iiint \frac{\mathbf{J}(\mathbf{r}')}{|\mathbf{r} - \mathbf{r}'|} \, d\tau' \tag{6-18}$$

which conveys the same information as the above component equations. Eq. (6-18) is called the *magnetic vector potential integral.* The magnetic flux density is obtained according to $\mathbf{B} = \nabla \times \mathbf{A}$, giving us a formal solution for the magnetostatic field in an unbounded homogeneous region.

Eq. (6-18) is viewed as the general solution for the magnetostatic field of electric currents. We shall consider the solutions for surface distributions of current and for line distributions of current to be special cases of the general solution, obtainable by appropriate limiting processes. For surface distributions $\mathbf{J}_s$, the volume integral reduces to a surface integral, and we have

$$\mathbf{A}(\mathbf{r}) = \frac{\mu}{4\pi} \iint \frac{\mathbf{J}_s(\mathbf{r}')}{|\mathbf{r} - \mathbf{r}'|} \, ds' \tag{6-19}$$

For filaments of current i, the volume integral reduces to a line integral, and we have

$$\mathbf{A}(\mathbf{r}) = \frac{\mu}{4\pi} \int \frac{i(\mathbf{r}')}{|\mathbf{r} - \mathbf{r}'|} \, d\mathbf{l}' \tag{6-20}$$

where the direction of the current is indicated by $d\mathbf{l}'$. There are no point distributions of electric current giving rise to a magnetostatic field, for these would violate the condition $\nabla \cdot \mathbf{J} = 0$. If we had a combination of volume distributions of electric current, surface distributions, and line distributions, the total magnetic vector potential would be simply the superposition of terms of the form of Eqs. (6-18) to (6-20).

Let us now look at the interpretation of the vector potential integrals. Integration is a summation process; so we can view the vector potential integrals as a summation over all current elements. A current element $i\mathbf{l}$ is formed by a current i extending over an incremental length $\mathbf{l}$, as defined in Sec. 2-4. Consider a current element, as shown in Fig. 6-1.

It is a consequence of the equation of continuity that charge must accumulate at the ends of such a current element. If the element carries a constant current, then the charge must be changing linearly with respect to time. The electric field associated with these charges must, therefore, also change with time, and the field is *not* magnetostatic. However, if the current element is a part of a total current distribution that satisfies $\nabla \cdot \mathbf{J} = 0$, then the charges associated with adjacent current elements cancel each other, and the electric field vanishes. This can be seen in the simple example of a number of current elements oriented into a loop, as shown in Fig. 6-2. Thus, even though the cur-

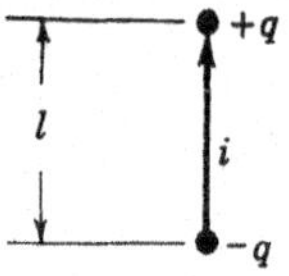

FIG. 6-1

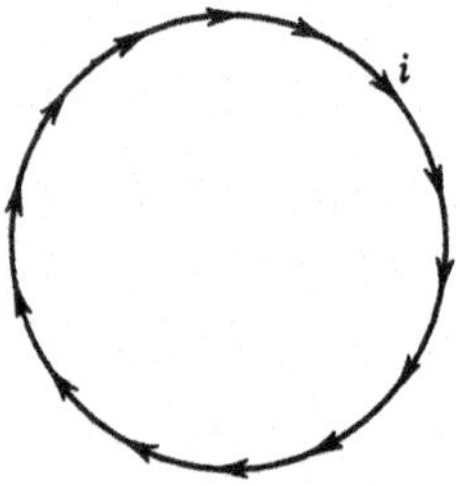

FIG. 6-2

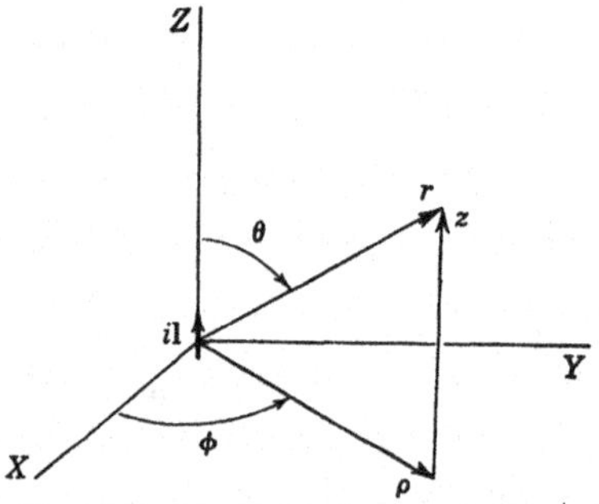

FIG. 6-1. A current element.
FIG. 6-2. A current loop formed of current elements.

FIG. 6-3. A z-directed current element at the coordinate origin.

rent element taken by itself does not produce a magnetostatic field, the integration over all differential current elements of a complete distribution of current does produce a magnetostatic field, so long as the total current is divergenceless.

Suppose we assume that our formulas, which were derived for divergenceless steady currents, are still valid even if the steady current is not divergenceless. (When we treat time-varying fields, we shall see that this is true.) Consider a current element of differential length, situated at the coordinate origin and pointing along the z axis. This is shown in Fig. 6-3. To evaluate the vector potential, we shall view the element as a line current and use Eq. (6-20). The current is taken to be constant over the element; so i can be removed from under the integral sign. The element is taken to be of incremental length; so the radius vector from all parts of the element to the field point is essentially the same, and this too can be removed from under the integral sign. We are left with just the integral over the line element, which is simply its length. The current points in the z direction, so there will be only an A_z. The magnetic vector potential associated with the current element is therefore specified by

$$A_z = \frac{\mu i l}{4\pi r} \tag{6-21}$$

Thus, **A** is a vector which always points in the direction of the current element, and its magnitude varies inversely with the distance from the element.

The magnetic field associated with the current element is given by $\mathbf{B} = \nabla \times \mathbf{A}$. Let us use cylindrical coordinates to evaluate this curl operation, in which case $r = \sqrt{\rho^2 + z^2}$. We find that there is only a ϕ component of **B**, given by

$$B_\phi = -\frac{\partial A_z}{\partial \rho} = \frac{\mu i l}{4\pi r^2}\left(\frac{\rho}{r}\right)$$

But $\rho/r = \sin\theta$ (see Fig. 6-3), and $H_\phi = B_\phi/\mu$; so the magnetic intensity is given by

$$H_\phi = \frac{il}{4\pi r^2}\sin\theta \tag{6-22}$$

Thus, **H** is entirely in the ϕ direction, varies with the sine of the angle between the direction of the current and the direction to the field point, and varies inversely with the square of the distance from the element to the field point. The magnitude of **H** is proportional to the moment of the current element. Remember that we have obtained this result by methods not yet justified; so we must accept it on faith. All that we can be sure of is that the field obtained by summing Eqs. (6-21) or (6-22) over all current elements of a divergenceless source will be correct.

We can put Eq. (6-21) into a form independent of our choice of coordinates by noting that **A** is in the direction of **l** and that the radius vector from the element to the field point is, in general, $\mathbf{r} - \mathbf{r}'$. Thus, for a current element located at the arbitrary point $\mathbf{r}'$, the vector potential at the point **r** is given by

$$\mathbf{A}(\mathbf{r}) = \frac{\mu i(\mathbf{r}')\mathbf{l}}{4\pi|\mathbf{r} - \mathbf{r}'|} \tag{6-23}$$

If we have a filament of current, summing over the **A**'s due to each differential length results in Eq. (6-20). Similar reasoning holds for volume and surface distributions of current, resulting in Eqs. (6-18) and (6-19). It is clear, in relation to line currents, that the vector potential integral is just a superposition of contributions from differential current elements; however, it is not so apparent that the same is true for volume and surface distributions. For a volume distribution of currents, we use Eq. (6-18), which is a summation over elements $\mathbf{J}\,d\tau$. The volume element $d\tau$ can be viewed as a cross section perpendicular to **J**, denoted by ds, times a length in the direction of **J**, denoted by dl. Then $\mathbf{J}\,d\tau = \mathbf{J}\,ds\,dl$. But

$J\,ds$ is the total current flowing through the volume element, and dl is the length over which it extends. Therefore, $J\,d\tau$ is simply the magnitude of the moment contained in $d\tau$, and the direction of the moment is that of $\mathbf{J}$. The differential current moments are thus $\mathbf{J}\,d\tau$, and when we use the vector potential integral we are summing over the contributions from all current elements. A similar argument holds for the case of surface current distributions.

Failure to understand the basic meaning of the vector potential integral formula might lead one to the incorrect conclusion that **A** and **J** can be expressed in terms of any components, say cylindrical or spherical. In most vector formulas, a division of the vectors into cylindrical or spherical components is permissible because all vectors happen to be at the same point in space. In the vector potential integral, however, we are evaluating **A** at one point, due to **J** at other points. The usual formulas for vector addition in terms of components apply to vectors at different points in space only if the unit coordinate vectors are all constant vectors, which is the case in rectangular coordinates. Since integration is fundamentally the same as addition, it follows that the vector potential integral can be separated into components of the same form only for rectangular components. There is still another way of looking at this question. This is by noting that **A** due to each current element points in the same direction as the current element itself. The ϕ direction, for example, at a given current element is not necessarily the same as the ϕ direction at the field point; so A_ϕ does not depend only on J_ϕ. If the current is given in cylindrical components, we should first express it in rectangular components and then apply the vector potential integral. In general, any equation which relates vectors at different points may be separated into rectangular component equations, but only in special cases is it permissible to separate such an equation into other coordinate components.

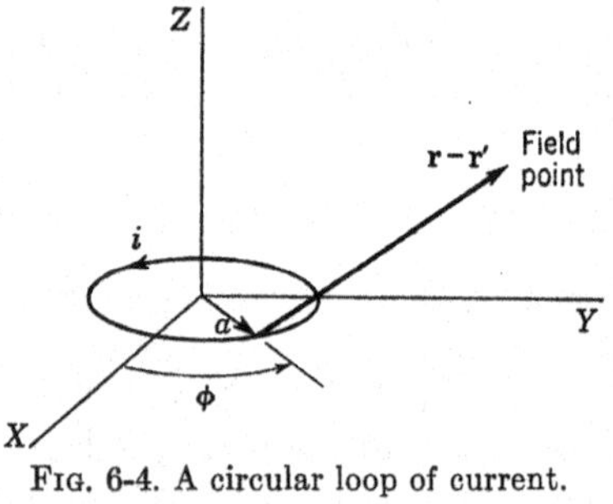

FIG. 6-4. A circular loop of current.

To illustrate the vector potential integral, consider a circular loop of constant current i and define coordinates as shown in Fig. 6-4. The source is a filament of current; so we use Eq. (6-20) with source coordinates specialized to ($\rho' = a,\ \phi',\ z' = 0$). The differential vector line element of source is

$$d\mathbf{l}' = \mathbf{u}_\phi a\,d\phi'$$

We should use rectangular components in the vector potential integral. The rectangular components of $\mathbf{u}_\phi$ are

$$\mathbf{u}_\phi = -\mathbf{u}_x \sin\phi' + \mathbf{u}_y \cos\phi'$$

We can now write the rectangular components of Eq. (6-20) as

$$A_x(\mathbf{r}) = \frac{-\mu a i}{4\pi} \int_0^{2\pi} \frac{\sin \phi'}{|\mathbf{r} - \mathbf{r}'|} d\phi'$$
$$A_y(\mathbf{r}) = \frac{\mu a i}{4\pi} \int_0^{2\pi} \frac{\cos \phi'}{|\mathbf{r} - \mathbf{r}'|} d\phi' \tag{6-24}$$

Note that $A_z = 0$, since there is no z-directed current. Depending upon the coordinate system chosen to represent the field point, we have

$$|\mathbf{r} - \mathbf{r}'| = \sqrt{(x - a \cos \phi')^2 + (y - a \sin \phi')^2 + z^2}$$
or
$$= \sqrt{\rho^2 + a^2 - 2\rho a \cos (\phi - \phi') + z^2} \tag{6-25}$$
or
$$= \sqrt{r^2 + a^2 - 2ra \sin \theta \cos (\phi - \phi')}$$

where the field coordinates are (x,y,z), (ρ,ϕ,z), or (r,θ,ϕ).

Before considering the further evaluation of Eqs. (6-24), let us formally complete the solution for the field, according to $\mathbf{B} = \nabla \times \mathbf{A}$. The curl operation can be carried out in any coordinate system, but for now let us arbitrarily pick rectangular coordinates. Since there is no A_z, the components of $\mathbf{B}$ are given by

$$B_x = -\frac{\partial A_y}{\partial z} \qquad B_y = \frac{\partial A_x}{\partial z} \qquad B_z = \frac{\partial A_y}{\partial x} - \frac{\partial A_x}{\partial y}$$

The differentiations can be performed under the integration signs of Eqs. (6-24), for they do not involve the integration variable. Only the denominator of the integrand is a function of the field coordinates. Using the first of Eqs. (6-25), we have

$$\frac{\partial}{\partial x}\left(\frac{1}{|\mathbf{r} - \mathbf{r}'|}\right) = \frac{a \cos \phi' - x}{|\mathbf{r} - \mathbf{r}'|^3}$$
$$\frac{\partial}{\partial y}\left(\frac{1}{|\mathbf{r} - \mathbf{r}'|}\right) = \frac{a \sin \phi' - y}{|\mathbf{r} - \mathbf{r}'|^3}$$
$$\frac{\partial}{\partial z}\left(\frac{1}{|\mathbf{r} - \mathbf{r}'|}\right) = \frac{-z}{|\mathbf{r} - \mathbf{r}'|^3}$$

and the formal solution for the field becomes

$$B_x = \frac{\mu a i z}{4\pi} \int_0^{2\pi} \frac{\cos \phi'}{|\mathbf{r} - \mathbf{r}'|^3} d\phi'$$
$$B_y = \frac{\mu a i z}{4\pi} \int_0^{2\pi} \frac{\sin \phi'}{|\mathbf{r} - \mathbf{r}'|^3} d\phi' \tag{6-26}$$
$$B_z = \frac{\mu a i}{4\pi} \int_0^{2\pi} \frac{\sin \phi'(a \sin \phi' - y) + \cos \phi'(a \cos \phi' - x)}{|\mathbf{r} - \mathbf{r}'|^3} d\phi'$$

Evaluation of the field in terms of tabulated functions will be considered after the following special case.

Along the axis of the loop (the z axis) the field can be evaluated quite simply. Specializing Eq. (6-25) to the z axis, we have

$$|\mathbf{r} - \mathbf{r}'| = \sqrt{a^2 + z^2}$$

which is no longer a function of ϕ' and which can now be removed from under the integration sign. The first two integrals of Eqs. (6-26) then vanish, for they are integrals over a complete cycle of a sinusoid. The last of Eqs. (6-26) becomes

$$\begin{aligned} B_z &= \frac{\mu i a^2}{4\pi(a^2 + z^2)^{3/2}} \int_0^{2\pi} (\sin^2 \phi' + \cos^2 \phi')\, d\phi' \\ &= \frac{\mu i a^2}{2(a^2 + z^2)^{3/2}} \end{aligned} \tag{6-27}$$

Thus, along the axis of the loop, **B** points in the axial direction and varies with the distance from the loop (z) according to Eq. (6-27).

The general solution can be expressed more concisely if we use cylindrical coordinates and components for the field point. The cylindrical components of **A** can be obtained by a straightforward transformation from Eqs. (6-24), but the following is a simpler procedure. There is no A_z; so there will be only ϕ and ρ components. We note that the problem has rotational symmetry about the z axis; so A_ρ and A_ϕ must be independent of ϕ. We therefore need evaluate A_ρ and A_ϕ only for one specific value of ϕ. If we pick the field point to lie in the $y = 0$ plane, that is, $\phi = 0$, we have

$$\begin{aligned} A_x(\rho,0,z) &= A_\rho(\rho,\phi,z) \\ A_y(\rho,0,z) &= A_\phi(\rho,\phi,z) \end{aligned} \tag{6-28}$$

In other words, the x and ρ components and the y and ϕ components coincide in this plane. However, in the plane $y = 0$, the x component of **A** vanishes, since the integrand of the first of Eqs. (6-24) is antisymmetrical about $\phi' = \pi$. Another way of looking at it is to note that, equidistant from a field point in the $y = 0$ plane, there are equal but oppositely directed x components of current. Thus, $A_\rho = 0$ everywhere, and we have only an A_ϕ. From Eqs. (6-24), (6-25), and (6-28), with $\phi = 0$, it follows that

$$A_\phi(\rho,\phi,z) = \frac{\mu a i}{4\pi} \int_0^{2\pi} \frac{\cos \phi'\, d\phi'}{\sqrt{\rho^2 + a^2 - 2a\rho \cos \phi' + z^2}} \tag{6-29}$$

To express Eq. (6-29) in terms of tabulated functions, we let

$$\phi' = \pi + 2\theta$$

obtaining $$A_\phi = \frac{\mu a i}{\pi} \int_0^{\pi/2} \frac{(2 \sin^2 \theta - 1)\, d\theta}{\sqrt{(a + \rho)^2 + z^2 - 4a\rho \sin^2 \theta}}$$

Defining $$\alpha = 2\sqrt{\frac{a\rho}{(a + \rho)^2 + z^2}} \tag{6-30}$$

we can rearrange the above equation into the form

$$A_\phi = \frac{\mu i}{\alpha\pi}\sqrt{\frac{a}{\rho}}\left[\left(1 - \frac{\alpha^2}{2}\right)K(\alpha) - E(\alpha)\right] \tag{6-31}$$

where $$K(\alpha) = \int_0^{\pi/2} \frac{d\theta}{\sqrt{1 - \alpha^2 \sin^2\theta}}$$

and $$E(\alpha) = \int_0^{\pi/2} \sqrt{1 - \alpha^2 \sin^2\theta}\, d\theta \tag{6-32}$$

are the complete elliptic integrals of the first and second kind, respectively. Tables of these functions are available.*

Making use of the relationships

$$\begin{aligned} \frac{\partial K}{\partial \alpha} &= \frac{E}{\alpha(1 - \alpha^2)} - \frac{K}{\alpha} \\ \frac{\partial E}{\partial \alpha} &= \frac{E}{\alpha} - \frac{K}{\alpha} \end{aligned} \tag{6-33}$$

we can obtain the field components by expanding $\mathbf{B} = \nabla \times \mathbf{A}$ in cylindrical components. This gives

$$\begin{aligned} B_\rho &= \frac{\mu i z}{2\pi\rho}\frac{1}{\sqrt{(a + \rho)^2 + z^2}}\left[-K(\alpha) + \frac{a^2 + \rho^2 + z^2}{(a - \rho)^2 + z^2}E(\alpha)\right] \\ B_z &= \frac{\mu i}{2\pi}\frac{1}{\sqrt{(a + \rho)^2 + z^2}}\left[K(\alpha) + \frac{a^2 - \rho^2 - z^2}{(a - \rho)^2 + z^2}E(\alpha)\right] \end{aligned} \tag{6-34}$$

Specialization of this solution to the z axis results in Eq. (6-27).

6-4. The Biot-Savart Law. We can use the analysis of the current element in the preceding section to construct field integrals giving $\mathbf{H}$ directly in terms of current. We have the result that $\mathbf{H}$ due to a z-directed current element at the origin (see Fig. 6-3) is given by Eq. (6-22). We can generalize this solution to apply to a current element at an arbitrary coordinate point by using the radius-vector notation. The current element for Eq. (6-22) is in the z direction; so θ is the angle between the direction of the current element $i\mathbf{l}$ and that of the radius vector from the element to the field point $(\mathbf{r} - \mathbf{r}')$. This calls to mind the cross or vector product of two vectors, defined by Eq. (2-14). Thus, by definition

$$\mathbf{l} \times (\mathbf{r} - \mathbf{r}') = \mathbf{u}_n l|\mathbf{r} - \mathbf{r}'| \sin\eta$$

where η is the angle between $\mathbf{l}$ and $(\mathbf{r} - \mathbf{r}')$, and $\mathbf{u}_n$ is a unit vector perpendicular to both $\mathbf{l}$ and $(\mathbf{r} - \mathbf{r}')$ according to the right-hand rule. Returning now to Eq. (6-22), we note that $\mathbf{H}$ is in the ϕ direction, which is perpendicular to both the z direction (that of $\mathbf{l}$) and the r direction (that

* E. Jahnke and F. Emde, "Tables of Functions," Dover Publications, New York, 1945, Sec. V.

of $\mathbf{r} - \mathbf{r}'$). Therefore, $\mathbf{H}$ is in the direction of $\mathbf{u}_n$. Combining all of this information, we can now write Eq. (6-22) in the form

$$\mathbf{H}(\mathbf{r}) = \frac{i(\mathbf{r}')\,\mathbf{l} \times (\mathbf{r} - \mathbf{r}')}{4\pi|\mathbf{r} - \mathbf{r}'|^3} \tag{6-35}$$

which is independent of the coordinate origin. This is the general form giving $\mathbf{H}$ at a point $\mathbf{r}$ due to a current $i\mathbf{l}$ at a point $\mathbf{r}'$. Eq. (6-35) is known as the *Biot-Savart law*. When $\mathbf{r}' = 0$ and $\mathbf{l}$ points in the z direction, Eq. (6-35) reduces to Eq. (6-22).

For filaments of current, Eq. (6-35) would apply to each incremental length of current. Thus, in Eq. (6-35), we replace $i\mathbf{l}$ by $i\,d\mathbf{l}$, and integrate over all filaments of current. This gives

$$\mathbf{H}(\mathbf{r}) = \frac{1}{4\pi} \int \frac{i(\mathbf{r}')\,d\mathbf{l}' \times (\mathbf{r} - \mathbf{r}')}{|\mathbf{r} - \mathbf{r}'|^3} \tag{6-36}$$

which is the general expression for line currents. The prime on $d\mathbf{l}'$ emphasizes that integration is with respect to the primed, or source, coordinates. For a single filament of current, i must be constant over the entire loop if the field is to be magnetostatic. In this case, i can be removed from under the integration, and we have

$$\mathbf{H}(\mathbf{r}) = \frac{i}{4\pi} \oint \frac{d\mathbf{l}' \times (\mathbf{r} - \mathbf{r}')}{|\mathbf{r} - \mathbf{r}'|^3} \tag{6-37}$$

If we have more than one loop of current, Eq. (6-37) can be applied to each loop, and the fields can be superimposed.

We might also want to integrate over volume distributions of current, or over surface distributions. For volume distributions, each volume element $d\tau$ contains a current moment $\mathbf{J}\,d\tau$, so integrating Eq. (6-35) over all volume elements gives

$$\mathbf{H}(\mathbf{r}) = \frac{1}{4\pi} \iiint \frac{\mathbf{J}(\mathbf{r}') \times (\mathbf{r} - \mathbf{r}')}{|\mathbf{r} - \mathbf{r}'|^3}\,d\tau' \tag{6-38}$$

For surface current distributions, the appropriate form would be

$$\mathbf{H}(\mathbf{r}) = \frac{1}{4\pi} \iint \frac{\mathbf{J}_s(\mathbf{r}') \times (\mathbf{r} - \mathbf{r}')}{|\mathbf{r} - \mathbf{r}'|^3}\,ds' \tag{6-39}$$

If we have a combination of volume, surface, and line distributions of current, the total magnetic field would, of course, be a superposition of terms of the form of Eqs. (6-36), (6-38), and (6-39). These field integrals we shall call *Biot-Savart integrals.*

To illustrate the application of a Biot-Savart integral, let us again consider the problem of a current loop. The situation is illustrated by

Fig. 6-4. Since we have a single filament of current, we use Eq. (6-37). This involves the use of the vector $\mathbf{r} - \mathbf{r}'$, as well as its magnitude. We have for the radius vector from the origin to the field point

$$\mathbf{r} = \mathbf{u}_x x + \mathbf{u}_y y + \mathbf{u}_z z$$

and for the radius vector from the origin to a representative current element

$$\mathbf{r}' = \mathbf{u}_\rho a$$

Also

$$\mathbf{u}_\rho = \mathbf{u}_x \cos \phi' + \mathbf{u}_y \sin \phi'$$

so the radius vector from a current element to the field point is

$$\mathbf{r} - \mathbf{r}' = \mathbf{u}_x(x - a \cos \phi') + \mathbf{u}_y(y - a \sin \phi') + \mathbf{u}_z z$$

We have the explicit expression for $d\mathbf{l}'$ given by

$$d\mathbf{l}' = \mathbf{u}_\phi a\, d\phi' = -\mathbf{u}_x a \sin \phi'\, d\phi' + \mathbf{u}_y a \cos \phi'\, d\phi'$$

The cross product appearing in the Biot-Savart law is now

$$\begin{aligned} d\mathbf{l}' \times (\mathbf{r} - \mathbf{r}') = {} & \mathbf{u}_x z a \cos \phi'\, d\phi' + \mathbf{u}_y z a \sin \phi'\, d\phi' \\ & - \mathbf{u}_z[(y - a \sin \phi') a \sin \phi'\, d\phi' + (x - a \cos \phi') a \cos \phi'\, d\phi'] \end{aligned}$$

We can separate Eq. (6-37) into its rectangular components, giving

$$\begin{aligned} H_x &= \frac{aiz}{4\pi} \int_0^{2\pi} \frac{\cos \phi'}{|\mathbf{r} - \mathbf{r}'|^3}\, d\phi' \\ H_y &= \frac{aiz}{4\pi} \int_0^{2\pi} \frac{\sin \phi'}{|\mathbf{r} - \mathbf{r}'|^3}\, d\phi' \\ H_z &= \frac{ai}{4\pi} \int_0^{2\pi} \frac{\sin \phi'(a \sin \phi' - y) + \cos \phi'(a \cos \phi' - x)}{|\mathbf{r} - \mathbf{r}|^3}\, d\phi' \end{aligned} \tag{6-40}$$

Letting $\mathbf{B} = \mu \mathbf{H}$, we have the same equations as Eqs. (6-26). Note that the Biot-Savart integrals give us the solution we require more directly than the vector potential method, for the differentiation is already performed. However, the vector potential method has certain advantages in problems involving the calculation of inductance and it can also be extended to the treatment of time-varying fields. For these reasons we emphasize the vector potential method.

6-5. Two-dimensional Current Distributions. Current distributions that are unvarying with respect to one direction in space are called two-dimensional distributions. If we choose the z direction as the direction of no variation, the current distribution is the same for all values of z; that is, it depends only on x and y. In practice, of course, we cannot realize these infinitely long currents, but we can use the concept to approximate currents that extend large distances in a given direction,

Just as in the three-dimensional case, the partial differential equations relating A_x, A_y, and A_z to μJ_x, μJ_y, and μJ_z, are the same as the equation relating V to q_v/ϵ in electrostatics. Thus, the solutions for the three rectangular components of **A** have the same form as the equation for V given by Eq. (5-67). We can combine the resulting three component equations into a single vector equation, as we did in Sec. 6-3 for the three-dimensional case. The vector equation is therefore

$$\mathbf{A}(\boldsymbol{\varrho}) = \frac{-\mu}{2\pi} \iint \mathbf{J}(\boldsymbol{\varrho}') \log |\boldsymbol{\varrho} - \boldsymbol{\varrho}'| \, ds' \tag{6-41}$$

where the notation means the same as the notation used in Sec. 5-7 for scalar potential. The magnetic field is then determined by $\mathbf{B} = \nabla \times \mathbf{A}$.

The above formula can be viewed as expressing the general relationship; two-dimensional surface distributions and line distributions can be viewed as special cases obtained by suitable limiting processes. If we have electric currents confined to a two-dimensional surface, they appear as a line in the cross-sectional plane, and Eq. (6-41) reduces to a line integral. Thus

$$\mathbf{A}(\boldsymbol{\varrho}) = \frac{-\mu}{2\pi} \int \mathbf{J}_s(\boldsymbol{\varrho}') \log |\boldsymbol{\varrho} - \boldsymbol{\varrho}'| \, dl' \tag{6-42}$$

would be the appropriate form for the **A** due to two-dimensional surface distributions. If the electric currents were confined to discrete filaments of current, the general integral would reduce to a summation. The line currents would have to be z-directed; otherwise charge would collect as it does in the case of the three-dimensional current element. Thus, for a number of z-directed filaments of current of infinite length, we have only a z component of **A** given by

$$A_z(\boldsymbol{\varrho}) = \frac{-\mu}{2\pi} \sum_{n=1}^{N} i_n(\boldsymbol{\varrho}_n') \log |\boldsymbol{\varrho} - \boldsymbol{\varrho}_n'| \tag{6-43}$$

The line current is the two-dimensional point source. If in a given problem there exists a combination of line, surface, and volume distributions of current, the total magnetic vector potential will be just a superposition of the contributions from each type.

The simplest two-dimensional problem is the single z-directed filament of current, for which we have already determined **H** in Sec. 4-6. Let us look at this problem again in terms of the two-dimensional magnetic vector potential. If the current has strength i and exists along the z axis (situated at the two-dimensional coordinate origin), we have from Eq. (6-43)

$$A_z = \frac{-\mu i}{2\pi} \log \rho \tag{6-44}$$

The magnetic field is given by $\mathbf{B} = \nabla \times \mathbf{A}$, $\mathbf{H} = \mathbf{B}/\mu$. All z derivatives are zero, so the curl operation gives only a ϕ component of $\mathbf{B}$ or $\mathbf{H}$. This is

$$H_\phi = \frac{-1}{\mu}\frac{\partial A_z}{\partial \rho} = \frac{i}{2\pi\rho} \tag{6-45}$$

which is precisely the solution we obtained in Sec. 4-6.

Using Eq. (6-45) as a starting point, we can develop two-dimensional Biot-Savart type of field integrals. Recasting Eq. (6-45) into vector notation, independent of coordinates, we have

$$\mathbf{H}(\boldsymbol{\varrho}) = \frac{\mathbf{i}(\boldsymbol{\varrho}') \times (\boldsymbol{\varrho} - \boldsymbol{\varrho}')}{2\pi|\boldsymbol{\varrho} - \boldsymbol{\varrho}'|^2} \tag{6-46}$$

If we have a continuous distribution of currents, we can divide it into differential line elements and apply Eq. (6-46) to each element. The field integral is then obtained by summing over the contributions from all line elements. The derivation of Eq. (6-46) and of the field integrals is left to the reader as an exercise.

A two-dimensional problem of practical interest is that of a pair of parallel line currents of infinite length, of equal magnitude but opposite direction. (This approximates the two-wire transmission line.) The picture of this situation in a plane perpendicular to the current filaments is shown in Fig. 6-5. Note that the geometry is the same as the geometry for parallel line charges, Sec. 5-7. We shall use the two-dimensional magnetic vector potential to obtain the solution.

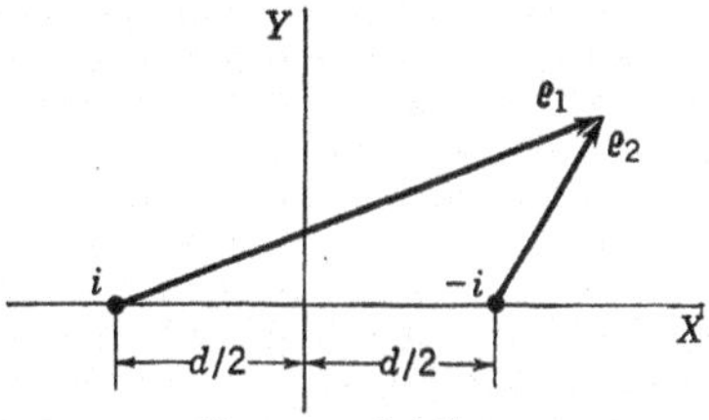

FIG. 6-5. Two parallel line currents.

Denoting the radial distance from the current i by ρ_1 and the radial distance from $-i$ by ρ_2, and applying Eq. (6-43), we have

$$A_z = \frac{\mu i}{2\pi}\log(\rho_2/\rho_1) \tag{6-47}$$

This has, of course, the same mathematical form as the equation for V due to parallel line charges. In terms of rectangular coordinates, the two radii are given by

$$\begin{aligned}\rho_1 &= \sqrt{\left(x + \frac{d}{2}\right)^2 + y^2}\\ \rho_2 &= \sqrt{\left(x - \frac{d}{2}\right)^2 + y^2}\end{aligned} \tag{6-48}$$

The magnetic field is then given by $\mathbf{B} = \nabla \times \mathbf{A}$, where there is only an

A_z. Thus, $B_z = 0$ and

$$B_x = \frac{\partial A_z}{\partial y} \qquad B_y = -\frac{\partial A_z}{\partial x} \tag{6-49}$$

where A_z is given above. We shall not bother to write down the explicit equations for **B**, for the manipulations are straightforward.

Let us now look at the behavior of the **B** lines, which by definition are lines in the direction of **B**. Since there is no B_z, these must lie entirely in planes transverse to the current filaments. Consider a differential line element of one such field line,

$$d\mathbf{l} = \mathbf{u}_x\, dx + \mathbf{u}_y\, dy$$

For this to be parallel to **B**, the ratio of the x component of $d\mathbf{l}$ to the x component of **B** must be the same as the ratio of the y component of $d\mathbf{l}$ to the y component of **B**. In equation form, this is

$$\frac{dx}{B_x} = \frac{dy}{B_y}$$

Substituting for **B** in terms of **A**, we have

$$\frac{\partial A_z}{\partial x}\, dx + \frac{\partial A_z}{\partial y}\, dy = 0 \tag{6-50}$$

Thus, the total differential of A_z is zero, which means that A_z = constant defines the **B** lines. Since A_z has the same mathematical form as V due to two equal but oppositely charged filaments, it follows that *the* **B** *lines from two line currents are coincident with the equipotential lines from two line charges.* Thus, a plot of the lines of **B** in this case would be identical with Fig. 5-15. The **E** lines associated with the above-mentioned two line charges are everywhere perpendicular to the equipotential lines. Therefore, if we have two filaments oppositely charged and carrying oppositely directed currents, then *the lines of* **B** *and those of* **E** *are everywhere mutually orthogonal.*

6-6. One-dimensional Current Distributions. A one-dimensional current distribution is a function of only one direction in space, being of infinite extent in the other two directions. We choose to designate the y and z directions as the directions of no variation; so the current distribution is a function of x only. We can use one-dimensional current distributions as an approximation to currents that extend for large distances transverse to the given direction of variation.

Again the partial differential equations relating rectangular components of **A** and μ**J** are the same as the equation relating V and q_v/ϵ, so we can obtain the solution by an interchange of symbols. The general solution, in terms of a volume distribution, has the same form as Eq. (5-77). In vector form, the magnetic vector potential is therefore

$$\mathbf{A}(x) = \frac{-\mu}{2}\int \mathbf{J}(x')|x - x'|\,dx' \tag{6-51}$$

where the notation is the same as the notation for scalar potentials. The magnetic field is, of course, given by $\mathbf{B} = \nabla \times \mathbf{A}$.

If the current is confined to surface distributions, that is, if it exists only in discrete sheets perpendicular to the x direction, the integration of Eq. (6-22) reduces to a summation. For example, if we had surface currents $\mathbf{J}_{s,1}, \mathbf{J}_{s,2}, \ldots, \mathbf{J}_{s,N}$, the formula for the magnetic vector potential would be

$$\mathbf{A}(x) = \frac{-\mu}{2}\sum_{n=1}^{N} \mathbf{J}_{s,n}(x_n')\,|x - x_n'| \tag{6-52}$$

Note that the current sheet is the one-dimensional point source; that is, it is defined by a discrete value of x.

To illustrate the one-dimensional field, let us consider a single sheet of current. We orient our coordinates so that the current flows in the y direction, the sheet being at $x = 0$. The vector potential then has only a y component, which, according to Eq. (6-52), is given by

$$A_y = \frac{-\mu}{2} J_s|x| \tag{6-53}$$

The magnetic field is now given by $\mathbf{B} = \nabla \times \mathbf{A}$ and $\mathbf{H} = \mathbf{B}/\mu$. Recalling that $|x|$ is shorthand for x, $x > 0$, and $-x$, $x < 0$, and noting that all derivatives except x derivatives must be zero, we have

$$H_z = -\frac{1}{\mu}\frac{\partial A_y}{\partial x} = \begin{cases} -\frac{1}{2}J_s & x > 0 \\ \frac{1}{2}J_s & x < 0 \end{cases} \tag{6-54}$$

as the only component of $\mathbf{H}$. Thus, the current sheet produces a constant $\mathbf{H}$ equal in magnitude to half the current strength. It is perpendicular to $\mathbf{J}_s$, pointing in opposite directions on opposite sides of the current sheet. This is shown in Fig. 6-6. Note that this solution could be obtained more directly by using symmetry conditions and the boundary condition for $\mathbf{H}$ at a current sheet, as called for in Prob. 4-21. We have used the one-dimensional magnetic vector potential merely to illustrate its application.

6-7. Current Dipoles. A loop of electric current of infinitesimal size is an elementary form of the *current dipole*, also called a *magnetic dipole*. The reason for the name "dipole" is that the magnetic field has the same functional form as the electric field due to a charge dipole. This we shall now show.

We have already treated the current loop of arbitrary size in Sec. 6-4, but we did not there proceed to the limit of a vanishingly small loop. We can, of course, take this limit in the general solutions, Eqs. (6-31) and (6-34), making use of the properties of the complete elliptic integrals. But let us start instead from the integral form of **A**, Eq. (6-29), so that we may show this limiting procedure in greater detail. It is convenient

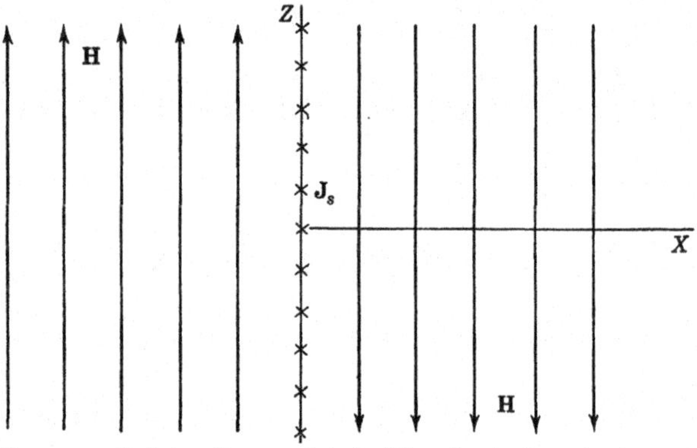

FIG. 6-6. The magnetic intensity associated with a sheet of constant current ($\mathbf{J}_s$ flows into the paper).

at this time to use the field point in spherical coordinates, so we can rewrite Eq. (6-29) as

$$A_\phi(r,\theta,\phi) = \frac{\mu a i}{4\pi} \int_0^{2\pi} \frac{\cos \phi' \, d\phi'}{\sqrt{r^2 + a^2 - 2ra \sin \theta \cos \phi'}} \tag{6-55}$$

We want to evaluate this as the loop becomes vanishingly small, that is, as $a \to 0$. The same procedure applies to the case of a finite-sized loop if we allow $a/r \to 0$, that is, if we wish to evaluate the field at distances large compared to the loop radius.

Eq. (6-55) can be rewritten as

$$A_\phi(r,\theta,\phi) = \frac{\mu a i}{4\pi r} \int_0^{2\pi} f\left(\frac{a}{r}\right) \cos \phi' \, d\phi' \tag{6-56}$$

where
$$f\left(\frac{a}{r}\right) = \left[1 - 2\frac{a}{r} \sin \theta \cos \phi' + \left(\frac{a}{r}\right)^2\right]^{-1/2}$$

Expanding $f(a/r)$ in a Maclaurin series, we have

$$f\left(\frac{a}{r}\right) = f(0) + f'(0)\left(\frac{a}{r}\right) + \frac{f''(0)}{2!}\left(\frac{a}{r}\right)^2 + \cdots$$
$$= 1 + \sin \theta \cos \phi' \left(\frac{a}{r}\right) + \cdots$$

Substituting this into Eq. (6-56) and integrating, we obtain

$$A_\phi(r,\theta,\phi) = \frac{\mu a i}{4\pi r}\left[0 + \pi \sin\theta\left(\frac{a}{r}\right) + \cdots\right]$$

When a/r is vanishingly small, all higher powers become negligible, and we have

$$A_\phi(r,\theta,\phi) = \frac{\mu i a^2}{4r^2}\sin\theta \tag{6-57}$$

This is the vector potential of a magnetic dipole. The field is obtained according to $\mathbf{B} = \mu\mathbf{H} = \nabla \times \mathbf{A}$, an operation which we perform in spherical coordinates. This gives

$$\mathbf{H} = \frac{ia^2}{4r^3}(\mathbf{u}_r 2\cos\theta + \mathbf{u}_\theta \sin\theta) \tag{6-58}$$

Note that this equation has the same functional form as Eq. (5-80) for **E** due to a charge dipole. The *magnetic moment* or *loop moment* of the current loop is defined as the current strength times the area of the loop. For a small circular loop, this moment is $\pi a^2 i$, so the magnetic field is proportional to the magnetic moment. A *point dipole* is formed according to a limiting process $a \to 0$, $i \to \infty$, such that $\pi a^2 i$ remains finite. The direction of the dipole moment is defined as the direction perpendicular to the plane of the loop, according to the right-hand rule. This is illustrated in Fig. 6-7. We denote the vector dipole moment by $i\mathbf{S}$, where **S** is a vector of magnitude equal to the area of the loop, and pointing in the direction of the axis of the loop.

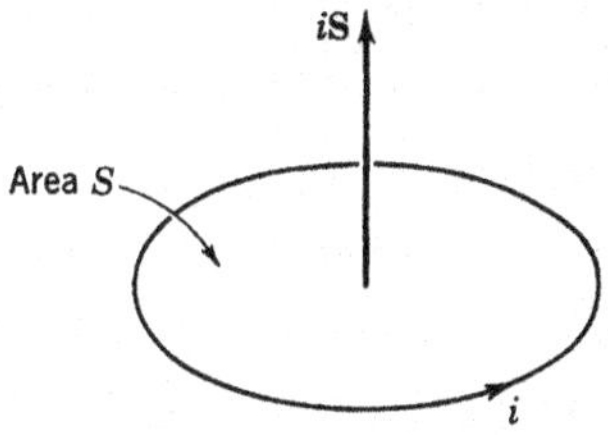

FIG. 6-7. The convention defining the vector moment of a current loop.

We now have a magnetic dipole that is mathematically analogous to the electric dipole of electrostatics. Starting from the magnetic dipole, we can construct higher multipoles having precisely the characteristics we found for the higher-order electrostatic multipoles. For example, a magnetic quadrupole can be obtained from two nearly coincident current loops oriented so that their dipole fields cancel. This is shown in Fig. 6-8. Just as the electric field of an electric quadrupole vanishes as r^{-4} as $r \to \infty$, so the magnetic field of the magnetic quadrupole vanishes as r^{-4} as $r \to \infty$, and so on for the higher multipoles. Thus, the only structural difference between the electrostatic field and the magneto-

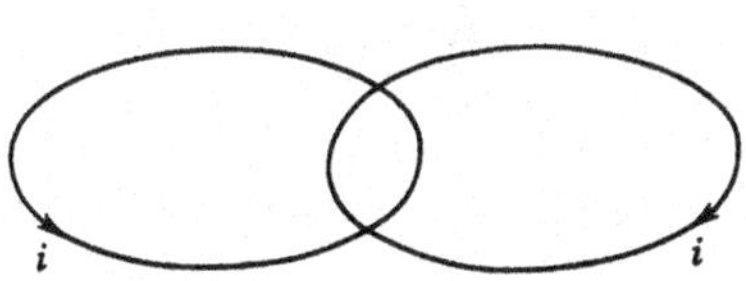

FIG. 6-8. A magnetic quadrupole.

static field is the point charge field of electrostatics. There is no magnetic "monopole" existing in nature.

Consider a current distribution of finite extent in a homogeneous medium of infinite extent. If the current has a net dipole moment, at great distances from the current the field reduces to the dipole field. If the net dipole moment is zero, but there exists a quadrupole moment, at great distances the field reduces to that of a magnetic quadrupole, and so on. In fact, the situation is entirely analogous to the electrostatic case, except for the absence of the monopole field.

Let us look again at the problem of a finite-sized loop of current and let us consider an alternative manner in which we might view it. Given a current loop, we divide it up into infinitesimal loops, as shown in Fig. 6-9. The sum of all the infinitesimal current loops is indeed the original large loop, for the currents in adjacent sides of the small loops cancel. We know precisely the field due to each infinitesimal loop, for this field has the form of Eq. (6-58). (Actually we have derived this result only for a circular loop, but in Sec. 7-7 we prove that the solution remains valid for infinitesimal loops of arbitrary shape.) By summing over all infinitesimal loops, we obtain the field of the original large loop. Note that each infinitesimal loop is a magnetic dipole; so *the field from a current loop is equivalent to the field from a surface distribution of magnetic dipoles.* Also, we can follow this procedure for any surface bounded by the current, so the loop of current is equivalent to a surface layer of magnetic dipoles over *any* surface bounded by the current.

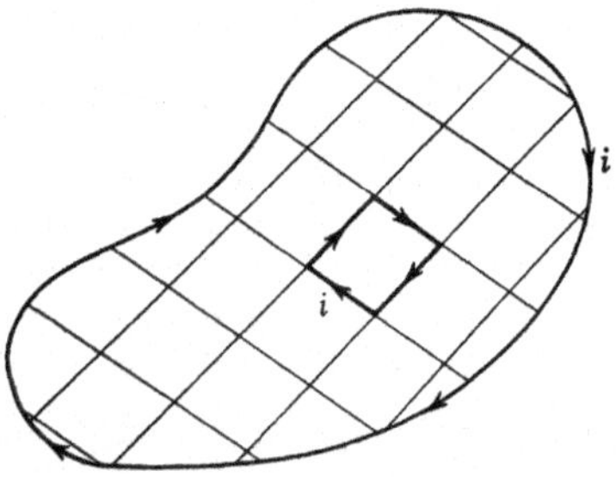

FIG. 6-9. A grid of infinitesimal current loops is equivalent to a single large loop.

6-8. The Solenoid. In Sec. 4-7, we obtained an approximate solution for a solenoid (coil). We wish now to consider the problem in a more precise manner, using the solution of this chapter. We shall, however, continue to make the working assumption that a solenoid is closely wound, so that the current effectively forms a sheet of current of density

$$J_s = \frac{Ni}{l} \tag{6-59}$$

where N is the number of turns, i is the current, and l is the length of the coil. For our analysis, we assume that there exists a cylinder of circumferentially directed surface current.

The solenoid is oriented with respect to the coordinate axes as shown in Fig. 6-10, the current sheet flowing in the ϕ direction. Each differential length of the solenoid can be viewed as a loop of current of magnitude $J_s\,dz'$, situated at the position $z = z'$. The field that each differential loop produces is that of Eqs. (6-26), with i replaced by $J_s\,dz'$ and z by $z - z'$. The total **B** produced by the coil can be found by integrating over the length of the solenoid. The resulting expressions for the field are rather complicated, so let us specialize our solution to the z axis.

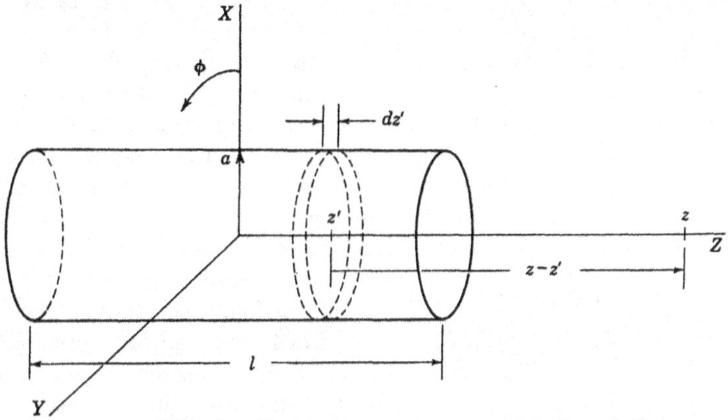

FIG. 6-10. The solenoid, showing parameters used in the solution.

For this case, replacing i by $J_s\,dz'$ and z by $z - z'$ in Eq. (6-27), we have from each differential loop of current

$$dB_z = \frac{\mu a^2 J_s\,dz'}{2[a^2 + (z - z')^2]^{3/2}}$$

along the z axis. Substituting for J_s from Eq. (6-59) and integrating over the length of the coil, we have along the axis

$$B_z = \frac{\mu a^2 N i}{2l} \int_{-l/2}^{l/2} \frac{dz'}{[a^2 + (z - z')^2]^{3/2}}$$

This can be integrated readily, and letting $\mathbf{H} = \mathbf{B}/\mu$, we obtain

$$H_z = \frac{Ni}{2l}\left[\frac{z + l/2}{\sqrt{a^2 + (z + l/2)^2}} - \frac{z - l/2}{\sqrt{a^2 + (z - l/2)^2}}\right] \qquad (6\text{-}60)$$

Since H_z is the only component of **H** along the axis of the solenoid, Eq. (6-60) gives the magnitude of **H** along the axis.

At the center of the solenoid, that is, at $z = 0$, Eq. (6-60) reduces to

$$H_z = \frac{Ni}{\sqrt{4a^2 + l^2}} \qquad (6\text{-}61)$$

If the length of the coil is much greater than its diameter, that is if $l \gg 2a$, then $H_z \approx Ni/l$, which is the solution we obtained for the ideal inductor in Sec. 4-7. At one end of the coil, that is, at $z = l/2$, Eq. (6-60) reduces to

$$H_z = \frac{Ni}{2\sqrt{a^2 + l^2}} \tag{6-62}$$

Thus, along the axis of the coil, *the magnitude of* **H** *at the end of a coil is exactly one-half of what it would be at the center of a coil twice as long,* if the ampere-turns per unit length are kept the same. If the coil is long and thin, then **H** at its ends is approximately half of the value of **H** at its center. Incidentally, we could have predicted this result from the superposition theorem, as follows. A coil can be considered to consist of two coils of half its length placed end to end. From symmetry considerations, we conclude that each of these half-coils must produce half of the total **H** that appears at the center of the original coil.

6-9. Reciprocity. The most familiar statement of reciprocity for magnetostatics deals with a pair of current loops, as follows. The magnetic flux through loop a due to loop b is equal to the magnetic flux through loop b due to loop a, if the currents in the two loops are equal. This is only one special interpretation of reciprocity for magnetostatics, a topic which we now consider.

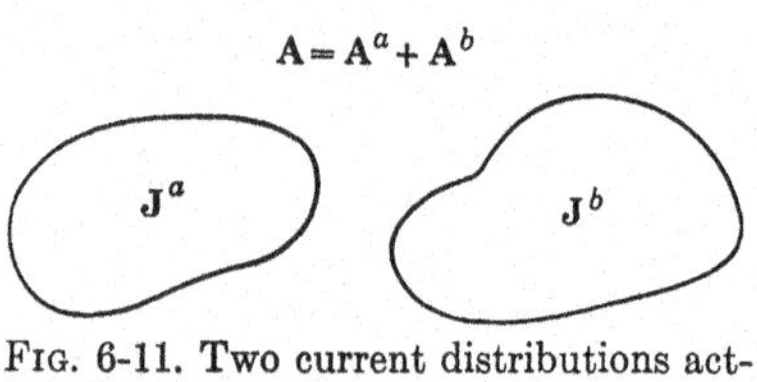

FIG. 6-11. Two current distributions acting simultaneously.

Suppose we have two sets of current, $\mathbf{J}^a$ and $\mathbf{J}^b$, acting simultaneously in a homogeneous medium of infinite extent. This is represented by Fig. 6-11. The partial potentials, $\mathbf{A}^a$ and $\mathbf{A}^b$, are those specified by Eq. (6-18) as applied to each set of current, that is

$$\mathbf{A}^a(\mathbf{r}) = \frac{\mu}{4\pi}\iiint \frac{\mathbf{J}^a(\mathbf{r}')}{|\mathbf{r} - \mathbf{r}'|}\,d\tau'$$

$$\mathbf{A}^b(\mathbf{r}) = \frac{\mu}{4\pi}\iiint \frac{\mathbf{J}^b(\mathbf{r}')}{|\mathbf{r} - \mathbf{r}'|}\,d\tau'$$

The total potential of the two sets of current acting together is, of course, the sum of the partial potentials. Taking the expression for $\mathbf{A}^b$, dot-multiplying it by $\mathbf{J}^a$, and integrating over all space, we have

$$\iiint \mathbf{J}^a \cdot \mathbf{A}^b\,d\tau = \frac{\mu}{4\pi}\iiint d\tau \iiint d\tau'\,\frac{\mathbf{J}^a(\mathbf{r}) \cdot \mathbf{J}^b(\mathbf{r}')}{|\mathbf{r} - \mathbf{r}'|}$$

The right-hand side is symmetrical in $\mathbf{r}$ and $\mathbf{r}'$, for $|\mathbf{r} - \mathbf{r}'| = |\mathbf{r}' - \mathbf{r}|$. Therefore, its value is unchanged if we interchange a and b; so

$$\iiint \mathbf{J}^a \cdot \mathbf{A}^b\,d\tau = \iiint \mathbf{J}^b \cdot \mathbf{A}^a\,d\tau \tag{6-63}$$

This is the reciprocity relationship for magnetostatics. We proved Eq. (6-63) only for steady currents in homogeneous media and their corresponding vector potential integrals. It can, however, also be proved for nonhomogeneous media (see Prob. 6-21).

The above reciprocity relationship seems to bear little resemblance to the reciprocity statement made in the first paragraph concerning flux. To specialize Eq. (6-63) to this case, consider both the a and b currents to be loops of constant current, i_a and i_b, respectively. The volume integrals reduce to line integrals, and we have

$$\iiint \mathbf{J}^a \cdot \mathbf{A}^b \, d\tau = i_a \oint \mathbf{A}^b \cdot d\mathbf{l}_a$$
$$\iiint \mathbf{J}^b \cdot \mathbf{A}^a \, d\tau = i_b \oint \mathbf{A}^a \cdot d\mathbf{l}_b$$

But, according to Eq. (6-7), the line integral of $\mathbf{A}$ about a closed contour is simply the magnetic flux linking the contour. Therefore, reciprocity for the case of two coils reduces to

$$i_a \psi_{b,a} = i_b \psi_{a,b} \tag{6-64}$$

where $\psi_{m,n}$ is the flux due to current m linking coil n.

In a manner analogous to the electrostatic case, we can use reciprocity as a mathematical tool to evaluate the vector potential or the field at a point. Suppose we choose the unit current element, $i\mathbf{l} = \mathbf{u}_l$, to be a mathematical "test" source. Let this current element be the source of the b field and let it be located at the point $\mathbf{r}'$. The reciprocity integral then reduces to

$$\iiint \mathbf{A}^a \cdot \mathbf{J}^b \, d\tau = \mathbf{u}_l \cdot \mathbf{A}^a(\mathbf{r}') = A_l{}^a(\mathbf{r}') \tag{6-65}$$

Thus, the unit current element measures the component of $\mathbf{A}$ (from an arbitrary a source) in the direction of the test current element. We can obtain any three components of $\mathbf{A}$ and thereby determine the vector itself, by orienting the test element successively in three noncoplanar directions. By orienting it in the x, y, and z directions successively, we can derive the formulas reciprocal to Eqs. (6-17). The formal mathematics of such a development parallels the development of Eq. (5-85) in the electrostatic case.

As an example, suppose that, using reciprocity, we solve for the vector potential from a current loop. We place a unit current element at a field point, as shown in Fig. 6-12. Denote the test element and field by b, and the original loop current and field by a. Using Eq. (6-65), we have for the reciprocity relationship

$$A_l{}^a = \iiint \mathbf{A}^b \cdot \mathbf{J}^a \, d\tau = i \oint \mathbf{A}^b \cdot d\mathbf{l}_a = i\psi_{b,a}$$

where $i = i_a$ is the loop current and $\psi_{b,a}$ is the flux due to the test b current linking the original a current loop. Thus, to obtain the various components of the **A** from the current loop, we need calculate only the flux from the test element which links the loop. Recall that **H** encircles a current element, with a null in the axial direction. It is evident that no flux links the loop for a z-directed test element; so $A_z^a = 0$. The symmetry is such that no net flux links the loop for a ρ-directed test element;

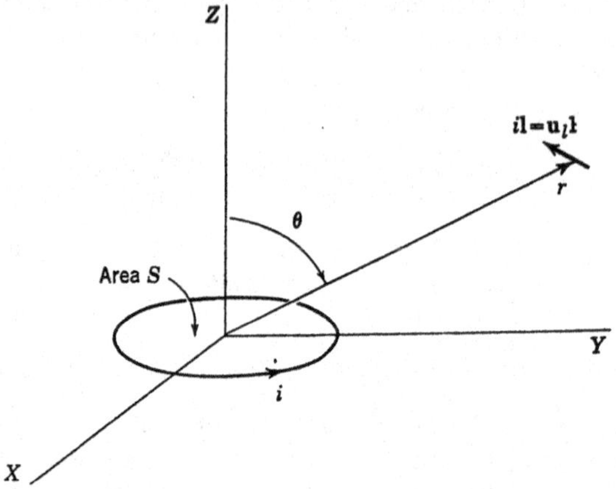

FIG. 6-12. Reciprocity used to evaluate the field of a current loop.

so $A_\rho^a = 0$. A ϕ-directed test element does produce a flux linking the loop; so

$$A_\phi^a = i\psi_{b,a} \text{ when } i_b\mathbf{l} = \mathbf{u}_\phi \tag{6-66}$$

This is the only cylindrical component of $\mathbf{A}^a$. For the sake of simplicity, let us take the test element distant from the loop, so that **H** due to the test element is approximately constant over the extent of the loop. According to Eq. (6-22), the field of the test element at the current loop is of magnitude

$$H^b = \frac{1}{4\pi r^2}$$

The component normal to the plane of the loop is

$$H_z^b = \frac{\sin\theta}{4\pi r^2}$$

The flux intercepted by the loop is the normal component of **B** times the area of the loop, or

$$\psi_{b,a} = \mu H_z^b S = \frac{\mu S \sin\theta}{4\pi r^2}$$

where S is the area of the loop. Finally, substituting this back into

Eq. (6-66), we have

$$A_\phi{}^a = \frac{\mu i S \sin \theta}{4\pi r^2} \tag{6-67}$$

which is identical with Eq. (6-57), with $\pi a^2 = S$.

If we use a unit magnetic dipole (current loop) as the test source instead of the current element, the reciprocity relationship can be used to measure (mathematically) the magnetic field directly. Taking the test source to be $i_b\mathbf{S} = \mathbf{u}_s$, we have

$$\begin{aligned} \iiint \mathbf{A}^a \cdot \mathbf{J}^b \, d\tau &= i_b \oint \mathbf{A}^a \cdot d\mathbf{l} = i_b \psi_{a,b} \\ &= i_b \mathbf{S} \cdot \mathbf{B}^a = B_s{}^a \end{aligned} \tag{6-68}$$

Therefore, according to Eq. (6-63), the component of **B** due to the original a sources in the direction of the test-loop moment is

$$B_s{}^a = \iiint \mathbf{A}^b \cdot \mathbf{J}^a \, d\tau \tag{6-69}$$

where $\mathbf{A}^b$ is the vector potential of the test loop. By orienting the test loop successively in the x, y, and z directions, we obtain the rectangular components of **B** and thereby determine the vector field. This procedure results in a formula identical with the Biot-Savart law, Eq. (6-38).

QUESTIONS FOR DISCUSSION

6-1. A field that is divergenceless is also called "solenoidal." What is the basis for this latter term?

6-2. What is the procedure for solving a problem in which both static charges and steady currents exist?

6-3. Given an arbitrary well-behaved vector field, can we always find a vector function whose curl is the original vector field?

6-4. Would you say that the lack of uniqueness of the magnetic vector potential is an advantage or a disadvantage in problem solving?

6-5. The general equation for the magnetic vector potential, Eq. (6-9), is a partial differential equation. Of what order is it? Is it linear? Is it homogeneous?

6-6. Would Eq. (6-14) separated into cylindrical components be a set of three scalar Poisson equations?

6-7. Could Eq. (6-18) be extended to nonhomogeneous regions by placing μ under the integral sign?

6-8. Which of Eqs. (6-2) are violated by the current element, Fig. 6-1?

6-9. Why is it basically necessary to use rectangular components when applying Eq. (6-18)?

6-10. Discuss the relationship of the Biot-Savart solution to the vector potential solution.

6-11. Could an infinitely long, closely wound solenoid be viewed as a two-dimensional current distribution?

6-12. Does A_z = constant define the **B** lines for any two-dimensional problem having only z-directed current?

6-13. Show that the x component of a one-dimensional divergenceless current distribution must be a constant (x being the direction of permissible variation). The integral of Eq. (6-51) then diverges; so the only permissible constant is zero. Thus, the current can have no x component.

6-14. Give some physical limitations to the concept of a point magnetic dipole.

6-15. The magnetic moment of a current distribution is defined as the moment of the point dipole giving the same distant field. What is the magnetic moment of a solenoid of N turns, radius a, and current i?

6-16. What are some of the limitations to the procedure of approximating a coil of wire by a current sheet?

6-17. Discuss the philosophy of using reciprocity to solve magnetostatic problems.

PROBLEMS

6-1. Show that the two vector potentials

$$\mathbf{A}_1 = \mathbf{u}_x \cos y + \mathbf{u}_y \sin x$$

and

$$\mathbf{A}_2 = \mathbf{u}_y(\sin x + x \sin y)$$

give the same **B** field. Show that this **B** is divergenceless. Using Eq. (6-9), show that both **A**'s result from the same current distribution in vacuum. Which one of the **A**'s satisfies Eq. (6-14)?

6-2. Evaluate $\nabla^2\mathbf{A}$ in cylindrical coordinates and components.

6-3. Evaluate $\nabla^2\mathbf{A}$ in spherical coordinates and components.

6-4. Determine the vector potential of a z-directed filament of constant current of length L, located along the z axis and centered at the origin.

6-5. Determine the **B** field from the results of Prob. 6-4.

6-6. Solve for the **H** field due to the current of Prob. 6-4 using the Biot-Savart integral.

6-7. Determine the vector potential of a square loop of constant current centered at the origin and lying in the $z = 0$ plane. (*Hint:* Use the results of Prob. 6-4.)

6-8. In the circular-loop problem, choose the field point to be in the $x = 0$ plane and write the equations corresponding to Eqs. (6-28). Using Eqs. (6-24), show that your equations give $A_\rho = 0$ and that the expression for A_ϕ reduces to Eq. (6-29).

6-9. Using Eqs. (6-31) and (6-33), derive Eqs. (6-34).

6-10. On a sphere of radius R, we have a surface distribution of current $\mathbf{J}_s = \mathbf{u}_\phi J_0$, where J_0 is a constant. Determine **B** at the center of the sphere. Determine **B** a large distance from the sphere.

6-11. Determine the vector potential for the two-dimensional problem of a z-directed current of width w in the x direction and of infinitesimal thickness in the y direction. Take the sheet to be centered at the origin.

6-12. Derive Eq. (6-46) and construct the appropriate two-dimensional Biot-Savart integral for volume distributions of current.

6-13. Set up the electric scalar potential integral for a ring of constant linear charge density at $\rho = a$, $z = 0$. Evaluate the potential in terms of complete elliptic integrals.

6-14. Set up the magnetic vector potential integral for a ring of z-directed current elements at $\rho = a$, $z = 0$, as shown in Fig. 6-13. Take the elements to be of incremental length l and sufficiently close together to be approximated by a sheet of surface current J_s. Show that the integral is the same as the integral in Prob. 6-13. Evaluate the magnetic vector potential in terms of complete elliptic integrals.

6-15. Given the one-dimensional current distribution

$$\mathbf{J} = \begin{cases} \mathbf{u}_z J_0 x & |x| < 1 \\ 0 & |x| > 1 \end{cases}$$

where J_0 is a constant, determine the magnetic intensity.

6-16. Given a disk of infinitesimal thickness and radius a, supporting a circumferential current $\mathbf{J}_s = \mathbf{u}_\phi J_0$, determine the magnetic vector potential in integral form. Use spherical coordinates with the z axis perpendicular to the center of the disk. Evaluate the potential for the case $r \gg a$.

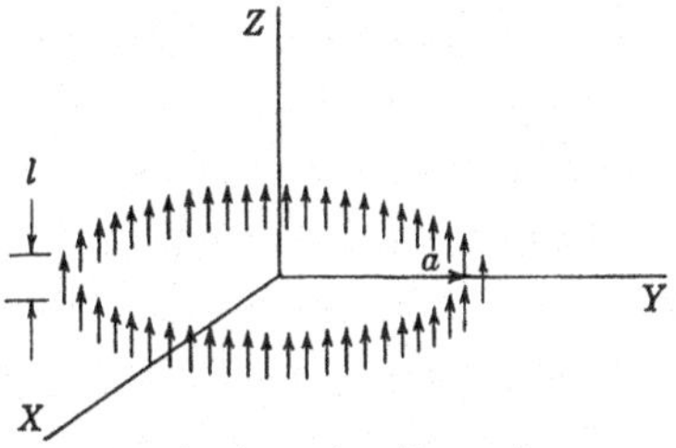

FIG. 6-13. A ring of z-directed current elements.

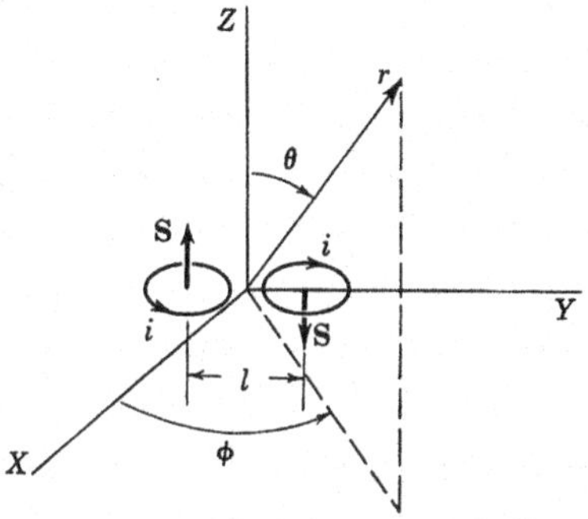

FIG. 6-14. Magnetic quadrupole for Prob. 6-17.

6-17. Determine the **B** field due to the magnetic quadrupole shown in Fig. 6-14. Take the limit $i \to \infty$, $S \to 0$, $l \to 0$, such that the product iSl remains finite.

6-18. Construct a graph of the ratio of H at the end of a solenoid to H at the center (along the axis) vs. the ratio $l/2a$ (length to diameter) for values up to $l/2a = 10$.

6-19. Solve Prob. 6-4 using reciprocity.

6-20. Solve Prob. 6-5 using reciprocity.

6-21. Prove that the reciprocity relationship, Eq. (6-63), is valid for nonhomogeneous media providing

$$\oint\!\!\oint (\mathbf{H}^a \times \mathbf{A}^b - \mathbf{H}^b \times \mathbf{A}^a) \cdot d\mathbf{s} = 0$$

over the surface surrounding the region of integration. (*Hint:* Start directly from the field equations as they apply to the a and b field.)

6-22. Prove the reciprocity relationship

$$\iiint \mathbf{H}^a \cdot \mathbf{J}^b \, d\tau = \iiint \mathbf{H}^b \cdot \mathbf{J}^a \, d\tau$$

providing

$$\oint\!\!\oint (\mathbf{H}^a \times \mathbf{H}^b) \cdot d\mathbf{s} = 0$$

over the surface surrounding the region of integration.

6-23. Using reciprocity, derive Eq. (6-18). Take the current element to be oriented successively in each of the three rectangular-coordinate directions, and obtain the magnetic vector potential integral from Eq. (6-65).

6-24. Using reciprocity, derive the Biot-Savart law, in the manner suggested in the last paragraph of Sec. 6-9.

CHAPTER 7

SOME FIELD CONCEPTS

7-1. On Fields in General. We can summarize the principal idea of Chap. 5 as follows. If we specify a vector function according to

$$\nabla \cdot \mathbf{E} = \frac{q_v}{\epsilon} \qquad \nabla \times \mathbf{E} = 0 \tag{7-1}$$

valid everywhere in space, then **E** can be constructed according to

$$\mathbf{E} = -\nabla V \tag{7-2}$$

where

$$V(\mathbf{r}) = \frac{1}{4\pi} \iiint \frac{q_v(\mathbf{r}')/\epsilon}{|\mathbf{r} - \mathbf{r}'|} \, d\tau' \tag{7-3}$$

Even though the concepts of electric intensity and charge density were used in deriving them, these formulas stand as a mathematical identity. It is intended that the above formulas express the general situation, covering by implication surface, line, and point distributions of sources. Mathematically, these singular types of source can be treated by using appropriate limiting operations. We shall show later in this section that the potential-integral solution is unique. In other words, there is a one-to-one correspondence between an irrotational field and its flow sources, if all space is included.

The principal idea of Chap. 6 can be summarized as follows: If we specify a vector function according to

$$\nabla \cdot \mathbf{B} = 0 \qquad \nabla \times \mathbf{B} = \mu \mathbf{J} \tag{7-4}$$

valid everywhere in space, then **B** can be constructed according to

$$\mathbf{B} = \nabla \times \mathbf{A} \tag{7-5}$$

where

$$\mathbf{A}(\mathbf{r}) = \frac{1}{4\pi} \iiint \frac{\mu \mathbf{J}(\mathbf{r}')}{|\mathbf{r} - \mathbf{r}'|} \, d\tau' \tag{7-6}$$

Again these equations stand as a mathematical identity, even though we used magnetic flux density and current density to derive them. The above formulas are meant to express the general case, covering by implication surface and line distributions of sources. These could, of course, be treated by using appropriate limiting operations. Again the potential-integral solution is unique, as we shall show shortly. That is, there is a one-to-one correspondence between a divergenceless field and its vortex sources, if all space is included.

Now suppose we have both flow sources and vortex sources of a vector function, that is,

$$\nabla \cdot \mathbf{C} = w \qquad \nabla \times \mathbf{C} = \mathbf{W} \tag{7-7}$$

are known everywhere in space. Since both divergence and curl are linear operations, we can employ superposition and define

$$\mathbf{C} = \mathbf{C}' + \mathbf{C}'' \tag{7-8}$$

where

$$\nabla \cdot \mathbf{C}' = w \qquad \nabla \times \mathbf{C}' = 0 \tag{7-9}$$

and

$$\nabla \cdot \mathbf{C}'' = 0 \qquad \nabla \times \mathbf{C}'' = \mathbf{W} \tag{7-10}$$

We can now construct a vector $\mathbf{C}'$ according to

$$\mathbf{C}' = -\nabla g \tag{7-11}$$

where

$$g(\mathbf{r}) = \frac{1}{4\pi} \iiint \frac{w(\mathbf{r}')}{|\mathbf{r} - \mathbf{r}'|} \, d\tau' \tag{7-12}$$

and a vector $\mathbf{C}''$ according to

$$\mathbf{C}'' = \nabla \times \mathbf{G} \tag{7-13}$$

where

$$\mathbf{G}(\mathbf{r}) = \frac{1}{4\pi} \iiint \frac{\mathbf{W}(\mathbf{r}')}{|\mathbf{r} - \mathbf{r}'|} \, d\tau' \tag{7-14}$$

The vector $\mathbf{C}$ is just the superposition of its irrotational part and its divergenceless part, so we construct

$$\mathbf{C} = -\nabla g + \nabla \times \mathbf{G} \tag{7-15}$$

with g and $\mathbf{G}$ given by Eqs. (7-10) and (7-14). Thus, *if we specify the flow sources and the vortex sources of a vector function everywhere, we can construct the vector function as the sum of the gradient of a scalar potential and the curl of a vector potential.* Again we assume that surface, line, and point sources are included according to appropriate limiting procedures. Note that the electric field expressed in terms of a scalar potential and the magnetic field expressed in terms of a vector potential are now just special cases of the above theorem.

We shall prove uniqueness for the general case, which includes electrostatics and magnetostatics as special cases. We start by assuming that there might be two solutions to Eqs. (7-7), which we denote by $\mathbf{C}$ and $\hat{\mathbf{C}}$. The difference field is defined to be $\delta\mathbf{C} = \mathbf{C} - \hat{\mathbf{C}}$. Subtracting Eqs. (7-7) written for the careted field from Eqs. (7-7) written for the uncareted field, we obtain

$$\nabla \cdot \delta\mathbf{C} = 0 \qquad \nabla \times \delta\mathbf{C} = 0 \tag{7-16}$$

since the sources of the two fields are everywhere the same. In view of the second of Eqs. (7-16), $\delta\mathbf{C}$ can be expressed in terms of a scalar potential, which we denote by δg. That is,

$$\delta\mathbf{C} = -\nabla\,\delta g$$

We now form the essentially positive quantity

$$\delta\mathbf{C}\cdot\delta\mathbf{C} = -\delta\mathbf{C}\cdot\nabla\,\delta g$$

In view of the vector identity $\nabla\cdot(a\mathbf{A}) = a\nabla\cdot\mathbf{A} + \mathbf{A}\cdot\nabla a$, we can write

$$(\delta C)^2 = -\nabla\cdot(\delta g\,\delta\mathbf{C}) + \delta g\nabla\cdot\delta\mathbf{C}$$

The last term of this equation is zero because of the first of Eqs. (7-16). Integrating the above equation throughout a region and applying the divergence theorem, we obtain

$$\iiint (\delta C)^2\,d\tau = -\oint\!\!\!\oint \delta g\,\delta\mathbf{C}\cdot d\mathbf{s} \qquad (7\text{-}17)$$

where the surface integral is over the bounding surface. We now choose the surface to be a sphere of large radius, so that all space is included in the volume integral as $r\rightarrow\infty$. We also choose δg to be the scalar potential with respect to infinity, that is $\delta g\rightarrow 0$ as $r\rightarrow\infty$. Assume that $\delta g\sim r^{-\alpha}$, then $\delta\mathbf{C} = -\nabla\,\delta g\sim r^{-\alpha-1}$. Since $ds = r^2\sin\theta\,d\theta\,d\phi\sim r^2$, the integrand of the surface integral of Eq. (7-17) varies as $r^{-2\alpha+1}$. Therefore, the surface integral of Eq. (7-17) vanishes if $\alpha > \frac{1}{2}$. This means that

$$\iiint(\delta C)^2\,d\tau = 0 \qquad (7\text{-}18)$$

when integrated over all space if $\delta\mathbf{C}$ vanishes at infinity more rapidly than $r^{-3/2}$. This is the case if, both $\mathbf{C}$ and $\hat{\mathbf{C}}$ vanish this rapidly. Since Eq. (7-18) can be valid only if $\delta\mathbf{C} = 0$ everywhere, it follows that $\mathbf{C} = \hat{\mathbf{C}}$ if the above condition is satisfied. We have already noted that the potential-integral solution vanishes at infinity at least as rapidly as r^{-2} if the sources are of finite extent. Therefore, *any other solution having the same sources everywhere and vanishing as rapidly at infinity as the potential-integral solution must be identically equal to the potential-integral solution.* The converse of this must also be true. *A field specified everywhere and vanishing at infinity more rapidly than $r^{-3/2}$ has associated with it a unique set of sources.* The uniqueness theorem can also be proven by expressing $\delta\mathbf{C}$ in terms of a vector potential, the conclusions being the same (see Prob. 7-2).

7-2. Surface Distributions of Sources. From an analytical standpoint, some of the most useful source distributions are surface layers. We have already treated surface distributions of electric charge, which are flow sources of **D**. The analysis in no way depended upon any special properties of electric flux density; so we may treat the result, Eq. (4-7), as a mathematical identity. Thus, if w_s is a surface distribution of flow sources of **C**, we have

$$\mathbf{u}_n \cdot (\mathbf{C}^{(1)} - \mathbf{C}^{(2)}) = C_n^{(1)} - C_n^{(2)} = w_s \tag{7-19}$$

where the notation is the same as the notation of Sec. 4-3. That is, $\mathbf{C}^{(1)}$ and $\mathbf{C}^{(2)}$ are the values of **C** on sides (1) and (2) of w_s, respectively, and $\mathbf{u}_n$ points into the region on side (1). In words, Eq. (7-19) states that *the normal component of the vector field is discontinuous across a surface distribution of flow sources, the magnitude of the discontinuity being equal to the surface density of the flow sources.*

We have also treated surface distributions of electric current, which are vortex sources of **H**. The analysis in no way depended upon any special properties of **H**, so we may treat the result, Eq. (4-22), as a mathematical identity. Thus, if $\mathbf{W}_s$ is a surface distribution of vortex sources of **C**, we have

$$\mathbf{u}_n \times (\mathbf{C}^{(1)} - \mathbf{C}^{(2)}) = \mathbf{W}_s \tag{7-20}$$

where the notation has the same meaning as in Eq. (7-19). In words, Eq. (7-20) states that *the tangential components of the vector field are discontinuous across a surface distribution of vortex sources, the magnitude of the discontinuity being equal to that of the surface density of vortex sources.*

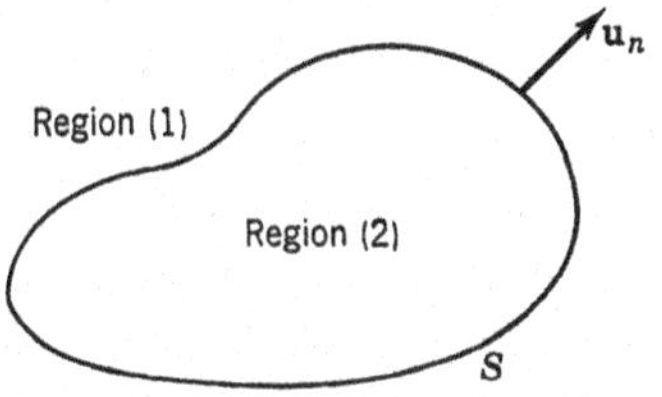

FIG. 7-1. A closed surface S.

Using the characteristics of surface distributions of sources, together with the uniqueness theorem of the preceding section, we have a powerful tool for establishing mathematical identities. Consider a closed surface S, as shown in Fig. 7-1. Denote the region external to S by (1) and the region internal to S by (2), and let $\mathbf{u}_n$ be the unit normal to S pointing into region (1). Specify that there exists in region (1) a source-free field $\mathbf{C}^{(1)}$ vanishing at infinity at least as rapidly as $1/r^2$ (which is the manner in which the potential integral solution vanishes). Specify that there exists in region (2) another source-free field $\mathbf{C}^{(2)}$. The total field must be produced by surface distributions of sources on S, given by Eqs. (7-19) and (7-20). It then follows from the uniqueness of the field that the potential-integral solution (involving the sources on S) is equal to $\mathbf{C}^{(1)}$ external to S and equal to $\mathbf{C}^{(2)}$ internal to S.

To illustrate the utility of the method outlined above, let us consider the following example. Let S be a sphere of radius r' with center at the coordinate origin. We specify that

$$\mathbf{C}^{(1)} = \mathbf{u}_r \frac{1}{r^2} \qquad \mathbf{C}^{(2)} = 0$$

shall define the field in regions (1) and (2), respectively. Applying Eqs. (7-19) and (7-20) at the surface of the sphere, we find the sources to be

$$w_s = \frac{1}{r'^2} \qquad \mathbf{W}_s = 0$$

There are flow sources only; so the field everywhere can be expressed in terms of a scalar potential. Taking a path along a radius vector from the origin, we have

$$g = \int_r^\infty C_r\,dr = \begin{cases} \dfrac{1}{r} & r > r' \\ \dfrac{1}{r'} & r < r' \end{cases}$$

as the scalar potential with respect to infinity. The potential integral

$$g(\mathbf{r}) = \frac{1}{4\pi} \oiint \frac{w_s(\mathbf{r}')}{|\mathbf{r} - \mathbf{r}'|}\,ds'$$

is also the scalar potential with respect to infinity, and must therefore equal the preceding expression. For our specific case, we have

$$g(r,\theta,\phi) = \frac{1}{4\pi} \int_0^{2\pi} d\phi' \int_0^\pi d\theta' \frac{\sin\theta'}{|\mathbf{r} - \mathbf{r}'|}$$

To keep the example as simple as possible, let us specialize the solution to the $\theta = 0$ axis, in which case

$$|\mathbf{r} - \mathbf{r}'| = \sqrt{r^2 + r'^2 - 2rr'\cos\theta'}$$

Then the integrand is not a function of ϕ', so the ϕ' integration gives 2π. Thus, we have

$$g(r,0,\phi) = \tfrac{1}{2} \int_0^\pi \frac{\sin\theta'\,d\theta'}{\sqrt{r^2 + r'^2 - 2rr'\cos\theta'}}$$

Because of the uniqueness of the solution, we can equate the two expressions for g and obtain the mathematical identity

$$\tfrac{1}{2} \int_0^\pi \frac{\sin\theta\,d\theta}{\sqrt{r^2 + r'^2 - 2rr'\cos\theta}} = \begin{cases} \dfrac{1}{r} & r > r' \\ \dfrac{1}{r'} & r < r' \end{cases} \tag{7-21}$$

We have chosen a simple illustration, but many very difficult integrals can be evaluated in this manner.

7-3. Electric Polarization. We have the two electric field vectors, **E** and **D**, fundamentally related to voltage and charge, respectively. We now define the *electric polarization* **P** as

$$\mathbf{P} = \mathbf{D} - \epsilon_0\mathbf{E} \tag{7-22}$$

Note that **P** exists only within matter, for in the absence of matter $\mathbf{D} = \epsilon_0\mathbf{E}$. Thus, we should expect **P** to be in some way related to the constitution of matter. The logic of the name "polarization" will be seen later in this section. From Eq. (7-22), it is evident that **P** has the same dimensions as **D**, that is, charge per area. The unit of **P** in the mksc system of units is therefore the *coulomb per square meter*.

In Chap. 5, we saw that a knowledge of the charge everywhere was sufficient to determine the electrostatic field in a homogeneous region of infinite extent (usually taken to be vacuum). We shall now show that a knowledge of both the charge and the polarization everywhere is sufficient to determine the electrostatic field if matter is present. Unfortunately, we usually do not know the polarization until after we have the solution to the problem.

Let us first consider the case for which the polarization is continuous and differentiable everywhere. The field equations for the electrostatic field are

$$\nabla \cdot \mathbf{D} = q_v \qquad \nabla \times \mathbf{E} = 0$$

Substituting for **D** from Eq. (7-22), and rearranging, we have

$$\nabla \cdot \mathbf{E} = \frac{1}{\epsilon_0}(q_v - \nabla \cdot \mathbf{P}) \qquad \nabla \times \mathbf{E} = 0 \tag{7-23}$$

Note that these equations have the same form as the equations for **E** from charges in vacuum, if $(q_v - \nabla \cdot \mathbf{P})$ is interpreted as the charge density. Thus, *when neutral matter becomes polarized, it acts like a distribution of charge existing in vacuum.* In accordance with the atomic picture of matter, this effect is interpreted as arising from a redistribution of the *bound* charged particles in the atoms or molecules of the matter (those incapable of moving freely from molecule to molecule). This is in contrast to q_v which is pictured as being due to the *free* charges in the matter (those able to move freely from molecule to molecule). We shall call the quantity $(q_v - \nabla \cdot \mathbf{P})$ the *absolute*-charge density, although other nomenclature has been used in the older literature.* Eqs. (7-23) are in a form showing all sources of **E**, there being only flow sources. Therefore, according to Sec. 7-1, the **E** field is given by the negative gradient of a scalar potential integral where the flow sources are $1/\epsilon_0(q_v - \nabla \cdot \mathbf{P})$. In other words, defining the absolute-charge density as

$$\hat{q}_v = q_v - \nabla \cdot \mathbf{P} \tag{7-24}$$

then

$$\mathbf{E} = -\nabla V \tag{7-25}$$

where

$$V(\mathbf{r}) = \frac{1}{4\pi\epsilon_0} \iiint \frac{\hat{q}_v(\mathbf{r}')}{|\mathbf{r} - \mathbf{r}'|}\, d\tau' \tag{7-26}$$

* Some authors call the flow sources of **D** true charge and the flow sources of $\epsilon_0\mathbf{E}$ free charge. See, for example, A. Abraham and R. Becker, "Classical Electricity and Magnetism," Hafner Publishing Company, New York, p. 74.

These equations must be amended or interpreted according to appropriate limiting procedures if **P** is a discontinuous function. Philosophically, this solution corresponds to an interpretation of matter entirely in terms of charge in vacuum, which agrees with modern atomic theory.

In most problems, matter is considered to have abrupt boundaries; so it is to be expected that **P** will be discontinuous at such boundaries. We shall now show that if the normal component of **P** is discontinuous at a boundary, so also is the normal component of **E**. Suppose that the surface S of Fig. 7-1 represents the boundary of a material object, that is, that there is vacuum in region (1), matter in region (2). In region (1) the electric flux density is given by

$$\mathbf{D}^{(1)} = \epsilon_0 \mathbf{E}^{(1)} \tag{7-27}$$

and in region (2) by

$$\mathbf{D}^{(2)} = \epsilon_0 \mathbf{E}^{(2)} + \mathbf{P} \tag{7-28}$$

At any material boundary, Eq. (4-7) must be satisfied. Thus, at the surface S

$$D_n^{(1)} - D_n^{(2)} = q_s \tag{7-29}$$

where q_s is a possible surface density of free charge on S. Substituting for $\mathbf{D}^{(1)}$ and $\mathbf{D}^{(2)}$ from the preceding equations and rearranging, we find that

$$E_n^{(1)} - E_n^{(2)} = \frac{1}{\epsilon_0}(q_s + P_n) \tag{7-30}$$

This is the same discontinuity in the normal component of **E** as would occur across a surface density of free charge in vacuum, equal in magnitude to $(q_s + P_n)$. Thus, *a discontinuity in the normal component of* **P** *at a boundary acts like a surface density of charge in vacuum.* Again we interpret this effect as arising from the bound charges in matter, in contrast to q_s which is due to the free charges in matter. We define the *absolute*-surface-charge density

$$\hat{q}_s = q_s + P_n \tag{7-31}$$

which acts as a flow source of $\epsilon_0\mathbf{E}$. The tangential components of **E** at any boundary must be continuous; so there are no surface vortex sources of **E**. According to the concepts of Sec. 7-2, we must include the flow sources of Eq. (7-31) in addition to those of Eq. (7-24) in the calculation of the potential integral. Thus, the form for the scalar potential showing explicitly the role played by material boundaries is

$$V(\mathbf{r}) = \frac{1}{4\pi\epsilon_0}\iiint \frac{\hat{q}_v(\mathbf{r}')}{|\mathbf{r} - \mathbf{r}'|}\,d\tau' + \frac{1}{4\pi\epsilon_0}\iint \frac{\hat{q}_s(\mathbf{r}')}{|\mathbf{r} - \mathbf{r}'|}\,ds' \tag{7-32}$$

The electric intensity is, of course, still given by Eq. (7-25). Even though we can view Eq. (7-26) as general, using suitable limiting procedures to take care of discontinuities in **P**, it is well to point out any surface densities of bound charge that appear at material surfaces. Otherwise these might easily be overlooked.

To recapitulate, we have determined all sources of **E** in terms of q_v and **P**. This gives us only flow sources, so the methods of solution of Chap. 5 apply directly. Once we obtain the electric intensity, the electric flux density is given by Eq. (7-27) external to matter and by Eq. (7-28) internal to matter. We could just as well have determined the sources of **D** in terms of q_v and **P**. This we shall do now, showing that **D** in general has both flow sources and vortex sources.

Again let us start out with the assumption that **P** is everywhere continuous and differentiable. Starting with the field equations, we substitute for **E** from Eq. (7-22) and we obtain

$$\nabla \cdot \mathbf{D} = q_v \qquad \nabla \times \mathbf{D} = \nabla \times \mathbf{P} \tag{7-33}$$

Thus, *polarized matter acts as a vortex source of* **D**. If there are no free charges present ($q_v = 0$), then **D** has only vortex sources and the equations take the same form as those for **B** due to electric currents in vacuum. In this case, the methods of Chap. 6 would apply directly for determining an "electric" vector potential in terms of $\nabla \times \mathbf{P}$. As we shall see in Sec. 7-5, the quantity $(1/\epsilon_0)\nabla \times \mathbf{P}$ could logically be called a "bound magnetic current."

At material boundaries, **P** is discontinuous, and we shall see that a distribution of surface vortex sources is then necessary. Again consider that Fig. 7-1 represents the boundary of a material object, that is, that there is vacuum in region (1) and matter in region (2). Eq. (7-29) holds always; so only the free-charge density on a surface is a flow source of **D**. However, for tangential field components, Eq. (3-29) must be satisfied; that is

$$\mathbf{u}_n \times (\mathbf{E}^{(1)} - \mathbf{E}^{(2)}) = 0 \tag{7-34}$$

at a material boundary. In other words, the tangential components of **E** are continuous. In region (1) Eq. (7-27) holds, and in region (2) Eq. (7-28). Substituting from these into Eq. (7-34), we arrive at

$$\mathbf{u}_n \times (\mathbf{D}^{(1)} - \mathbf{D}^{(2)}) = \mathbf{P} \times \mathbf{u}_n \tag{7-35}$$

The term on the right-hand side must therefore be the surface distribution of vortex sources of **D** appearing on the material boundary. These sources must also be included in the vector potential for **D** according to the analysis in Sec. 7-2.

The mathematics used in determining **E** and **D** from **P** is, of course, the same as the mathematics used in Chaps. 5 and 6, so few examples will be needed. However, it is worthwhile to show now that the field due to each element of **P** is identical with the field due to a charge dipole, justifying the name "polarization" for **P**.

Consider a volume element $\Delta\tau$ within which there is a uniform **P**, as shown in Fig. 7-2. Outside $\Delta\tau$ we have vacuum. $\nabla \cdot \mathbf{P}$ is zero everywhere; so we have no volume distribution of flow sources of **E**. A component of **P** normal to the boundary exists only over the top and bottom of the block; so surface flow sources appear only at the top and bottom. According to Eq. (7-30), these are $+P$ over the top and $-P$ over the bottom. The total bound charge on the top is then $P\,\Delta A$, and on the bottom $-P\,\Delta A$, where ΔA is the cross-sectional area of $\Delta\tau$. At any point distant from $\Delta\tau$, these charges appear to be point charges. They are separated by the height of $\Delta\tau$, which we shall call Δh, so these charges form a dipole of moment $P\,\Delta A\,\Delta h$. But $\Delta A\,\Delta h$ is simply $\Delta\tau$, and the direction of the dipole is that of **P**, so the vector dipole moment is $\mathbf{P}\,\Delta\tau$. Therefore, the scalar potential for the element of Fig. 7-2 is given by Eq. (5-81) with ql replaced by $\mathbf{P}\,\Delta\tau$. Thus

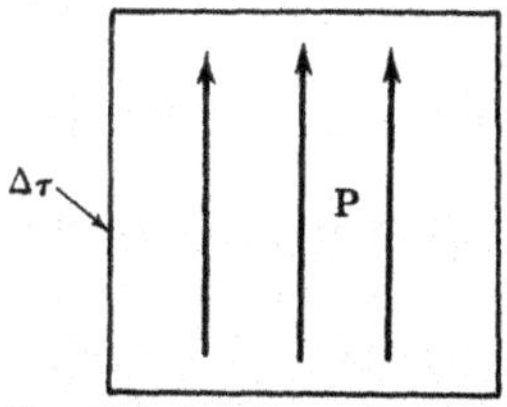

FIG. 7-2. A volume element of uniform polarization.

$$V(\mathbf{r}) = \frac{\mathbf{P}(\mathbf{r}') \cdot (\mathbf{r} - \mathbf{r}')}{4\pi\epsilon_0 |\mathbf{r} - \mathbf{r}'|^3}\,\Delta\tau' \qquad (7\text{-}36)$$

and $\mathbf{E} = -\nabla V$. As $\Delta\tau \to 0$, the volume element becomes a true point dipole, so we arrive at the conclusion that *each volume element of* **P** *acts like a point charge dipole.* The **E** lines associated with an element of **P** are, of course, just those of Fig. 5-19. To obtain the total field associated with a distribution of **P**, we could sum over all differential volume elements. However, the procedure of summing over a volume distribution of dipole moments is generally more difficult than determining the sources of **E** or **D**; so we use the latter method.

A couple of other simple examples can be given immediately because we have solved the analogous problems in Chaps. 5 and 6. If we had a long, thin cylinder of polarized matter with **P** uniform and directed along the axis, we would have, effectively, two point charges separated by the length of the cylinder. Thus, the **E** field would be the field of two point charges of equal magnitude and opposite sign, one at each end of the cylinder. A plot of the equipotential lines would be equivalent to Fig. 5-9*d*, and the **E** lines would be perpendicular to these equipotential lines. If we had a very short cylinder, that is, a disk of polarized matter with **P** uniform and directed along the axis, we could use the alternative analysis in terms of vortex sources. **P** internal to the disk has no curl; so we would have only a surface vortex source of **D** according to Eq. (7-35). This source would be in the circumferential (or ϕ) direction about the disk's edge, and have a surface density P. Thus, the total strength would be Ph, where h is the thickness of the disk. We have now precisely the same problem as that of calculating **H** due to a loop of current. The solution uses Eqs. (6-40), with **H** replaced by **D** and i replaced by Ph. For example, from Eq. (6-27) it follows that along the axis of the disk

$$D_z = \frac{Pha^2}{2(a^2 + z^2)^{3/2}} \tag{7-37}$$

where a is the radius of the disk and z is the distance along the axis from the disk to the field point.

The atomic picture of electric polarization is as follows. In the unpolarized state, a neutral molecule has coincident "centers" of negative and positive charge. This is suggested by Fig. 7-3a. When a

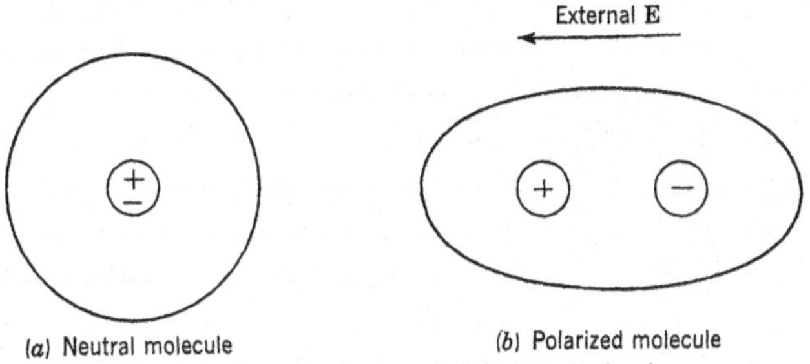

FIG. 7-3. Atomic picture of electric polarization.

neutral molecule is acted upon by an external electric field, the bound atomic particles are on the average displaced. This produces a displacement of the center of negative charge with respect to the center of positive charge, as suggested by Fig. 7-3b. The result is an "atomic dipole." From the macroscopic viewpoint, each volume element of matter contains many such atomic dipoles, giving rise to a density of dipole moments. It is this density of dipole moments that we call the polarization **P** of the material. In regions where **P** has no divergence, the net charge in each volume element remains zero. We can imagine that this results from displacement of equal amounts of charge into and out of the volume element. A divergence of **P** exists when more bound charge enters than leaves the volume element, or vice versa. The most common occurrence of this is at a material boundary, where it gives rise to the surface density of bound charge mentioned earlier. To illustrate this, consider the polarized dielectric slab of Fig. 7-4. In the interior of the slab, there are equal amounts of negative and positive bound charge; that is, there is no net charge. At the two surfaces, however, an excess of one polarity over the other gives the effect of a surface density of charge. In Fig. 7-4, the neutral interior of the dielectric lies between the two dashed surfaces, and the surface densities of bound charge lie between the dashed surfaces and the dielectric boundary. As a matter of fact, *for a neutral, linear, homogeneous, and isotropic dielectric body, all bound charge must appear as a surface charge density.* This follows from the divergence of Eq. (7-22), since $\nabla \cdot \mathbf{D} = \epsilon \nabla \cdot \mathbf{E} = 0$. A volume density of bound charge can occur only when one or more of the above conditions are violated.

In linear media, **D** is proportional to **E**; therefore **P** must be proportional to **E**. The *electric susceptibility* χ_e is defined according to

$$\mathbf{P} = \chi_e \epsilon_0 \mathbf{E} \tag{7-38}$$

Since $\mathbf{D} = \epsilon \mathbf{E}$ in linear media, it follows from Eq. (7-22) that the capacitivity (or permittivity) is related to the susceptibility according to

$$\epsilon = \epsilon_0(1 + \chi_e) \tag{7-39}$$

Thus, a knowledge of χ_e is equivalent to a knowledge of ϵ. It is evident from the above equations that χ_e is dimensionless.

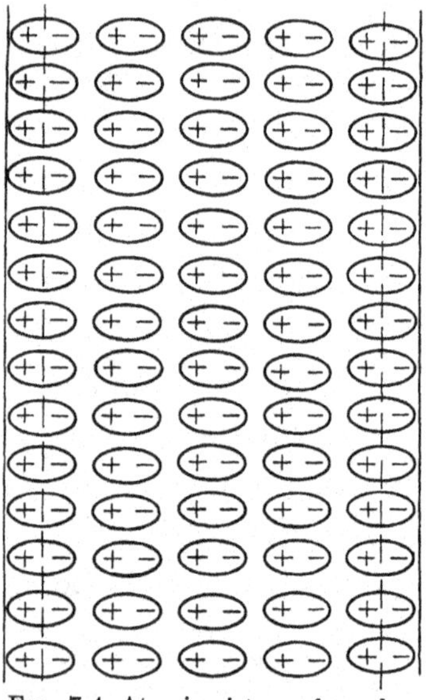

FIG. 7-4. Atomic picture of a polarized dielectric slab.

7-4. Magnetization. We have the two magnetic field vectors, **B** and **H**, fundamentally related to flux and current, respectively. We now define the *magnetization* **M** as

$$\mathbf{M} = \frac{1}{\mu_0}\mathbf{B} - \mathbf{H} \tag{7-40}$$

Note that **M** exists only within matter, for in the absence of matter $\mathbf{B} = \mu_0 \mathbf{H}$. Thus, **M** must be related to the constitution of matter. If we compare Eq. (7-40) with the equation for electric polarization, Eq. (7-22), the following question comes to mind. Why not define magnetization as $(\mathbf{B} - \mu_0 \mathbf{H})$, which has the same mathematical form as electric polarization? Actually, we could make such a definition, which would give us a "magnetic polarization" equal to μ_0 (a constant) times the magnetization. However, it is conventional to relate **M** to infinitesimal electric-current loops rather than to "magnetic charge dipoles," and the definition of Eq. (7-40) is better suited to this purpose. It is evident from Eq. (7-40) that **M** has the same dimensions as **H**, that is, current per length. In the mksc system, the unit of magnetization is therefore the *ampere per meter.*

In Chap. 6, we saw that a knowledge of the electric current everywhere was sufficient to determine the magnetostatic field in a homogeneous region of infinite extent (usually taken to be vacuum). We shall now show that a knowledge of both the current and the magnetization everywhere is sufficient to determine the magnetostatic field if matter is present. Just as in the case of electric polarization, we often do not know the magnetization until after we have the solution to the problem. However, in magnetostatics, there exists quite commonly the phenomenon of permanent magnetism. In this case a magnetization exists independent of any "impressed field," and magnetization is then viewed as the primary source of the field.

We shall start with the assumption that the magnetization is continuous and differentiable everywhere. The field equations for the magnetostatic field are

$$\nabla \cdot \mathbf{B} = 0 \qquad \nabla \times \mathbf{H} = \mathbf{J}$$

Substituting for **H** from Eq. (7-40) and rearranging, we have

$$\nabla \cdot \mathbf{B} = 0 \qquad \nabla \times \mathbf{B} = \mu_0(\mathbf{J} + \nabla \times \mathbf{M}) \tag{7-41}$$

These equations have the same form as the equations for **B** due to currents in free space, if $(\mathbf{J} + \nabla \times \mathbf{M})$ is interpreted as the current density. Thus, *when matter becomes magnetized, it acts like a current existing in vacuum.* According to the atomic theory of matter, this effect is interpreted as resulting from atomic magnetic moments of the particles constituting the matter. Accordingly, the effective current $\nabla \times \mathbf{M}$ is called a *bound*-current density, in contrast with **J** which is called the *free*-current density. The quantity $(\mathbf{J} + \nabla \times \mathbf{M})$ could then properly be called the *absolute*-current density. Eqs. (7-41) are written in a form showing all sources of **B**, there being only vortex sources. Therefore, according to Sec. 7-1, the **B** field is given by the curl of a vector potential integral, where the vortex sources are $\mu_0(\mathbf{J} + \nabla \times \mathbf{M})$. In other words, defining the absolute-current density

$$\hat{\mathbf{J}} = \mathbf{J} + \nabla \times \mathbf{M} \tag{7-42}$$

we have

$$\mathbf{B} = \nabla \times \mathbf{A} \tag{7-43}$$

where

$$\mathbf{A}(\mathbf{r}) = \frac{\mu_0}{4\pi} \iiint \frac{\hat{\mathbf{J}}(\mathbf{r}')}{|\mathbf{r} - \mathbf{r}'|}\, d\tau' \tag{7-44}$$

Let us not lose sight of the fact that these equations are explicit only if **M** is everywhere continuous and differentiable. If **M** is discontinuous, we must treat Eq. (7-44) according to appropriate limiting operations or else amend it as we do below. Philosophically, the above solution interprets all matter in terms of electric currents or magnetic dipoles in vacuum, which is consistent with modern atomic concepts.

In most problems, matter is considered to have abrupt boundaries, and

M is often discontinuous at such boundaries. We shall now show that if the tangential components of **M** are discontinuous at a boundary, so also will be the tangential components of **B**. Again we take Fig. 7-1 to represent a material boundary, with vacuum in region (1) and matter in region (2). The magnetic intensity in region (1) is given by

$$\mathbf{H}^{(1)} = \frac{1}{\mu_0}\mathbf{B}^{(1)} \tag{7-45}$$

and in region (2) by

$$\mathbf{H}^{(2)} = \frac{1}{\mu_0}\mathbf{B}^{(2)} - \mathbf{M} \tag{7-46}$$

At any boundary, Eq. (4-22) must be valid. Thus, at the surface S

$$\mathbf{u}_n \times (\mathbf{H}^{(1)} - \mathbf{H}^{(2)}) = \mathbf{J}_s \tag{7-47}$$

where $\mathbf{J}_s$ is a possible surface distribution of free current on S. Substituting from Eqs. (7-45) and (7-46) into Eq. (7-47) and rearranging, we obtain

$$\mathbf{u}_n \times (\mathbf{B}^{(1)} - \mathbf{B}^{(2)}) = \mu_0(\mathbf{J}_s + \mathbf{M} \times \mathbf{u}_n) \tag{7-48}$$

This is the same discontinuity in the tangential components of **B** as would occur across a surface distribution of free current in vacuum, equal in magnitude to $(\mathbf{J}_s + \mathbf{M} \times \mathbf{u}_n)$. Thus, *the discontinuities in the tangential components of* **M** *at a boundary act like a surface distribution of current in vacuum.* Again we interpret the bound current $(\mathbf{M} \times \mathbf{u}_n)$ as being due to the atomic currents in matter. We define the *absolute* surface current distribution

$$\hat{\mathbf{J}}_s = \mathbf{J}_s + \mathbf{M} \times \mathbf{u}_n \tag{7-49}$$

which acts as a vortex source of $\mathbf{B}/\mu_0$. The normal component of **B** at any boundary must be continuous, so there is no surface flow source of **B**. According to the analysis made in Sec. 7-2, we must now add the vortex sources of Eq. (7-49) to those of Eq. (7-42) in calculating the potential-integral solution. Thus, a more general form for the magnetic vector potential is

$$\mathbf{A}(\mathbf{r}) = \frac{\mu_0}{4\pi}\iiint \frac{\hat{\mathbf{J}}(\mathbf{r}')}{|\mathbf{r} - \mathbf{r}'|}\,d\tau' + \frac{\mu_0}{4\pi}\iint \frac{\hat{\mathbf{J}}_s(\mathbf{r}')}{|\mathbf{r} - \mathbf{r}'|}\,ds' \tag{7-50}$$

which shows explicitly the role played by material boundaries. The magnetic flux density is then found according to $\mathbf{B} = \nabla \times \mathbf{A}$. We can, of course, view Eq. (7-44) as the general case, with Eq. (7-50) obtainable by suitable limiting processes. But we should always be careful not to overlook the surface discontinuities that exist at material boundaries.

We have obtained a solution by determining the sources of **B**. There is the alternative method of determining the sources of **H**. For a distribution of **M** continuous and differentiable everywhere, we can substitute for **B** from Eq. (7-40) into the field equations and obtain

$$\nabla \cdot \mathbf{H} = -\nabla \cdot \mathbf{M} \qquad \nabla \times \mathbf{H} = \mathbf{J} \tag{7-51}$$

Thus, *magnetized matter acts as a flow source of* **H**. If there are no free currents present, the **H** has only flow sources, and the equations have the same form as those for **D** due to electric charge in vacuum. Thus, the methods of Chap. 5 apply directly for determining a magnetic scalar potential in terms of $(-\nabla \cdot \mathbf{M})$. The use of flow sources of **H** is found to be quite convenient for analysis, and the quantity $(-\nabla \cdot \mathbf{M})$ has been given the name *bound magnetic pole density.* As we shall see in Sec. 7-5, the quantity $(-\mu_0\nabla \cdot \mathbf{M})$ could also logically be called a "bound magnetic charge."*

At material boundaries, the discontinuity in **M** gives rise to a surface density of flow sources of **H**. Again consider that Fig. 7-1 represents a boundary of matter, with vacuum in region (1) and matter in region (2). Eq. (7-47) is always valid; so the only vortex sources of **H** on S are due to free current. However, for the normal field component, Eq. (3-28) must be satisfied; that is

$$\mathbf{u}_n \cdot (\mathbf{B}^{(1)} - \mathbf{B}^{(2)}) = B_n{}^{(1)} - B_n{}^{(2)} = 0 \tag{7-52}$$

at a material boundary. In region (1) Eq. (7-45) is valid, and in region (2) Eq. (7-46). Substituting these equations into Eq. (7-52), we obtain

$$H_n{}^{(1)} - H_n{}^{(2)} = M_n \tag{7-53}$$

Thus, M_n is the surface density of flow sources of **H** appearing on the material boundary. In other words, *the surface density of bound magnetic poles on the boundary is* M_n. These sources must be included in the scalar potential integral according to Sec. 7-2.

The equations for **H** in terms of **M** when **J** is zero, Eqs. (7-51), take the same mathematical form as the equations for **E** in terms of **P** when q_v is zero, Eqs. (7-23). Similarly, the equations for **B** in terms of **M** when **J** is zero, Eqs. (7-41), take the same mathematical form as the equations for **D** in terms of **P** when q_v is zero, Eqs. (7-33). Therefore, all the examples given for the field due to electric polarization are valid for the corresponding case of magnetization. For example, a small block of uniformly magnetized material acts as a dipole source. Thus, the **H** field is derivable from a scalar potential, which we shall denote by U, and which we shall name *magnetic scalar potential.* In Eq. (7-36), we need

* Some authors have used the names "magnetic pole" and "magnetic charge" interchangeably, and this has been the cause of considerable misinterpretation. In this text, the magnetic charge density is defined as μ_0 times the magnetic pole density.

merely replace V by U, $\mathbf{P}$ by $\mu_0\mathbf{M}$, and ϵ_0 by μ_0, to obtain

$$U(\mathbf{r}) = \frac{\mathbf{M}(\mathbf{r}') \cdot (\mathbf{r} - \mathbf{r}')}{4\pi|\mathbf{r} - \mathbf{r}'|^3} \Delta\tau' \tag{7-54}$$

The magnetic intensity is then given by $\mathbf{H} = -\nabla U$, and a picture of the $\mathbf{H}$ lines would be given by Fig. 5-19. Thus, *each volume element of* $\mathbf{M}$ *acts like a point magnetic dipole.* To obtain the total field associated with a distribution of $\mathbf{M}$, we could sum over all differential volume elements. However, it is usually more convenient to determine the sources of $\mathbf{H}$ or $\mathbf{B}$ and then use the potential-integral solution, than to sum over a volume distribution of dipole moments.

We saw in Sec. 6-7 that the field due to an infinitesimal loop of current is identical in form with the field due to a charge dipole. Thus, each volume element of magnetization can be viewed as a loop of current rather than as a charge dipole. This is a more common view. To obtain the current-loop analysis of an element of $\mathbf{M}$, we need merely look at the problem in terms of the $\mathbf{B}$ vector. If we have uniform magnetization in the volume element $\Delta\tau$ (consider Fig. 7-2 with $\mathbf{P}$ replaced by $\mathbf{M}$), the curl of $\mathbf{M}$ is zero everywhere, but we have a discontinuity in $\mathbf{M}$ at the surface of $\Delta\tau$. Applying Eq. (7-49) to the surface of $\Delta\tau$, we obtain a surface density of vortex source of magnitude $\hat{J}_s = M$ about the sides of the block in a direction perpendicular to $\mathbf{M}$ according to the right-hand rule (thumb in the direction of $\mathbf{M}$, fingers in the direction of $\hat{\mathbf{J}}_s$). The total strength of this bound current is $\hat{J}_s\,\Delta h$, where Δh is the height of $\Delta\tau$. The magnitude of the moment of this loop is the current strength times the area enclosed, or $\hat{J}_s\,\Delta h\,\Delta A$, where ΔA is the cross-sectional area of $\Delta\tau$. But $\Delta\tau = \Delta h\,\Delta A$, so the magnitude of the loop moment is $\hat{J}_s\,\Delta\tau = M\,\Delta\tau$. The direction of the moment is that of $\mathbf{M}$ so the vector loop moment is $\mathbf{M}\,\Delta\tau$. The formulas of Sec. 6-7 now apply if we replace the loop moment $\pi a^2 i$ by $M\,\Delta\tau$. Thus, if $\Delta\tau$ is situated at the coordinate axis with $\mathbf{M}$ in the z direction, from Eq. (6-58) it follows that

$$\mathbf{H} = \frac{M\,\Delta\tau}{4\pi r^3}(\mathbf{u}_r 2\cos\theta + \mathbf{u}_\theta \sin\theta) \tag{7-55}$$

It is left as an exercise for the reader to show that this is identical with the result obtained from $\mathbf{H} = -\nabla U$ with U given by Eq. (7-54).

From the atomic viewpoint, the principal source of atomic magnetic dipoles is an intrinsic magnetic moment possessed by electrons. This can be pictured as an electron "spin," or the rotation of the electron about its axis. In this sense it is a current of atomic origin. Another possible source of an atomic magnetic dipole is the orbital motion of electrons in the atoms. This again can be pictured as a current of atomic origin. In the unmagnetized state, matter is pictured as having atomic dipoles of

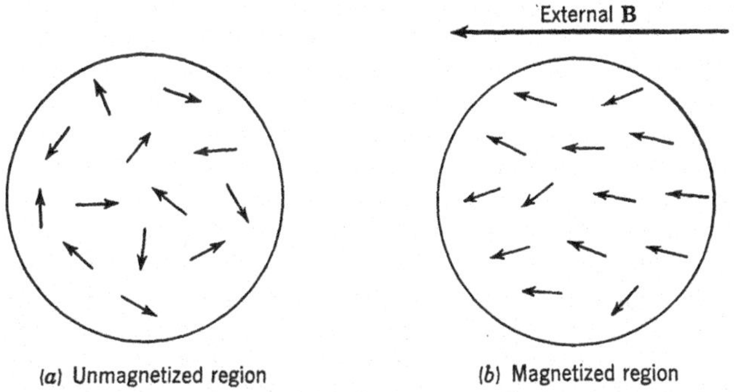

FIG. 7-5. Atomic picture of magnetization.

random orientation, giving no net effect. This interpretation is suggested by Fig. 7-5*a*. When the atomic dipoles are acted upon by an external magnetic field, they tend to align themselves parallel to the field, as suggested by Fig. 7-5*b*. From the macroscopic viewpoint, a volume element of matter contains many such atomic dipoles, giving rise to a density of dipole moments. It is this density of dipole moments that we call the magnetization **M** of the material. The above picture is, of course, a greatly simplified and incomplete description of a very complicated phenomenon. In regions where **M** has no curl, the net current in each volume element is zero. This we can picture as due to a cancelation of components of atomic currents. When curl **M** exists, there is incomplete cancelation of atomic currents, and a net bound-current density results. The most common occurrence of this is at a material boundary, resulting in the bound surface current. To illustrate, let us picture the atomic dipoles in terms of microscopic current loops. A magnetized slab of magnetic material is represented in Fig. 7-6. In each interior volume element of macroscopic size, there is zero net current. But at the surface, which is represented by the region between the dashed line and the material boundary in Fig. 7-6, a net bound current results. In fact, *for a current-free, linear,*

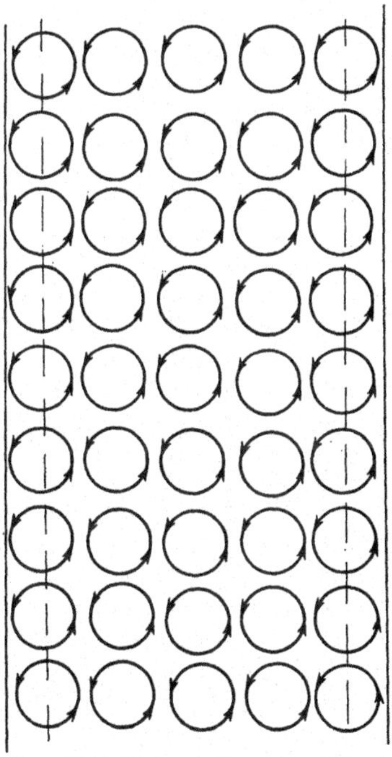

FIG. 7-6. Picture of a magnetized slab in terms of atomic current loops.

homogeneous, isotropic body, all bound current appears as a surface current distribution. This follows from the curl of Eq. (7-40), for then

$$\nabla \times \mathbf{H} = \mu_0^{-1} \nabla \times \mathbf{B} = 0$$

Violation of one or more of the above conditions can, of course, result in a volume distribution of bound current.

Magnetism existing as a primary field source, that is, in the absence of an "applied" field, is known as *permanent magnetism.* A knowledge of the magnetization **M** of a permanent magnet is sufficient to determine its field, according to the concepts of this section. As an example, let us consider the so-called "bar magnet." This consists of a cylinder of

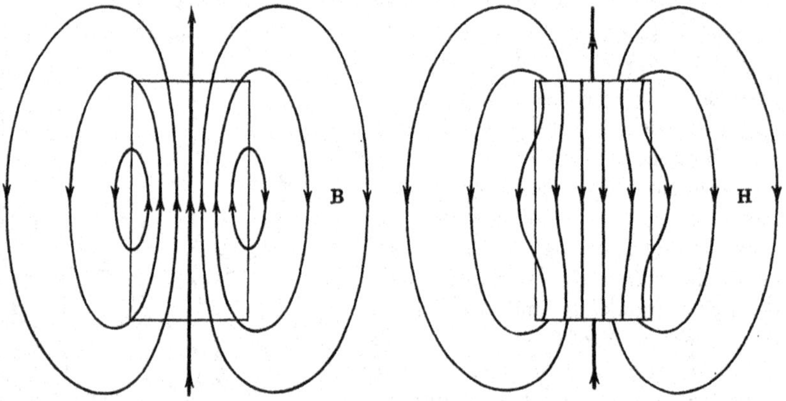

(a) **B** lines (b) **H** lines

FIG. 7-7. The field associated with an ideal bar magnet.

ferromagnetic metal magnetized so that **M** is approximately of constant magnitude and directed along the cylinder axis. Following the usual procedure for determining the vortex sources of **B**, we find only a surface density, $\hat{J}_s = M$, in the circumferential direction about the cylinder. This produces exactly the same **B** as a cylinder of conduction currents, of the same surface density, in free space. A closely wound solenoid of N turns carrying a constant current i forms effectively a sheet of current of linear density $J_s = Ni/l$, where l is the length of the solenoid. Therefore, *a bar magnet of constant axial magnetization* **M** *produces the same* **B** *field as a solenoid in free space, which is of the same size and shape as the bar magnet and of ampere-turns* $Ni = Ml$. The **B** lines associated with a bar magnet are sketched in Fig. 7-7*a*. We have available the alternative analysis of determining the flow sources of **H**. These sources are found in the usual manner to be two "plates" of surface pole density covering the end faces of the bar magnet, of density M over one end and of density $-M$ over the other end. Thus, *a bar magnet of constant* *axial magnetization* **M** *produces the same* **H** *field as two plates of uniform surface-pole density M in free space, of the same size and shape as the end faces of the bar magnet.* The field produced by these two plates is spatially identical with the **E** field produced by two plates of uniform surface-charge density. The **H** lines associated with a bar magnet are sketched in Fig. 7-7*b*. Note that outside the magnet the **B** and **H** lines have the same form, for $\mathbf{B} = \mu_0\mathbf{H}$, but inside the magnet the **B** and **H** lines are quite different. Two limiting cases of the bar magnet would be (*a*) a long needle-shaped magnet, and (*b*) a short disk-shaped magnet. In the first case, we would have two point poles for the **H** sources, and the field in the vicinity of one end of the magnet would have the same form as the field due to a point charge. In the second case, we would have a loop of vortex source for **B**, and the field would have the same form as the field due to a loop of current.

In linear media, **B** is proportional to **H** and, therefore, **M** also must be proportional to **H**. The *magnetic susceptibility* χ_m is defined according to

$$\mathbf{M} = \chi_m \mathbf{H} \tag{7-56}$$

Since $\mathbf{B} = \mu\mathbf{H}$ in linear media, it follows from Eq. (7-40) that the inductivity (or permeability) is related to the magnetic susceptibility according to

$$\mu = \mu_0(1 + \chi_m) \tag{7-57}$$

Thus, a knowledge of χ_m is equivalent to a knowledge of μ. It is evident from the above equations that χ_m is dimensionless. For most media, $\mu \approx \mu_0$ and $\chi_m \approx 0$. For ferromagnetic media, $\mu \gg \mu_0$, but such media are quite nonlinear.

7-5. The Static Equivalence Principle. So far, we have always considered the field everywhere in terms of its sources everywhere. In many cases, we are concerned with the field only in certain regions of space and we may not even know the sources outside these regions. Given a region of space, *any number of distributions of sources outside this region may produce the same field within the region.* We have already had many examples of this. For instance, a uniformly charged sphere produces the same field (that of a point charge) external to the sphere, regardless of its diameter. Also, flow sources may produce exactly the same field in a given region as vortex sources. For example, the field from a charge dipole is equivalent to the field from a current loop. When we are interested only in the field within a given region, it makes no difference to us whether or not we know the "actual" sources; we need only possible sources. Two sets of sources producing the same field within a given region of space are said to be *equivalent within that region.*

Let us first look at magnetostatics. The fundamental equations for magnetostatic fields in linear media are Eqs. (6-2), so long as the mathematical problem is to represent the physical problem everywhere in space. If we are interested in the magnetostatic field only within a given region of space, there is no mathematical reason why we should not consider this field to be produced, at least in part, by flow sources of **B** external to the given region. It is necessary only that there be no flow sources of **B** internal to the region. Thus, we can introduce into our mathematical problem flow sources of **B** to be used as equivalent sources. The *generalized equations of magnetostatics* then become

$$\nabla \times \mathbf{H} = \mathbf{J} \qquad \nabla \cdot \mathbf{B} = m_v \tag{7-58}$$

where m_v is called the *magnetic charge density*. It follows from the second of Eqs. (7-58) that m_v has the dimensions of magnetic flux per volume. The unit of magnetic charge density in the mksc system is therefore the *weber per cubic meter*. The volume integral of m_v we call simply the *magnetic charge* m, that is

$$m = \iiint m_v \, d\tau \tag{7-59}$$

Since the dimensions of m_v are those of magnetic flux per volume, it follows from Eq. (7-59) that the dimensions of m are the same as those of magnetic flux. The unit of magnetic charge in the mksc system is thus the *weber*. We may have occasion to use the concepts of surface densities of magnetic charge m_s, line distributions m_l, and point distributions m. These are defined in a manner corresponding to the electric charge concepts, Sec. 2-1. The integral forms of Eqs. (7-58) are

$$\oint \mathbf{H} \cdot d\mathbf{l} = i \qquad \oiint \mathbf{B} \cdot d\mathbf{s} = m \tag{7-60}$$

where i is the electric current through any surface bounded by the contour to which the line integral is applied and where m is the magnetic charge enclosed by the surface over which the surface integral is taken.

If we have a surface distribution of electric current $\mathbf{J}_s$, then according to Sec. 7-2,

$$\mathbf{u}_n \times (\mathbf{H}^{(1)} - \mathbf{H}^{(2)}) = \mathbf{J}_s \tag{7-61}$$

where $\mathbf{u}_n$ points into region (1). If we have a surface distribution of magnetic charge m_s, then

$$B_n^{(1)} - B_n^{(2)} = m_s \tag{7-62}$$

where the positive direction of the normal is into region (1). Note that Eq. (7-61) is the usual relationship for **H** in actual media, whereas Eq. (7-62) applies to the physical world only where $m_s = 0$.

We have shown the uniqueness of a vector field if we specify all its sources everywhere. We have a slightly different case when we specify the flow sources of **B** and the vortex sources of **H**. If the medium is linear and homogeneous everywhere, that is, if $\mathbf{B} = \mu\mathbf{H}$ with μ a constant, then the flow sources of **H** are m_v/μ, and the vortex sources of **B** are $\mu\mathbf{J}$; so we know all sources of both **B** and **H**. The uniqueness theorem as stated in Sec. 7-1 then applies directly. However, it is not necessary to be so restrictive concerning uniqueness. It is necessary merely that the medium be linear; that is, μ may be a function of position (see Prob. 7-23). The magnetostatic field is unique (*a*) if the electric current and magnetic charge are specified everywhere, (*b*) if the medium is linear everywhere, and (*c*) if the field vanishes at infinite distances at least as rapidly as r^{-2}.

We are now prepared to formulate the equivalence theorem for magnetostatics. Let us first take a special case of the general equivalence

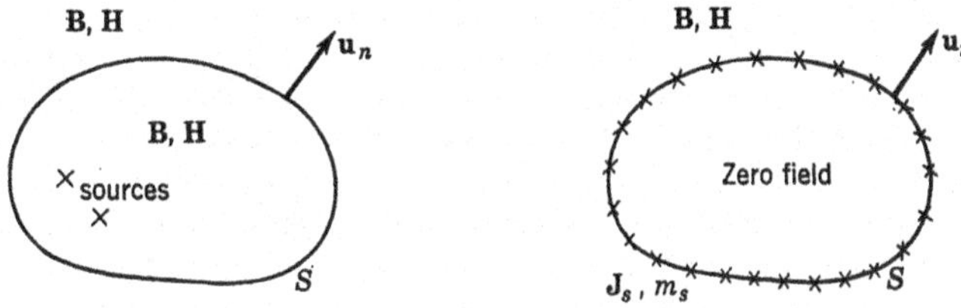

(*a*) Original problem (*b*) Equivalent problem

FIG. 7-8. A special case of the equivalence theorem.

theorem. Suppose we have a set of sources (say steady currents and magnetized matter) of finite extent, producing a magnetostatic field **B**, **H** throughout space. We choose an arbitrary closed surface S such that there is only vacuum external to S, as suggested by Fig. 7-8*a*. A second problem, equivalent to the original problem outside S, is now set up as follows. External to S we specify the field to be the same as in the original problem. Internal to S we specify the field to be zero. Both of these choices are possible solutions to the source-free magnetostatic field equations in vacuum (the external field is the field of the original problem; the internal field is the null field). To support such a field, there must be surface sources on S according to Eqs. (7-61) and (7-62). For our particular case, these reduce to

$$\mathbf{J}_s = \mathbf{u}_n \times \mathbf{H} \qquad m_s = B_n \tag{7-63}$$

where $\mathbf{u}_n$ points outward from S, and **H** and **B** are those of the original problem. According to our concepts of uniqueness, the sources specified by Eqs. (7-63), existing on S and acting in vacuum, must produce the original field external to S and zero field internal to S. This is illustrated by Fig. 7-8*b*.

The field produced by the equivalent sources of Fig. 7-8*b* can now be expressed in terms of the potential integrals. We have already defined the scalar magnetic potential integral U to be the potential integral of the flow sources of $\mathbf{H}$, so

$$U(\mathbf{r}) = \frac{1}{4\pi\mu_0} \oint\!\!\oint \frac{m_s(\mathbf{r}')}{|\mathbf{r} - \mathbf{r}'|}\, ds'$$

where m_s is given by Eqs. (7-63) and the integration is carried out over S. The magnetic vector potential $\mathbf{A}$ is

$$\mathbf{A}(\mathbf{r}) = \frac{\mu_0}{4\pi} \oint\!\!\oint \frac{\mathbf{J}_s(\mathbf{r}')}{|\mathbf{r} - \mathbf{r}'|}\, ds'$$

where $\mathbf{J}_s$ is given by Eqs. (7-63) and the integration is carried out over S. The field of the equivalent problem is then given by

$$\mu_0\mathbf{H} = \mathbf{B} = -\mu_0\nabla U + \nabla \times \mathbf{A}$$

where U and $\mathbf{A}$ are given above. Substituting from Eqs. (7-63) and combining the above equations, we obtain the formula

$$4\pi\mathbf{B}(\mathbf{r}) = -\nabla \oint\!\!\oint \frac{\mathbf{B}(\mathbf{r}') \cdot d\mathbf{s}'}{|\mathbf{r} - \mathbf{r}'|} - \nabla \times \oint\!\!\oint \frac{\mathbf{B}(\mathbf{r}') \times d\mathbf{s}'}{|\mathbf{r} - \mathbf{r}'|} \tag{7-64}$$

which is valid for $\mathbf{r}$ outside S if all original sources are within S. Eq. (7-64) is, of course, a mathematical identity, and can be proved in a mathematical manner (see Prob. 7-24).

Before considering the general equivalence principle, let us look at the analogous electrostatic case. The fundamental equations of electrostatic fields in linear media are Eqs. (5-1), so long as the mathematical problem is to represent the physical problem everywhere. If we are interested in the electrostatic field only within a given region of space, then there is no mathematical reason why we cannot consider this field to be produced, at least in part, by vortex sources of $\mathbf{E}$ external to the given region. Thus, we can introduce into our equations vortex sources of $\mathbf{E}$ to serve as equivalent sources. The *generalized equations of electrostatics* then become

$$\nabla \times \mathbf{E} = -\mathbf{K} \qquad \nabla \cdot \mathbf{D} = q_v \tag{7-65}$$

where $\mathbf{K}$ is called the *magnetic-current density.* So far as electrostatics is concerned, there is no apparent reason for the minus sign in the first of Eqs. (7-65), and we can view this sign merely as a convention. However, when general time-varying fields are considered, it is found that $\mathbf{K}$ can be viewed as the rate of flow of magnetic charge. If $\mathbf{K}$ and m_v are to satisfy the same continuity equation as $\mathbf{J}$ and q_v, then the minus sign is necessary. It follows from the first of Eqs. (7-65) that $\mathbf{K}$ has the dimensions of voltage per area. The unit of volume distributions of magnetic-current density in the mksc system is therefore the *volt per square meter.* The surface integral of the normal component of **K** we call simply the *magnetic current k;* that is

$$k = \iint \mathbf{K} \cdot d\mathbf{s} \tag{7-66}$$

Since **K** has the dimensions of voltage per area, k must have the same dimensions as voltage. The unit of magnetic current in the mksc system is, therefore, the *volt.* We may have occasion to use the concepts of surface distributions of magnetic current $\mathbf{K}_s$ and of line distributions k. These are defined as analogous to the corresponding electrical concepts, Sec. 2-4. The integral forms of Eqs. (7-65) are

$$\oint \mathbf{E} \cdot d\mathbf{l} = -k \qquad \oint\!\!\oint \mathbf{D} \cdot d\mathbf{s} = q \tag{7-67}$$

where k is the magnetic current through any surface bounded by the contour to which the line integral is applied and where q is the electric charge enclosed by the surface over which the surface integral is taken.

The minus sign appearing in the first of Eqs. (7-67) indicates that the lines of **E** tend to encircle a magnetic current in the direction opposite

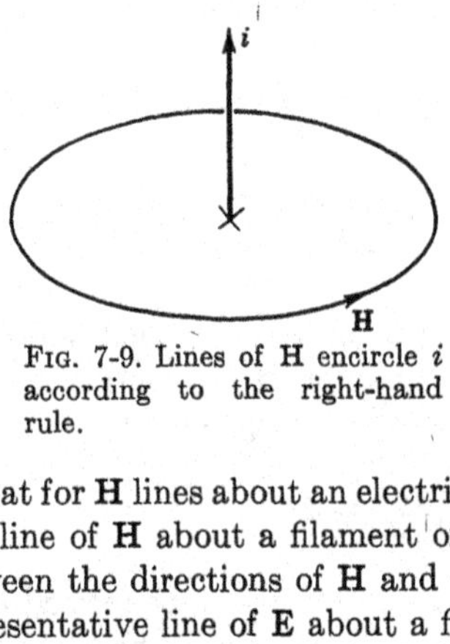

FIG. 7-9. Lines of **H** encircle i according to the right-hand rule.

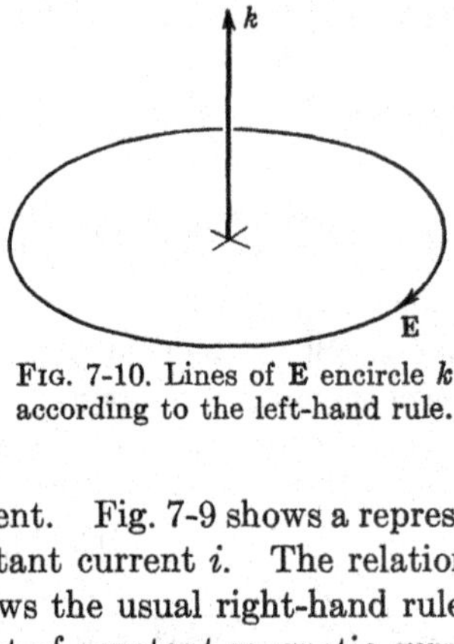

FIG. 7-10. Lines of **E** encircle k according to the left-hand rule.

to that for **H** lines about an electric current. Fig. 7-9 shows a representative line of **H** about a filament of constant current i. The relationship between the directions of **H** and i follows the usual right-hand rule. A representative line of **E** about a filament of constant magnetic current k is shown in Fig. 7-10. The relationship between the directions of **E** and k therefore obeys the *left-hand* rule. (The fingers of the left hand point in the direction of **E**, the thumb in the direction of k or **K**.)

If we have a surface distribution of magnetic current $\mathbf{K}_s$, then in the usual manner

$$\mathbf{u}_n \times (\mathbf{E}^{(1)} - \mathbf{E}^{(2)}) = -\mathbf{K}_s \tag{7-68}$$

where $\mathbf{u}_n$ points into region (1). The minus sign in Eq. (7-68) expresses the fact that $-\mathbf{K}$ are the vortex sources of **E**. If we have a surface distribution of electric charge q_s, then

$$D_n^{(1)} - D_n^{(2)} = q_s \tag{7-69}$$

which is the same relationship as that which holds for actual media. Note that Eq. (7-68) reduces to the form applicable to actual media where $\mathbf{K}_s = 0$.

All of the generalized equations for electrostatics have the same mathematical form as the generalized equations for magnetostatics. Therefore, conclusions reached for generalized electrostatics apply directly to generalized magnetostatics, and vice versa. If the medium is linear and homogeneous everywhere, that is, $\mathbf{D} = \epsilon\mathbf{E}$ with ϵ constant, the uniqueness theorem of Sec. 7-1 applies directly. If the medium is linear but not homogeneous, that is, if ϵ is a function of position, the more general uniqueness theorem mentioned for magnetostatics applies (see Prob. 7-23).

The special case of the equivalence theorem, already formulated for magnetostatics, applies to electrostatics if we make the appropriate changes of symbols and of terminology. The original problem (analogous to Fig. 7-8*a*) consists of sources producing an electrostatic field **E**, **D** throughout space, with only vacuum external to the surface S. Following the same reasoning as before, we arrive at the conclusion that the surface sources

$$\mathbf{K}_s = \mathbf{E} \times \mathbf{u}_n \qquad q_s = D_n \tag{7-70}$$

with vacuum everywhere, produce exactly the same field **E**, **D** external to S, and zero field internal to S. This would be illustrated by a figure analogous to Fig. 7-8*b*.

We can obtain the field produced by the above equivalent sources by means of the potential integrals. The scalar potential integral associated with the q_s is the usual electric scalar potential V, given by

$$V(\mathbf{r}) = \frac{1}{4\pi\epsilon_0} \oiint \frac{q_s(\mathbf{r}')}{|\mathbf{r} - \mathbf{r}'|} \, ds'$$

with q_s given by Eq. (7-70). The vector potential can be formed in the usual manner, as described in Sec. 7-1. A special symbol has been given to the vector potential that involves vortex sources of **E**. Thus, the *electric vector potential* **F** is defined according to

$$\mathbf{F}(\mathbf{r}) = \frac{\epsilon}{4\pi} \iiint \frac{\mathbf{K}(\mathbf{r}')}{|\mathbf{r} - \mathbf{r}'|} \, d\tau' \tag{7-71}$$

where this equation is intended to include symbolically the cases of surface and line distributions of magnetic current. Note that, strictly speaking, **F** is the vector potential of $-\mathbf{D}$ because of the minus sign in

Eqs. (7-65). The **D** associated with **F** is given by $\mathbf{D} = -\nabla \times \mathbf{F}$, in contrast with the magnetic case where $\mathbf{B} = \nabla \times \mathbf{A}$. Returning to our equivalent problem, the electric vector potential associated with the equivalent magnetic currents is

$$\mathbf{F}(\mathbf{r}) = \frac{\epsilon_0}{4\pi} \oiint \frac{\mathbf{K}_s(\mathbf{r}')}{|\mathbf{r} - \mathbf{r}'|} \, ds'$$

with $\mathbf{K}_s$ given by Eq. (7-70). The field produced by the equivalent sources is then

$$\epsilon_0 \mathbf{E} = \mathbf{D} = -\epsilon_0 \nabla V - \nabla \times \mathbf{F}$$

with V and **F** given above.

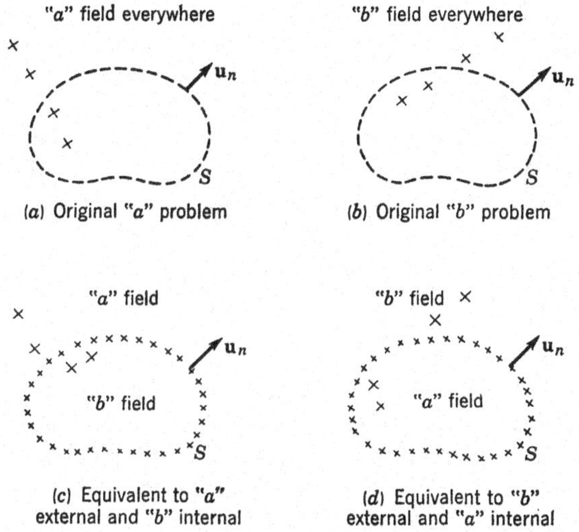

FIG. 7-11. The general form of the equivalence principle.

We shall now proceed to the general form of the static equivalence principle. To include both electrostatics and magnetostatics, we shall use the term "field" to represent either **E**, **D**, or **B**, **H** and we shall use the term "source" to represent either **K**, q_v, or **J**, m_v. Consider two distinct problems involving sources and matter in an unbounded region, as suggested by Figs. 7-11*a* and *b*. In each problem we define the same mathematical surface S. We can set up a problem equivalent to a external to S and b internal to S as follows. External to S, we place the a sources and matter that are external to S in the a problem, and specify the field to be the original a field. Internal to S, we place the b sources and matter that are internal to S in the b problem, and specify the field to be the original b field. To support this postulated field, we must also have surface distributions of flow sources and vortex sources on S in the usual manner. In an electrostatic problem, according to Eqs. (7-68) and (7-69), these are

$$\begin{aligned} \mathbf{K}_s &= (\mathbf{E}^a - \mathbf{E}^b) \times \mathbf{u}_n \\ q_s &= D_n{}^a - D_n{}^b \end{aligned} \tag{7-72}$$

where $\mathbf{u}_n$ points outward from S, the $\mathbf{E}^a$, $\mathbf{D}^a$ are the field vectors over S in the original a problem; the $\mathbf{E}^b$, $\mathbf{D}^b$ are the field vectors over S in the original b problem. In a magnetostatic problem, the surface sources according to Eqs. (7-61) and (7-62) are

$$\begin{aligned} \mathbf{J}_s &= \mathbf{u}_n \times (\mathbf{H}^a - \mathbf{H}^b) \\ m_s &= B_n{}^a - B_n{}^b \end{aligned} \tag{7-73}$$

where the notation is the same as the notation for the electrostatic case. It follows from the uniqueness theorem that the surface sources defined above, plus those original sources that have been retained, do actually produce the assumed field. This equivalent problem that we have been discussing is illustrated by Fig. 7-11*c*. We can also set up the alternative case for which the problem is equivalent to b externally and a internally. This follows the same line of reasoning with a and b interchanged; so the surface sources would then be given by Eqs. (7-72) or (7-73) with a replaced by b and b by a. This is illustrated by Fig. 7-11*d*. Note that the simplified case treated earlier, Fig. 7-8, is the special case for which all of the a sources are within S and the b sources are everywhere zero.

Let us give a couple of examples to illustrate the use of the equivalence principle before discussing it further. Suppose that we had a bar magnet of the type discussed in Sec. 7-4 and that we did not know how to determine its field in terms of its magnetization. Exploring its field with a compass, or perhaps with iron filings, we note that the field lines appear to be leaving one end of the magnet and entering the other end. We conclude that, at least approximately, the field is normal to the ends of the magnet and that it is essentially zero over the sides of the magnet. We then proceed to apply the special case of the equivalence theorem, Fig. 7-8, to a surface coinciding with that of the magnet. Applying Eqs. (7-63), we find that, according to our assumed field, the equivalent sources consist of two plates of magnetic charge, situated over that part of S covered by the poles of the magnet. If we assume that the magnetic charge density is uniform and if we calculate the field, we arrive at precisely the solution illustrated by Fig. 7-7*b*. Actually, this exact equality is an accident, for the field produced by two charged plates is not zero over the sides of S and is not perpendicular over the top and bottom (see Fig. 7-7*b*). However, this example does serve to illustrate how the equivalence principle, coupled with fortunate guessing, can lead to approximate solutions.

Another example, this time mathematically exact, but physically unimportant, is as follows. A point charge q, situated at the coordinate origin, produces a field

$$\mathbf{D} = \frac{q}{4\pi r^2}\,\mathbf{u}_r$$

Suppose we apply the equivalence principle to an infinite plane passing through the charge. Taking the plane to be horizontal, we wish to keep the same field above the plane and to have zero field below the plane. Since the surface cuts through the source, we must retain that part which is above the plane, that is, half of the point charge. Applying

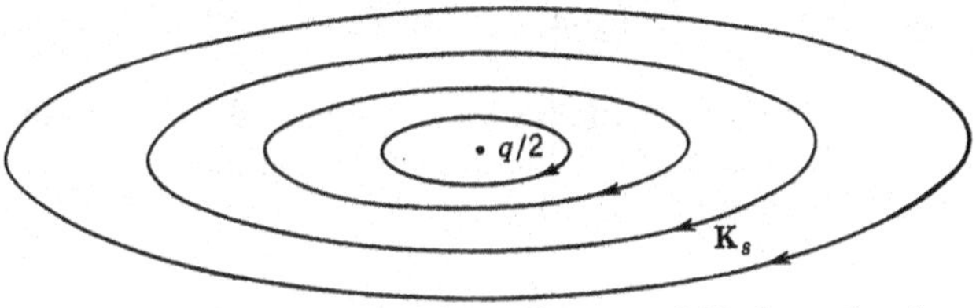

FIG. 7-12. This source distribution produces the same field above the plane of $\mathbf{K}_s$ as does a point charge q.

Eqs. (7-68) and (7-69) to the plane, for the assumed conditions, we find that $q_s = 0$ and

$$\mathbf{K}_s = \frac{-q}{4\pi\epsilon r^2}\,\mathbf{u}_\phi \tag{7-74}$$

Therefore, this ϕ-directed magnetic-current sheet plus a point charge of strength $q/2$ at the origin produces the same field as a point charge q above the sheet and zero field below. The source distribution for this case is shown in Fig. 7-12.

Suppose we now apply superposition to the situation illustrated by Fig. 7-12. We know that the point charge $q/2$, acting alone, produces half the field of a point charge q. This is just half the total field that we know to exist above the plane of the current. Therefore, the magnetic-current sheet, acting alone, must produce a like field above its plane. Below its plane, the magnetic-current sheet must produce a field exactly the negative of the point charge, since in this region the total field is zero. We therefore know the field produced by the magnetic-current sheet alone. It is

$$\mathbf{D} = \begin{cases} \dfrac{q}{8\pi r^2}\,\mathbf{u}_r & \text{above the sheet} \\[2ex] \dfrac{-q}{8\pi r^2}\,\mathbf{u}_r & \text{below the sheet} \end{cases} \tag{7-75}$$

If we compare Eqs. (7-65) with Eqs. (7-58), it is evident that an electric current numerically equal to the negative of a magnetic current produces an $\mathbf{H}$ field numerically equal to the $\mathbf{E}$ field of the magnetic current. In

Eq. (7-74), let $q/4\pi\epsilon = A$, an amplitude factor. Then, substituting for q in Eq. (7-75), we have $D = \epsilon A/2r^2$, or $E = A/2r^2$. Therefore, an infinite plane sheet of electric currents given by

$$\mathbf{J}_s = \frac{A}{r^2}\mathbf{u}_\phi \tag{7-76}$$

produces a field

$$\mathbf{H} = \begin{cases} \dfrac{A}{2r^2}\mathbf{u}_r & \text{above the sheet} \\ \dfrac{-A}{2r^2}\mathbf{u}_r & \text{below the sheet} \end{cases} \tag{7-77}$$

Fig. 7-13 shows a sketch of the field associated with the current sheet of Eq. (7-76). Note that this result can be obtained more directly by using the uniqueness concepts of Sec. 7-2.

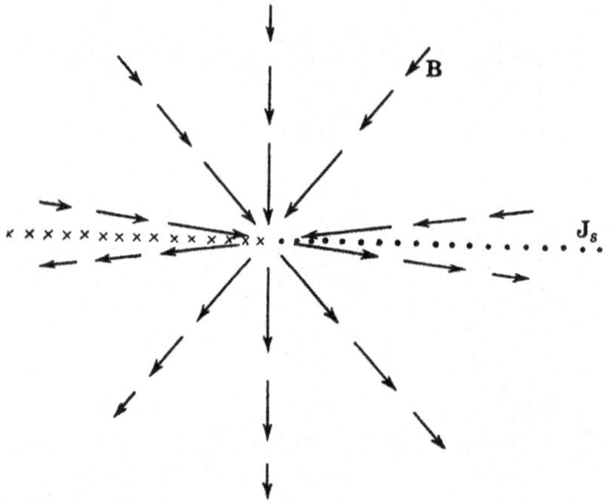

FIG. 7-13. The **B** associated with the electric current sheet of Eq. (7-76).

The treatment of problems involving polarized and magnetized matter, as discussed in Secs. 7-3 and 7-4, can be included within the framework of the equivalence principle. If we write the generalized equations of electrostatics for vacuum in a form showing explicitly the sources of **E**, we have

$$\nabla \cdot \mathbf{E} = \frac{1}{\epsilon_0}\hat{q}_v \qquad \nabla \times \mathbf{E} = -\hat{\mathbf{K}} \tag{7-78}$$

where the carets indicate equivalent sources rather than "actual" sources. Comparing Eqs. (7-78) with Eqs. (7-23), we see that

$$\hat{q}_v = q_v - \nabla \cdot \mathbf{P} \qquad \hat{\mathbf{K}} = 0 \tag{7-79}$$

can be considered to be equivalent sources for polarized media. Such sources acting in vacuum would give the correct **E** everywhere, but **D** would be correct only external to matter. In other words, Eqs. (7-79) would give equivalent sources for **D** only in regions external to polarized matter. Note that $\hat{q}_v$ of Eqs. (7-79) is the same as $\hat{q}_v$ of Eq. (7-24); that is, the equivalent charge in this case is identical with the absolute charge defined in Sec. 7-3. If we write the generalized equations of electrostatics for vacuum in a form showing explicitly the sources of **D**, we have

$$\nabla \cdot \mathbf{D} = \hat{q}_v \qquad \nabla \times \mathbf{D} = -\epsilon_0 \hat{\mathbf{K}} \tag{7-80}$$

Comparing these with Eqs. (7-33), we see that

$$\hat{q}_v = q_v \qquad \hat{\mathbf{K}} = -\frac{1}{\epsilon_0} \nabla \times \mathbf{P} \tag{7-81}$$

can be considered to be equivalent sources for polarized media. In this case, the sources of Eqs. (7-81) acting in vacuum would give the correct **D** everywhere, but **E** would be correct only external to matter. Thus, the sources of Eqs. (7-81) would be equivalent sources for **E** only in the region external to polarized matter.

The analogous procedure applies to magnetostatics. Writing the generalized equations of magnetostatics for vacuum showing explicitly the sources of **B**, we have

$$\nabla \cdot \mathbf{B} = \hat{m}_v \qquad \nabla \times \mathbf{B} = \mu_0 \hat{\mathbf{J}} \tag{7-82}$$

the carets again indicating equivalent sources. Comparing these with Eqs. (7-41), we can set

$$\hat{m}_v = 0 \qquad \hat{\mathbf{J}} = \mathbf{J} + \nabla \times \mathbf{M} \tag{7-83}$$

and obtain equivalent sources for **B** everywhere. However, only in the region external to magnetized matter would these sources acting in vacuum be equivalent for **H**. Note that the equivalent current in this case is equal to the absolute current defined by Eq. (7-42). If we write the generalized equations of magnetostatics for vacuum showing the sources of **H**, we have

$$\nabla \cdot \mathbf{H} = \frac{1}{\mu_0} \hat{m}_v \qquad \nabla \times \mathbf{H} = \hat{\mathbf{J}} \tag{7-84}$$

Comparing these with Eqs. (7-51), we can set

$$\hat{m}_v = -\mu_0 \nabla \cdot \mathbf{M} \qquad \hat{\mathbf{J}} = \mathbf{J} \tag{7-85}$$

and obtain equivalent sources for **H** everywhere. However, only in regions external to magnetized matter would these sources acting in vacuum be equivalent for **B**.

7-6. Duality. The concept of duality is merely a formal recognition that, if two problems have equations of the same mathematical form, they have solutions of the same mathematical form. Two equations having the same mathematical form are called *dual equations.* Quantities which occupy the same mathematical position in dual equations are called *dual quantities.*

To illustrate the duality concept, let us consider the static equivalence principle. The generalized equations of electrostatics are summarized

TABLE 7-1. DUAL EQUATIONS FOR STATIC FIELDS

Generalized electrostatics	*Generalized magnetostatics*
$\nabla \cdot \mathbf{D} = q_v$	$\nabla \cdot \mathbf{B} = m_v$
$\nabla \times \mathbf{E} = -\mathbf{K}$	$\nabla \times \mathbf{H} = \mathbf{J}$
$\mathbf{D} = \epsilon \mathbf{E}$	$\mathbf{B} = \mu \mathbf{H}$
$\mathbf{E} = -\nabla V - \frac{1}{\epsilon} \nabla \times \mathbf{F}$	$\mathbf{H} = -\nabla U + \frac{1}{\mu} \nabla \times \mathbf{A}$
$V = \frac{1}{4\pi\epsilon} \iiint \frac{q_v \, d\tau'}{\lvert \mathbf{r} - \mathbf{r}' \rvert}$	$U = \frac{1}{4\pi\mu} \iiint \frac{m_v \, d\tau'}{\lvert \mathbf{r} - \mathbf{r}' \rvert}$
$\mathbf{F} = \frac{\epsilon}{4\pi} \iiint \frac{\mathbf{K} \, d\tau'}{\lvert \mathbf{r} - \mathbf{r}' \rvert}$	$\mathbf{A} = \frac{\mu}{4\pi} \iiint \frac{\mathbf{J} \, d\tau'}{\lvert \mathbf{r} - \mathbf{r}' \rvert}$

in the first column of Table 7-1. In the second column are listed the dual equations for generalized magnetostatics. By inspection, we can now form a table of dual quantities, Table 7-2. If we take the equations for electrostatics in Table 7-1 and replace symbols of the first column of Table 7-2 by their duals in the second column, we obtain the equations for magnetostatics in Table 7-1. Similarly, any solution to the equations of the first column of Table 7-1 can be converted to a solution of the equations of the second column, and vice versa, by an interchange of symbols according to Table 7-2.

TABLE 7-2. DUAL QUANTITIES FOR STATIC FIELDS

Generalized electrostatics	*Generalized magnetostatics*
$\mathbf{E}$	$\mathbf{H}$
$\mathbf{D}$	$\mathbf{B}$
q	m
$-\mathbf{K}$	$\mathbf{J}$
ϵ	μ
V	U
$-\mathbf{F}$	$\mathbf{A}$

Whenever quantities satisfy dual relationships, it is possible to visualize them according to *dual concepts.* A dual concept need have no relationship to "reality." It is merely a recognition that, since the equations for dual quantities are the same, a mental picture which holds for one quantity will hold also for its dual quantity. These mental pictures are of value because they aid in our thinking and serve as a guide to the mathematics. In fact, it was the duality relationship of the electromagnetic field equations to equations for an elastic, incompressible fluid that first enabled Maxwell to formulate the field equations. Maxwell was so impressed by this duality of equations that it led him to postulate the existence of an ideal fluid called "ether" permeating all space. The concept of an ether has since been discarded, but it served the early investigators of electromagnetic theory well as a guide to their thinking. A dual concept that can be of help to us is one between elec-

TABLE 7-3. SUMMARY OF THE USE OF THE POTENTIAL INTEGRAL SOLUTION FOR STATIC FIELDS

Given $\nabla \cdot \mathbf{C} = w$, $\nabla \times \mathbf{C} = \mathbf{W}$ everywhere, then

$$\mathbf{C} = -\nabla g + \nabla \times \mathbf{G}$$

where

$$g = \frac{1}{4\pi} \iiint \frac{w \, d\tau'}{|\mathbf{r} - \mathbf{r}'|} \qquad \mathbf{G} = \frac{1}{4\pi} \iiint \frac{\mathbf{W} \, d\tau'}{|\mathbf{r} - \mathbf{r}'|}$$

	C	w	W	w_s	$\mathbf{W}_s$	g	G
Electric charge (homogeneous media)	E	$\frac{q_v}{\epsilon}$	0	$\frac{q_s}{\epsilon}$	0	V	0
	D	q_v	0	q_s	0	ϵV	0
Electric current (homogeneous media)	B	0	$\mu\mathbf{J}$	0	$\mu\mathbf{J}_s$	0	A
	H	0	J	0	$\mathbf{J}_s$	0	$\frac{1}{\mu}\mathbf{A}$
Polarized matter (no free charge)	E	$\frac{-1}{\epsilon_0}\nabla \cdot \mathbf{P}$	0	$\frac{1}{\epsilon_0}P_n$	0	V	0
	D	0	$\nabla \times \mathbf{P}$	0	$\mathbf{P} \times \mathbf{u}_n$	0	$-\mathbf{F}$
Magnetized matter (no free current)	B	0	$\mu_0\nabla \times \mathbf{M}$	0	$\mu_0\mathbf{M} \times \mathbf{u}_n$	0	A
	H	$-\nabla \cdot \mathbf{M}$	0	M_n	0	U	0
Generalized statics (homogeneous media)	E	$\frac{q_v}{\epsilon}$	$-\mathbf{K}$	$\frac{q_s}{\epsilon}$	$-\mathbf{K}_s$	V	$-\frac{1}{\epsilon}\mathbf{F}$
	D	q_v	$-\epsilon\mathbf{K}$	q_s	$-\epsilon\mathbf{K}_s$	ϵV	$-\mathbf{F}$
	B	m_v	$\mu\mathbf{J}$	m_s	$\mu\mathbf{J}_s$	μU	A
	H	$\frac{m_v}{\mu}$	J	$\frac{m_s}{\mu}$	$\mathbf{J}_s$	U	$\frac{1}{\mu}\mathbf{A}$

tric and magnetic charge and current. We already have a good mental picture of electric charge and current. This is because we have used them so often. The same picture holds for magnetic charge and current, for these quantities satisfy equations of the same form. Thus, we can visualize collections of magnetic charge, and we can visualize the motion of magnetic charge as giving rise to magnetic current.

Actually, all of the solutions we have been able to carry out up to now follow one general pattern of procedure, that of determining all sources of a given vector and then applying the potential integrals. Table 7-3 summarizes the notation used for the various cases we have considered.

The solution of one problem by the potential-integral method can be translated readily into the solution of other problems, through appropriate interchange of symbols. For example, we have already seen that a disk of uniformly polarized matter and a disk of uniformly magnetized matter present essentially the same problem. This problem is equivalent also to the problem of two charged plates and to the problem of a current loop.

7-7. Dipole Layers and Current Loops. At the end of Sec. 6-7 we showed how a current loop might be considered in terms of a grid of infinitesimal loops. Each infinitesimal loop behaves like a point dipole, so we conclude that a loop of current is equivalent to a surface layer of magnetic dipoles. We now wish to reconsider this situation in the light of the concepts of this chapter.

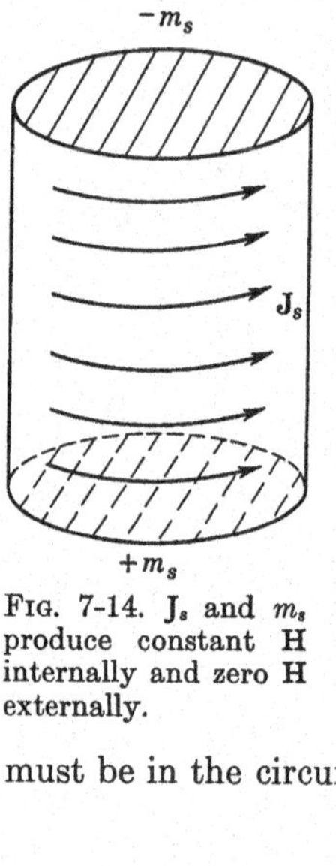

FIG. 7-14. $\mathbf{J}_s$ and m_s produce constant H internally and zero H externally.

By means of the equivalence principle, let us determine the sources that produce the following **H** field. Internal to a given cylinder of finite length, we wish the field to be a constant **H** in the direction of the cylinder axis, which is a valid solution to the static field equations. External to the given cylinder, we wish the field to vanish, that is, to be the null solution to the field equations. In order to determine the sources over the cylinder that are required to produce this field, we apply Eqs. (7-73). Since **H** internally is constant and in the axial direction and since **H** externally is zero, it follows that $\mathbf{J}_s$ must be in the circumferential direction and of magnitude

$$J_s = H \tag{7-86}$$

J_s exists only over the sides of the cylinder. Since $\mathbf{B} = \mu\mathbf{H}$ internally is constant and in the axial direction and since **B** is zero externally, it follows that m_s must exist over the ends of the cylinder and that its magnitude must be

$$|m_s| = \mu H \tag{7-87}$$

The magnetic charge must be positive on the end of the cylinder where the **H** lines begin and must be negative on the end where they terminate. It follows from the uniqueness theorem that the above-determined sources produce the assumed field. Note that this conceptual model, illustrated by Fig. 7-14, gives precisely the field of the ideal inductor postulated in Sec. 4-7.

The field from the sources of Fig. 7-14 is produced by the combined action of both the current and the charge. Part of the field is due to the current, and part is due to the charge. Since superposition applies, it follows that, external to the cylinder, the field produced by the charges alone must be exactly the negative of that produced by the currents alone. This must be so since the sum of the two fields is zero. If we replace the positive magnetic charge by negative charge, and vice versa, *this distribution produces exactly the same magnetic field external to the cylinder as does the electric current.* Eliminating H between Eqs. (7-86) and (7-87), we find that the magnetic charge is related to the current sheet according to

$$|m_s| = \mu J_s \tag{7-88}$$

A sketch of the **H** lines associated with the cylinder of currents would be Fig. 7-7*a*; a sketch of the **H** lines associated with the two charged plates would be Fig. 7-7*b*. (Our present problem is thus seen to be dual to the problem of the ideal bar magnet.) Note that the external fields are the same in both cases, but the internal fields are not. However, addition of the internal field from the currents to the negative of the internal field from the charges must give the constant $H = J_s$ in the axial direction.

The above analysis is independent of the shape of the cross section of the cylinder. It requires only that the surface current be constant and directed perpendicular to the cylinder axis. To solve for the field from the cylinder of currents, we could use the vector potential integral. To solve for the field from the charged plates, we could use the scalar potential integral. Since the two solutions give the same magnetic field external to the cylinder, this shows that either a vector potential or a scalar potential can be used in the external region. This we already know, for external to the cylinder both $\nabla \times \mathbf{H} = 0$ and $\nabla \cdot \mathbf{H} = 0$. Instead of using the concept of equivalent sources, we could find these potentials by purely mathematical methods. However, most of us can remember concepts more readily than formal mathematics, so we choose to use the equivalent magnetic-charge picture.

If we now let the length of the cylinder become infinitesimal, we obtain a loop of current from the cylinder of current and a dipole layer of charge from the charged plates. Thus, *the field from a loop of constant current is the same as the field from a uniform dipole layer of charge.* Letting l represent the infinitesimal length of the cylinder, the total current in the loop is given by

$$i = J_s l \tag{7-89}$$

The dipole moment per unit area of the dipole layer is by definition $m_s \mathbf{l}$, where $\mathbf{l}$ points from negative to positive charge. Using Eq. (7-88) and Eq. (7-89), it follows that the surface dipole-moment density equivalent to the current loop is

$$m_s l = \mu i \tag{7-90}$$

Since we know how to analyze current loops (by the vector potential integral), we can now treat uniform dipole layers. If we had treated dipole layers previously, we could have used those results for the equivalent current loop. Incidentally, if we assume that the electric charge on the plates of a parallel-plate capacitor is distributed uniformly, we have the problem dual to that of the two layers of magnetic charge. We therefore conclude that *the electric field external to a thin parallel-plate capacitor is approximately the same as the magnetic field from a current loop coincident with the edge of the capacitor.*

Suppose we now go a step further and let the cross section of the cylinder, as well as the length, become infinitesimal. The two plates of magnetic charge now become point charges of strength

$$m = m_s S \tag{7-91}$$

where S is the cross-sectional area of the cylinder. These point charges are separated by the infinitesimal distance l, so we have a point dipole of strength ml. Substituting from Eq. (7-91) into Eq. (7-90), we have

$$ml = \mu i S \tag{7-92}$$

which gives the *charge-dipole moment* ml in terms of the equivalent *current-dipole moment* iS. The two produce numerically the same field only if the charge-dipole moment is μ times the current-dipole moment. It was because we wanted to relate magnetization $\mathbf{M}$ to elementary current loops rather than to charge dipoles that we defined $\mu_0\mathbf{M}$ mathematically analogous to polarization $\mathbf{P}$, which we related to charge dipoles. At any rate, we have once again reached the conclusion that a small loop of current produces the same field as a charge dipole. Our previous analysis, Sec. 6-7, was actually done for a circular loop, whereas our present analysis is valid for loops of arbitrary shape.

Since the magnetic charge dipole is the problem dual to the electric charge dipole, we already have the solution to it. Substituting dual quantities from Table 7-2 into Eq. (5-81), we have the magnetic scalar potential U associated with the magnetic charge dipole given by

$$U(\mathbf{r}) = \frac{m\mathbf{l}(\mathbf{r}') \cdot (\mathbf{r} - \mathbf{r}')}{4\pi\mu|\mathbf{r} - \mathbf{r}'|^3} \tag{7-93}$$

If the dipole is z-directed and situated at the coordinate origin, we have, from $\mathbf{H} = -\nabla U$ or from Eq. (5-80) by duality, the field given by

$$\mathbf{H} = \frac{ml}{4\pi\mu r^3}(\mathbf{u}_r 2\cos\theta + \mathbf{u}_\theta \sin\theta) \tag{7-94}$$

The field from a small current loop is identical with that from the charge dipole if Eq. (7-92) is satisfied. Therefore, substituting from Eq. (7-92) into Eq. (7-93), we have a magnetic scalar potential associated with a small current loop given by

$$U(\mathbf{r}) = \frac{i\mathbf{S}(\mathbf{r}') \cdot (\mathbf{r} - \mathbf{r}')}{4\pi|\mathbf{r} - \mathbf{r}'|^3} \tag{7-95}$$

where the direction of $\mathbf{S}$ is assigned according to Fig. 6-7. The field from a small loop located at the coordinate origin and lying in the x-y plane is given by Eq. (7-94) with ml replaced according to Eq. (7-92). Thus

$$\mathbf{H} = \frac{iS}{4\pi r^3} (\mathbf{u}_r 2 \cos \theta + \mathbf{u}_\theta \sin \theta) \tag{7-96}$$

which is identical with Eq. (6-58) with $S = \pi a^2$. We could also obtain a scalar magnetic potential associated with a finite-sized loop by applying Eq. (7-95) to each differential element of cross-sectional area and then integrating. However, in this case it is simpler to apply the vector potential integral.

Let us calculate the magnetic field due to a double layer of magnetic charge that is equivalent to a circular current loop and then compare this with the result obtained for the current loop itself (Sec. 6-3). To keep the mathematics simple, we shall consider the field only along the axis of the disk. First, consider a disk of single-layer magnetic charge m_s. We have already treated the dual problem, a disk of electric charge, in Sec. 5-2. Applying duality, according to Table 7-2, to Eq. (5-15), we have

$$B_z^+(0,0,z) = \frac{m_s z}{2}\left(\frac{1}{|z|} - \frac{1}{\sqrt{a^2 + z^2}}\right)$$

This is the only component of $\mathbf{B}$ along the axis of the disk (z axis). The superscript "+" is placed on B_z^+ to indicate that it is the $\mathbf{B}$ due to the disk of positive charge alone. The $\mathbf{B}$ produced by a disk of negative charge will be the negative of the $\mathbf{B}$ produced by a positive disk. But in the dipole layer, the negative disk is an incremental distance l below the positive disk. Therefore, the $\mathbf{B}$ from the negative disk B_z^- will be the negative of the above equation evaluated at $z + l$. Since l is an incremental distance, this is given by

$$B_z^- = -\left(B_z^+ + l\frac{\partial}{\partial z} B_z^+\right)$$

The $\mathbf{B}$ from the entire dipole layer is just the sum of B_z^+ and B_z^-, or

$$B_z(0,0,z) = -\frac{m_s l}{2}\frac{\partial}{\partial z}\left(\frac{z}{|z|} - \frac{z}{\sqrt{a + z^2}}\right)$$

Performing the indicated differentiation, we obtain

$$B_z(0,0,z) = \frac{m_s l a^2}{2(a^2 + z^2)^{3/2}} \tag{7-97}$$

But if this disk of dipole charge is to be numerically equivalent to a loop of current i of the same size, Eq. (7-90) must be satisfied. Substituting $m_s l = \mu i$ into Eq. (7-97), we have the identical answer that we obtained by the vector potential integral, Eq. (6-27).

QUESTIONS FOR DISCUSSION

7-1. Given all sources of a vector field, are its scalar and vector potentials unique?

7-2. Does the uniqueness theorem of Sec. 7-1 apply to two-dimensional problems?

7-3. Why is the scalar potential used in the example of Sec. 7-2 unique?

7-4. The vector **P** plays the dual role of a "field" quantity and a "source" quantity. Discuss this statement.

7-5. Discuss the philosophical difference between bound charge and free charge.

7-6. In view of the discussion of the last paragraph of Sec. 5-9, why would you expect a polarized dielectric to give a dipole field at distant points?

7-7. Since bound charge and free charge are fundamentally the same physical entity, why do we distinguish between them?

7-8. What is the relationship between the susceptibility χ_e and the dielectric constant ϵ_r?

7-9. Discuss the reason for the difference in definition between **P** and **M**.

7-10. What is the philosophical difference between bound current and free current?

7-11. Why is the concept of "bound magnetic poles" in magnetostatics more useful than the concept of "bound magnetic current" in electrostatics?

7-12. Discuss the difference between the **B** lines and the **H** lines of magnetized matter.

7-13. Why introduce the concepts of magnetic charge and magnetic current if they do not exist in nature?

7-14. Discuss the difference between "magnetic charge" and "magnetic poles."

7-15. The generalized equations for magnetostatics can be written as $u = i$, $\psi = m$. Similarly, the generalized equations for electrostatics can be written as $v = -k$, $\psi^e = q$. Explain this notation.

7-16. What are the mksc units of m_s, m_l, and $\mathbf{K}_s$?

7-17. Do the equivalent sources of Figs. 7-11*c* and *d* in general act in vacuum?

7-18. What is the definition of the term "equivalent sources"?

7-19. Discuss the concept of duality.

7-20. Make tables of duals corresponding to Tables 7-1 and 7-2 for the concepts of polarization and magnetization.

7-21. Discuss the use of Table 7-3.

7-22. A pair of charged plates and a cylinder of currents satisfying Eq. (7-88) produce the same field external to the cylinder. Do they produce the same field internal to the cylinder? If not, what is the relationship between the two internal fields?

7-23. Discuss the significance of the factor μ in Eq. (7-92), relating the charge dipole to the current dipole. What would be the relationship between a "magnetic pole dipole" and a current dipole?

7-24. What is the basis for the substitution $m_s l = \mu i$ mentioned in the last sentence of Sec. 7-7?

PROBLEMS

7-1. Given the potentials

$$g = re^{-r} \cos \theta \qquad \mathbf{G} = \mathbf{u}_\phi re^{-r} \sin \theta$$

determine the field **C**, its flow sources w, and its vortex sources **W** according to the concepts of Sec. 7-1.

7-2. Prove the uniqueness theorem of Sec. 7-1 by expressing the difference field $\delta\mathbf{C}$ in terms of a vector potential $\delta\mathbf{G}$ according to

$$\delta\mathbf{C} = \nabla \times \delta\mathbf{G}$$

7-3. Using the uniqueness theorem already proved in Sec. 7-1, show that the vector potential **G** is unique if, in addition to Eq. (7-7), we specify that

$$\nabla \cdot \mathbf{G} = 0$$

and that **G** vanishes at infinity more rapidly than $r^{-3/2}$. Show also that the scalar potential g is unique if it vanishes at infinity more rapidly than $r^{-1/2}$.

7-4. Using the results of Prob. 7-1 and the uniqueness theorem of Prob. 7-3, write down some definite integrals that we have evaluated.

7-5. Assume that external to a sphere of radius a there exists a field

$$\mathbf{C}^{(1)} = K_1 r^{-3}(\mathbf{u}_r\, 2 \cos \theta + \mathbf{u}_\theta \sin \theta)$$

and that internal to the sphere there exists

$$\mathbf{C}^{(2)} = \mathbf{u}_z K_2$$

where K_1 and K_2 are constants. Show that there is no volume distribution of sources for the field: $\mathbf{C} = \mathbf{C}^{(1)}$, $r > a$; $\mathbf{C} = \mathbf{C}^{(2)}$, $r < a$. Determine the relationship between K_1 and K_2 if there are only flow sources on the sphere $r = a$. Determine these flow sources in terms of K_1.

7-6. For the field of Prob. 7-5, determine the relationship between K_1 and K_2 if there are only vortex sources on the sphere $r = a$. Determine the vortex sources in terms of K_1.

7-7. Using the scalar potential of the field of Prob. 7-5, show that

$$\frac{3}{8\pi} \int_0^{2\pi} d\phi' \int_0^{\pi} d\theta' \frac{\sin 2\theta'}{\sqrt{r^2 + a^2 - 2ra \cos \psi}} = \begin{cases} \dfrac{a \cos \theta}{r^2} & r > a \\ \dfrac{r \cos \theta}{a^2} & r < a \end{cases}$$

where $\qquad \cos \psi = \cos \theta \cos \theta' + \sin \theta \sin \theta' \cos (\phi - \phi')$

Specialize this equation to the axis $\theta = 0$, showing

$$\frac{3}{4} \int_0^{\pi} \frac{\sin 2\theta\, d\theta}{\sqrt{r^2 + a^2 - 2ra \cos \theta}} = \begin{cases} \dfrac{a}{r^2} & r > a \\ \dfrac{r}{a^2} & r < a \end{cases}$$

7-8. Given a dielectric sphere of radius a, with uniform polarization

$$\mathbf{P} = \mathbf{u}_z P_0$$

where P_0 is a constant, determine the bound charge distribution. Using the results of Prob. 7-5, determine the **E** field both internal and external to the dielectric. From a knowledge of **P** and **E**, find the **D** field both internal and external to the dielectric. Note that the polarized sphere gives the dipole field at all points external to the sphere. Sketch the **E** lines and the **D** lines.

7-9. Obtain Eq. (7-32) from Eq. (7-26) using a limiting process.

7-10. Consider a polarized body in vacuum. Using Eq. (7-36) to represent the field from each element of the body, obtain an integral for the scalar potential. Show that this integral can be transformed mathematically into Eq. (7-32).

7-11. Given a magnetized sphere of radius a, with uniform magnetization

$$\mathbf{M} = \mathbf{u}_z M_0$$

where M_0 is a constant, determine the bound current distribution. Using the results of Prob. 7-6, determine the **B** field both internal and external to the sphere. From a knowledge of **B** and **M**, find the **H** field both internal and external to the sphere. Note that the magnetized sphere gives the dipole field at all points external to the sphere. Sketch the **B** and **H** lines. Compare this result with that of Prob. 7-8.

7-12. Obtain Eq. (7-50) from Eq. (7-44) using a limiting process.

7-13. Consider a toroidal body of magnetized matter with a narrow slot cut as shown in Fig. 7-15. The cross section of the toroid is circular. If the magnetization

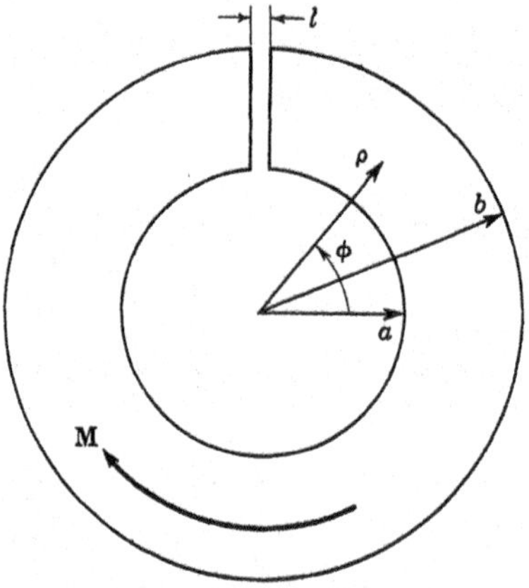

FIG. 7-15. Magnetized toroid for Prob. 7-13.

is given by

$$\mathbf{M} = \frac{-\mathbf{u}_\phi M_0}{\rho}$$

where M_0 is a constant, determine the bound pole distribution. Evaluate approximately the **H** field in the slot, and state your approximations.

7-14. Develop the general equation for the magnetic scalar potential analogous to Eq. (7-32) assuming no electric currents. Show mathematically that, *external to matter*, the two formulas: $\mathbf{H} = -\nabla U$ with U as derived above, and $\mathbf{H} = \mu_0^{-1}\nabla \times \mathbf{A}$ with **A** given by Eq. (7-50), are identical. Show that the formulas $\mathbf{H} = -\nabla U$ and $\mathbf{H} = \mu_0^{-1}\nabla \times \mathbf{A} - \mathbf{M}$ are identical everywhere.

7-15. Given a bar magnet of circular cross section of radius a, of length l, and with constant axial magnetization, derive an expression for the **B** field at the midpoint of the pole face. Plot a curve of B/M at this point vs. l/a for $0 \leq l/a \leq 10$.

7-16. Determine the equivalent sources on an infinite cylinder $\rho = a$ which give zero magnetic field $\rho < a$ and which give the magnetic dipole field $\rho > a$. Take the dipole field to the field from a z-directed moment iS at the coordinate origin.

7-17. Determine the equivalent sources on an infinite cylinder $\rho = a$ which give zero electric field $\rho < a$ and which give the point-charge field $\rho > a$. Take the point-charge field to be the field from a charge q at the coordinate origin.

7-18. Consider the infinite plane $z = 0$, and find the equivalent sources which in the region $z < 0$ give zero electric field and in the region $z > 0$ give the field of a charge q at $z = -a$. From symmetry considerations, show that the equivalent charge alone over the plane $z = 0$ gives in the region $z < 0$ the field of a point charge $q/2$ at $z = a$, and in the region $z > 0$ gives the field of a point charge $q/2$ at $z = -a$. Show that the equivalent magnetic current alone over the plane $z = 0$ gives in the region $z < 0$ the field of a point charge $-q/2$ at $z = a$, and in the region $z > 0$ gives the field of a point charge $q/2$ at $z = -a$. Extension of this line of reasoning can lead to the "image theory" of the next chapter.

7-19. Find the sources on the sphere $r = a$ necessary to terminate the field of the z-directed magnetic dipole at $r = 0$. In other words, we want the dipole field $r < a$ and zero field $r > a$. Note that the dipole itself must be present. What is the relationship of these "terminating" sources to the equivalent sources giving a dipole field $r > a$ and zero field $r < a$?

7-20. State the electric problem dual to Prob. 7-16, and write down the answer using the duality relationships of Table 7-2.

7-21. Obtain Eq. (6-60) using the equivalent magnetic charge according to the concepts of Sec. 7-7. Use the equation dual to Eq. (5-15) for the field from the disks of equivalent magnetic charge.

7-22. Establish Eq. (7-97) by considering each surface element of the dipole layer of magnetic charge as a point dipole of incremental moment and by integrating over all surface elements.

7-23. Show that the magnetostatic field is unique if the medium is linear, that is, if

$$\mathbf{B} = \mu \mathbf{H}$$

with μ at most a function of position, and if the flow sources of **B** and the vortex sources of **H** are specified, providing **B** and **H** vanish more rapidly than $r^{-3/2}$ at infinity. Give the corresponding statement of uniqueness for the electrostatic field.

7-24. Prove Eq. (7-64) mathematically, using vector identities.

CHAPTER 8

THE BOUNDARY-VALUE PROBLEM

8-1. Introduction. A boundary-value problem is a problem in which the field within a given region is to be found in terms of certain values of the field over the boundary of the region. We have already encountered the boundary-value concept in the equivalence principle of Sec. 7-5. There we saw that equivalent sources could be found in terms of the field over the bounding surface. A comprehensive treatment of boundary-value problems would require much more time than we are willing to spend in this book. We shall therefore concentrate on methods whereby boundary-value problems may be reduced to equivalent problems solvable by the techniques of Chaps. 5 and 6.

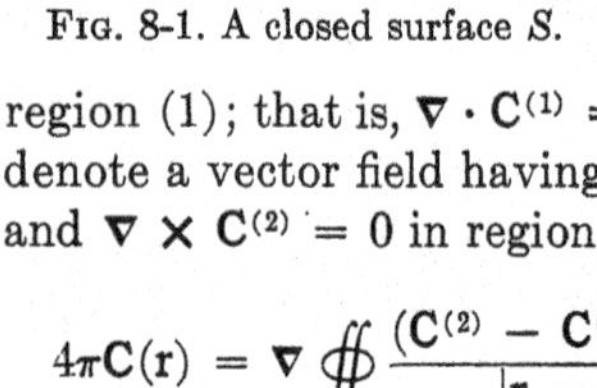

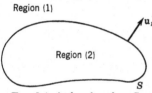

FIG. 8-1. A closed surface S.

Let us start our discussion using some of the concepts of Chap. 7. Consider a closed surface S dividing all space into two regions, as shown in Fig. 8-1. Let $\mathbf{C}^{(1)}$ denote a vector field having no sources in region (1); that is, $\nabla \cdot \mathbf{C}^{(1)} = 0$ and $\nabla \times \mathbf{C}^{(1)} = 0$ in region (1). Let $\mathbf{C}^{(2)}$ denote a vector field having no sources in region (2); that is, $\nabla \cdot \mathbf{C}^{(2)} = 0$ and $\nabla \times \mathbf{C}^{(2)} = 0$ in region (2). We can then derive the identity

$$4\pi\mathbf{C}(\mathbf{r}) = \nabla \oiint \frac{(\mathbf{C}^{(2)} - \mathbf{C}^{(1)}) \cdot d\mathbf{s}'}{|\mathbf{r} - \mathbf{r}'|} + \nabla \times \oiint \frac{(\mathbf{C}^{(2)} - \mathbf{C}^{(1)}) \times d\mathbf{s}'}{|\mathbf{r} - \mathbf{r}'|} \qquad (8\text{-}1)$$

where
$$\mathbf{C}(\mathbf{r}) = \begin{cases} \mathbf{C}^{(1)} & \mathbf{r} \text{ in region (1)} \\ \mathbf{C}^{(2)} & \mathbf{r} \text{ in region (2)} \end{cases}$$

This is established as follows: (*a*) assume the field $\mathbf{C} = \mathbf{C}^{(1)}$ external to S and $\mathbf{C} = \mathbf{C}^{(2)}$ internal to S, (*b*) determine the necessary surface sources on S in terms of $\mathbf{C}^{(1)}$ and $\mathbf{C}^{(2)}$, (*c*) form the potential integrals from these surface sources, and (*d*) determine the field from the potential integrals. Eq. (8-1) can be readily extended to include the case for which $\mathbf{C}^{(1)}$ and $\mathbf{C}^{(2)}$ have sources in their respective regions by adding the appropriate potential integrals for these sources (see Sec. 8-2).

In a boundary-value problem, we are interested in the field only in one region, say region (1) of Fig. 8-1. We can immediately obtain a solution for $\mathbf{C}$ external to S by choosing $\mathbf{C}^{(2)}$ in Eq. (8-1) to be the null field. This gives us Eq. (7-64), which is a formal solution to the boundary-value problem. However, we can choose $\mathbf{C}^{(2)}$ to be *any* source-free field in region (2), so we can write an infinity of formal solutions for region (1). All of these solutions give the same field in region (1). In view of this arbitrariness in the choice of $\mathbf{C}^{(2)}$, we might ask the question: Can $\mathbf{C}^{(2)}$ be chosen so that one of the integrals in Eq. (8-1) vanishes? If $\mathbf{u}_n \cdot \mathbf{C}^{(2)} = \mathbf{u}_n \cdot \mathbf{C}^{(1)}$ over S, only the second term of Eq. (8-1) would remain, giving a formula for $\mathbf{C}^{(1)}$ in region (1) in terms only of its tangential components over S. If $\mathbf{u}_n \times \mathbf{C}^{(2)} = \mathbf{u}_n \times \mathbf{C}^{(1)}$ over S, only the first term of Eq. (8-1) would remain, giving a formula for $\mathbf{C}^{(1)}$ in region (1) in terms only of its normal component. We then have the problem of determining the required $\mathbf{C}^{(2)}$ field. This actually involves the solution of a boundary-value problem for the required $\mathbf{C}^{(2)}$. Nevertheless, our question can be answered affirmatively in most cases. It is usually possible to determine a field in terms only of its normal component over the boundary, or only of its tangential components, or even of its normal component over part of the boundary and of its tangential components over the rest. We shall consider this topic further in the next section.

In some problems we have two or more different media in the region of interest. According to the concepts of Sec. 7-2, the normal component of a vector field is continuous across any boundary unless there is a surface distribution of flow sources. The tangential components of a vector field are continuous across any boundary unless there is a surface distribution of vortex sources. The usual procedure used in treating discontinuities in media is to obtain solutions in each medium and then "match" the fields over the boundaries between media.

In any region for which there are no vortex sources, the field can be expressed in terms of a scalar potential. In many cases it is easier to solve the boundary-value problem for the scalar potential than for the field vector. In any region for which there are no flow sources, the field can be expressed in terms of a vector potential. Thus, we could solve the boundary-value problem for the vector potential instead of for the field vector. However, for our purposes it is more convenient to solve for the field vector than for the vector potential.

8-2. Uniqueness. The usual type of boundary-value problem is to determine the field within a source-free region, given certain values of the field over the boundary of the region. We shall now formalize this problem, determining the boundary conditions necessary for a unique solution. A closed region will be used for the discussion, but the results will apply also to an open region (one extending to infinity) if the field vanishes at least as rapidly as r^{-2} as $r \to \infty$.

Consider a region enclosed by a surface S, for example, region (2) of

Fig. 8-1. Within S, we have no sources, that is,

$$\left.\begin{aligned} \nabla \cdot \mathbf{C} &= 0 \\ \nabla \times \mathbf{C} &= 0 \end{aligned}\right\} \qquad \text{within } S \tag{8-2}$$

This field can be thought of as produced by sources external to S. The vector **C** might be any of the various field quantities **E**, **D**, **B**, **H**, or **J**. We consider any two solutions to Eqs. (8-2), denoted $\mathbf{C}'$ and $\mathbf{C}''$, and form the difference field,

$$\delta\mathbf{C} = \mathbf{C}' - \mathbf{C}''$$

Since both $\mathbf{C}'$ and $\mathbf{C}''$ are solutions to Eqs. (8-2), so also is $\delta\mathbf{C}$, that is

$$\left.\begin{aligned} \nabla \cdot \delta\mathbf{C} &= 0 \\ \nabla \times \delta\mathbf{C} &= 0 \end{aligned}\right\} \qquad \text{within } S \tag{8-3}$$

Those conditions for which $\delta\mathbf{C} = 0$ everywhere within S are the conditions for uniqueness of solution, for when those conditions are met $\mathbf{C}' = \mathbf{C}''$ and the two solutions are the same.

In view of the second of Eqs. (8-2), for each solution there exists a scalar potential g such that

$$\mathbf{C}' = -\nabla g' \qquad \mathbf{C}'' = -\nabla g''$$

Defining the difference potential

$$\delta g = g' - g''$$

we have

$$\delta\mathbf{C} = -\nabla\, \delta g$$

We now form the essentially positive quantity

$$\delta\mathbf{C} \cdot \delta\mathbf{C} = -\delta\mathbf{C} \cdot \nabla\, \delta g$$

and we note the vector identity $\nabla \cdot (a\mathbf{A}) = a\nabla \cdot \mathbf{A} + \mathbf{A} \cdot \nabla a$. Using this identity and the first of Eqs. (8-3), we can reduce the above equation to

$$(\delta C)^2 = -\nabla \cdot (\delta g\, \delta\mathbf{C})$$

This equation can now be integrated throughout the region enclosed by S, and the divergence theorem can be applied to the right-hand side *if* $(\delta g\, \delta\mathbf{C})$ is well-behaved within S. This gives

$$\iiint (\delta C)^2\, d\tau = -\oint\!\!\oint \delta g\, \delta\mathbf{C} \cdot d\mathbf{s}$$

When the right-hand term vanishes, this equation can be satisfied only if $\delta\mathbf{C} = 0$ everywhere within S, that is, if $\mathbf{C}' = \mathbf{C}''$. Therefore, the field is unique (*a*) if

$$\oint\!\!\oint \delta g \, \delta \mathbf{C} \cdot d\mathbf{s} = 0 \tag{8-4}$$

and (*b*) if $\delta g \, \delta \mathbf{C}$ is well-behaved within S.

We shall assume that condition (*b*) is met and give some cases for which uniqueness is obtained. First of all, Eq. (8-4) is satisfied if $\delta \mathbf{C} \cdot \mathbf{u}_n = \delta C_n$ is zero over S, that is, if the normal components of $\mathbf{C}'$ and $\mathbf{C}''$ are equal over S. Second, Eq. (8-4) is satisfied if δg is zero over S, that is, if g' and g'' are equal over S. Still more generally, δC_n could be zero over part of S and δg zero over the rest of S; Eq. (8-4) would then be satisfied. These possibilities can be summarized: *A well-behaved field having a well-behaved scalar potential is unique in a source-free region if its normal component is specified over the bounding surface or if its scalar potential is specified over the bounding surface or if its normal component is specified over part of the bounding surface and its potential over the rest.* Other possibilities exist, one of which is given in Sec. 8-3. An example of a problem in which the scalar potential is not well-behaved although all the other conditions are met, is given in Prob. 8-1. Uniqueness of the field is not obtained in this case.

An alternative uniqueness proof makes use of the vector potential. In view of the first of Eqs. (8-2), there exist vector potentials such that

$$\mathbf{C}' = \nabla \times \mathbf{G}' \qquad \mathbf{C}'' = \nabla \times \mathbf{G}''$$

Defining the difference potential

$$\delta \mathbf{G} = \mathbf{G}' - \mathbf{G}''$$

we have

$$\delta \mathbf{C} = \nabla \times \delta \mathbf{G}$$

We then form the essentially positive quantity

$$\delta \mathbf{C} \cdot \delta \mathbf{C} = \delta \mathbf{C} \cdot \nabla \times \delta \mathbf{G}$$

and we note the vector identity $\nabla \cdot (\mathbf{A} \times \mathbf{B}) = \mathbf{B} \cdot \nabla \times \mathbf{A} - \mathbf{A} \cdot \nabla \times \mathbf{B}$. Using this identity and the second of Eqs. (8-3), we can reduce the above equation to

$$(\delta C)^2 = \nabla \cdot (\delta \mathbf{G} \times \delta \mathbf{C})$$

This is now integrated throughout the region enclosed by S. *If* $(\delta \mathbf{G} \times \delta \mathbf{C})$ is well-behaved, we can apply the divergence theorem to the right-hand side, obtaining

$$\iiint (\delta C)^2 \, d\tau = \oint\!\!\oint (\delta \mathbf{G} \times \delta \mathbf{C}) \cdot d\mathbf{s}$$

When the right-hand term vanishes, this equation can be satisfied only if $\delta \mathbf{C} = 0$ everywhere within S, that is, if $\mathbf{C}' = \mathbf{C}''$. Therefore, the field is unique (*a*) if

$$\oint\!\!\oint (\delta \mathbf{G} \times \delta \mathbf{C}) \cdot d\mathbf{s} = 0 \tag{8-5}$$

and (*b*) if $\delta\mathbf{G} \times \delta\mathbf{C}$ is well-behaved within S.

Assuming that condition (*b*) is met, we can give some cases for which uniqueness is obtained. We can write the integrand of Eq. (8-5) as

$$\mathbf{u}_n \cdot \delta\mathbf{G} \times \delta\mathbf{C} = \delta\mathbf{C} \cdot \mathbf{u}_n \times \delta\mathbf{G} = \delta\mathbf{G} \cdot \delta\mathbf{C} \times \mathbf{u}_n$$

according to a vector identity. The integrand vanishes identically if either $\mathbf{u}_n \times \delta\mathbf{G}$ or $\mathbf{u}_n \times \delta\mathbf{C}$ vanish at each point on S, that is, if the tangential components of $\mathbf{G}'$ and $\mathbf{G}''$ or of $\mathbf{C}'$ and $\mathbf{C}''$ are equal. Thus, *a well-behaved field having a well-behaved vector potential is unique in a source-free region if its tangential components are specified over the bounding surface or if the tangential components of its vector potential are specified over the bounding surface or if the former are specified over part of the surface and the latter over the rest.* Other possibilities for uniqueness exist. An example of a case for which the vector potential is not well-behaved, although all the other conditions are met, is given in Prob. 8-3. Uniqueness of the field is not obtained in this problem.

We have a slightly different problem when sources exist within the region of interest. In this case, our starting equations are

$$\left.\begin{aligned} \nabla \cdot \mathbf{C} &= w \\ \nabla \times \mathbf{C} &= \mathbf{W} \end{aligned}\right\} \qquad \text{within } S \tag{8-6}$$

A possible solution to Eqs. (8-6) is the potential integral solution of Sec. 7-1. This is the solution which would hold if there were no sources outside S. The total solution can be visualized as the sum of the field produced by sources within S, which we take to be the above-mentioned potential integral solution and which we call a *particular solution*, plus the field produced by the sources outside S, which we call a *complementary solution*. Thus, let

$$\mathbf{C} = \mathbf{C}_{ps} + \mathbf{C}_{cs} \tag{8-7}$$

where $\mathbf{C}_{ps}$ and $\mathbf{C}_{cs}$ denote the particular solution and the complementary solution, respectively. Both $\mathbf{C}$ and $\mathbf{C}_{ps}$ satisfy Eqs. (8-6) within S; so $\mathbf{C}_{cs}$ satisfies the homogeneous equations

$$\left.\begin{aligned} \nabla \cdot \mathbf{C}_{cs} &= 0 \\ \nabla \times \mathbf{C}_{cs} &= 0 \end{aligned}\right\} \qquad \text{within } S \tag{8-8}$$

All of the preceding discussion on $\mathbf{C}$ in a source-free region now applies directly to $\mathbf{C}_{cs}$ in a region containing sources. A specification of boundary conditions on $\mathbf{C}$ over S is equivalent to a specification of boundary conditions on $\mathbf{C}_{cs}$ over S, since $\mathbf{C}_{ps}$ is a known solution.

8-3. Electrostatics. The fundamental equations for the electrostatic field are

$$\nabla \times \mathbf{E} = 0 \qquad \mathbf{D} = \epsilon \mathbf{E}$$
$$\nabla \cdot \mathbf{D} = q_v \qquad \mathbf{E} = -\nabla V \tag{8-9}$$

Thus, **E** is a field having no vortex sources. If the region is homogeneous, **D** also has no vortex sources. We shall now look at these equations according to boundary-value problem concepts.

In electrostatics, matter is classified as either dielectric ($\sigma = 0$), or conductor ($\sigma \neq 0$). Of course, all matter has some degree of conductivity, but that for which σ is very small can be approximated by considering σ to be zero. Inside a conductor, the electrostatic field **E** must be zero; otherwise $\mathbf{J} = \sigma \mathbf{E}$ would exist, violating the assumed static condition. Since $\mathbf{D} = \epsilon \mathbf{E}$, **D** also is zero. Thus, *the electrostatic* **E** *and* **D** *fields are zero internal to a conductor.* Since $\mathbf{E} = -\nabla V$ and **E** is zero, it follows that *the electrostatic potential V is constant internal to a conductor.*

Some light will be shed upon the above statements by the following development. In linear matter

$$\nabla \cdot \mathbf{J} = \nabla \cdot (\sigma \mathbf{E}) = -\frac{\partial q_v}{\partial t}$$

and

$$\nabla \cdot \mathbf{D} = \nabla \cdot (\epsilon \mathbf{E}) = q_v$$

If the medium is homogeneous, σ and ϵ can be removed from under the divergence operations, and the above equations combine to

$$\frac{\partial q_v}{\partial t} + \frac{\sigma}{\epsilon} q_v = 0$$

The solution to this equation is

$$q_v(t) = q_v(0)e^{-(\sigma/\epsilon)t} \tag{8-10}$$

Thus, if any charge exists internal to matter at some time $t = 0$, it decays exponentially to zero. By the time $t = \epsilon/\sigma$ the charge will have decayed to $1/e$ (37 per cent) of its initial value. The time $t = \epsilon/\sigma$ is called the *relaxation time* of the material. In good conductors, the relaxation time is of the order of 10^{-16} second; so a nonzero charge density could exist only for a small fraction of a microsecond. On the other hand, in good dielectrics the relaxation time is of the order of 10^6 seconds; so a nonzero charge density could exist for days. This is a justification for approximating good dielectrics by perfect dielectrics for purposes of analysis. When a conductor is placed in an electrostatic field, its atomic charges redistribute themselves almost instantaneously, preserving the equipotential characteristics of the conductor.

We must now look at the various boundary conditions found in electrostatics. First of all, no vortex sources of **E** exist in nature, so the *tangen-*

tial components of **E** *are continuous across any material boundary.* This is mathematically expressed by $E_t^{(1)} = E_t^{(2)}$, the superscripts denoting regions (1) and (2). The equivalent behavior of **D** at a boundary is $D_t^{(1)}/\epsilon_1 = D_t^{(2)}/\epsilon_2$, where ϵ_1 and ϵ_2 denote ϵ in regions (1) and (2), respectively. In nature, charge is the flow source of **D**; so *the normal component of* **D** *is discontinuous across a material boundary if a surface layer of charge exists.* The precise nature of this discontinuity we know to be

$$D_n^{(1)} - D_n^{(2)} = q_s$$

where the unit normal points into region (1). The equivalent behavior of **E** at a boundary is $\epsilon_1 E_n^{(1)} - \epsilon_2 E_n^{(2)} = q_s$. These boundary conditions are summarized in Col. I of Table 8-1.

We often deal with the scalar potential when solving boundary-value problems; so we should also have the equivalent boundary conditions on V. The change in V across a surface is given by $dV = -\mathbf{E} \cdot d\mathbf{l}$. Since **E** remains finite even if a surface layer of charge exists, *the scalar potential V is continuous across any material boundary.* This is equivalent to continuity of the tangential components of **E**. Since $D_n = \epsilon E_n = -\epsilon\, \partial V/\partial n$, it follows from $D_n^{(1)} - D_n^{(2)} = q_s$ that *the normal derivative of V may be discontinuous across a material boundary.* These boundary conditions also are summarized in Col. I of Table 8-1.

TABLE 8-1. BOUNDARY CONDITIONS FOR THE ELECTROSTATIC FIELD
The direction of the normal is into region (1).

Field quantity	(I) General	(II) No surface charge	(III) (1) dielectric, (2) conductor
E	$E_t^{(1)} = E_t^{(2)}$ $\epsilon_1 E_n^{(1)} - \epsilon_2 E_n^{(2)} = q_s$	$E_t^{(1)} = E_t^{(2)}$ $\epsilon_1 E_n^{(1)} = \epsilon_2 E_n^{(2)}$	$E_t^{(1)} = 0$ $\epsilon E_n^{(1)} = q_s$
D	$D_n^{(1)} - D_n^{(2)} = q_s$ $\dfrac{D_t^{(1)}}{\epsilon_1} = \dfrac{D_t^{(2)}}{\epsilon_2}$	$D_n^{(1)} = D_n^{(2)}$ $\dfrac{D_t^{(1)}}{\epsilon_1} = \dfrac{D_t^{(2)}}{\epsilon_2}$	$D_n^{(1)} = q_s$ $D_t^{(1)} = 0$
V	$V^{(1)} = V^{(2)}$ $\epsilon_1 \dfrac{\partial V^{(1)}}{\partial n} - \epsilon_2 \dfrac{\partial V^{(2)}}{\partial n} = -q_s$	$V^{(1)} = V^{(2)}$ $\epsilon_1 \dfrac{\partial V^{(1)}}{\partial n} = \epsilon_2 \dfrac{\partial V^{(2)}}{\partial n}$	$V^{(1)} = \text{constant}$ $\epsilon_1 \dfrac{\partial V^{(1)}}{\partial n} = -q_s$

We shall now consider two special cases of the general boundary conditions. At a dielectric-to-dielectric interface, no surface charge results from application of a field. In other words, if the boundary is initially uncharged, no charge can accumulate due to the action of the field. Thus, the boundary conditions for an uncharged dielectric-to-dielectric boundary are obtained from the general relations by letting $q_s = 0$. These are summarized in Col. II of Table 8-1. At a dielectric-to-conductor interface, a surface charge can exist as a result of establishing a field, for the conductor permits a flow of charge. In fact, *any net charge on a conductor appears only as a surface charge*, for no net charge can exist internal to a conductor. Also, a conductor is an equipotential surface, with **E** and **D** zero within the conductor. The boundary conditions for a dielectric-to-conductor boundary are then obtained from the general relations by setting **E** and **D** in the conductor equal to zero and V equal to a constant. Note that *the normal component of* **D** *is equal to the surface-charge density on the conductor.* These boundary conditions are summarized in Col. III of Table 8-1.

The boundary conditions needed to obtain a unique solution have been discussed in Sec. 8-2. There is, however, an additional condition applicable to electrostatics that we have not mentioned. Letting the field $\mathbf{C} = \mathbf{E}$ and the potential $g = V$, suppose that part of the boundary surface encloses a conductor. Since the conductor is an equipotential surface, Eq. (8-4) taken over the conductor is of the form

$$\oiint \delta V\, \delta\mathbf{E} \cdot d\mathbf{s} = \delta V \oiint (\mathbf{E}' - \mathbf{E}'') \cdot d\mathbf{s} = \delta V \frac{1}{\epsilon} (q' - q'')$$

where q' and q'' are the total charges on the conductor for the $\mathbf{E}'$ and $\mathbf{E}''$ fields. This vanishes if $q' = q''$. Therefore, *a specification of the total charge on conducting bodies is sufficient to ensure uniqueness of the field.* This is, of course, in addition to the usual boundary conditions over those surfaces that do not coincide with conductors.

Given sufficient boundary conditions, the solution is unique; so any procedure used to obtain this solution is justified. The general procedure is to take known solutions to the field equations, or linear combinations of them, and attempt to satisfy boundary conditions. At this stage of development, our supply of known solutions is somewhat limited, but it is nevertheless sufficient for quite a number of problems. For example,

$$\mathbf{E} = \frac{\mathbf{u}_r K_1}{r^2} \qquad V = \frac{K_1}{r} + K_2$$

is a known solution which we recognize as being associated with a point charge at the origin. But now we view it merely as a function which satisfies the field equations, regardless of K_1 and K_2. It is these constants (parameters not dependent on the coordinates) that we may adjust to satisfy boundary conditions. Other known solutions are those associated with a line charge, plane charge, charge dipole, or any of the problems that we have already solved.

We can immediately reinterpret all our solutions involving static charges in a homogeneous dielectric in terms of conductors in a homogeneous dielectric. About any distribution of static charge there exist equipotential surfaces (see, for example, Fig. 5-9). *Any equipotential surface may be considered to be the surface of a conductor*, for the boundary conditions are sufficient to ensure uniqueness. Since each distribution of static charge has an infinity of equipotential surfaces, each solution is that of an infinity of boundary-value problems involving conductors in a homogeneous dielectric.

Facility in solving boundary-value problems comes only with practice. We shall give a few examples here, others being given as problems at the end of the chapter.

A. Charged conducting sphere. The equipotential surfaces due to a point charge q are concentric spheres about q. It follows from the above discussion that a single conducting sphere of any size having a total charge q produces the same field external to the sphere as does a point charge q located at its center. Therefore, the field is

$$\begin{aligned} \mathbf{E} &= \frac{q}{4\pi\epsilon r^2}\,\mathbf{u}_r && \text{external to the sphere} \\ \mathbf{E} &= 0 && \text{internal to the sphere} \end{aligned} \tag{8-11}$$

The charge must exist on the sphere as a surface distribution, obtainable from $D_n = q_s$. Letting R denote the radius of the sphere, we have

$$q_s = \frac{q}{4\pi R^2} \tag{8-12}$$

Thus, the charge is uniformly distributed over the surface of the sphere. Instead of the total charge on the sphere, we might know the potential of the sphere with respect to infinity (or distant objects). Denoting the potential of the sphere by V_0, we can obtain the total charge on the sphere according to $V_0 = q/4\pi\epsilon R$ and then obtain

$$\mathbf{E} = \frac{V_0 R}{r^2}\,\mathbf{u}_r \qquad \text{external to the sphere} \tag{8-13}$$

We say that a point charge at the origin is equivalent to a charged sphere in the region external to the sphere.

B. Coaxial conducting cylinders. For this problem we consider two conducting cylinders of infinite length, as suggested by Fig. 8-2. We note first that the equipotential surfaces of the line charge field

$$V = -K_1 \log \rho + K_2 \qquad \mathbf{E} = \frac{\mathbf{u}_\rho K_1}{\rho} \tag{8-14}$$

are concentric cylinders; so Eqs. (8-14) have the desired functional form of the solution. Suppose we specify the potentials on the conducting cylinders, say V_1 on R_1 and V_2 on R_2. To satisfy these conditions, the first of Eqs. (8-14) for $\rho = R_1$ and $\rho = R_2$ becomes

$$V_1 = -K_1 \log R_1 + K_2 \qquad V_2 = -K_1 \log R_2 + K_2$$

These two equations can now be solved for K_1 and K_2 in terms of V_1 and V_2, giving

$$K_1 = \frac{V_1 - V_2}{\log (R_2/R_1)} \qquad K_2 = \frac{V_1 \log R_2 - V_2 \log R_1}{\log (R_2/R_1)} \tag{8-15}$$

These constants, substituted back into Eqs. (8-14), give the desired solution. Note that **E** depends only on the potential difference between the two cylinders.

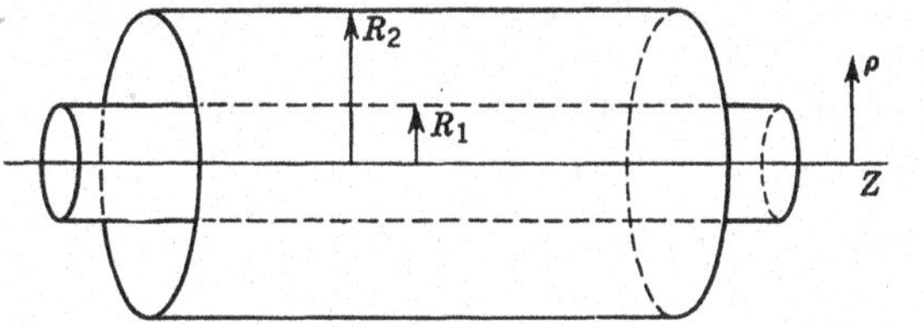

FIG. 8-2. Coaxial cylinders.

C. Parallel conducting plates, two dielectrics. Fig. 8-3 represents the one-dimensional problem of a pair of parallel conducting plates between which are two dielectrics. We recall that the equipotential surfaces of the infinite charged sheet were parallel planes; so we should expect to use this form of solution. However, we now have a discontinuity in the media at the interface of the two dielectrics; so we would expect to find some sort of discontinuity in the field quantities at this surface. The

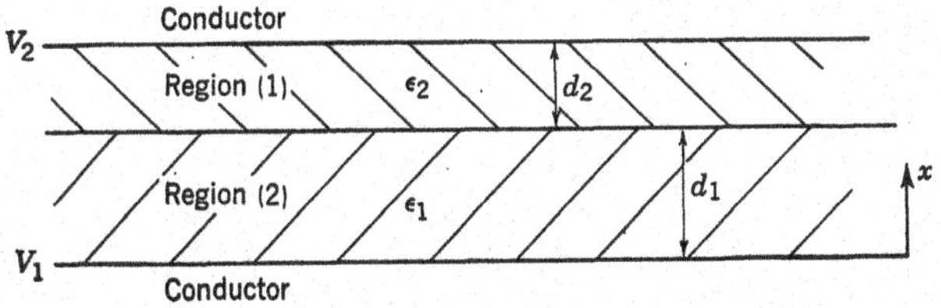

FIG. 8-3. Parallel plates and dielectric slabs.

usual procedure is to consider solutions for each region separately and then to satisfy the boundary conditions at the interface. Thus we take

$$\mathbf{E}^{(1)} = \mathbf{u}_x K_1 \qquad \mathbf{E}^{(2)} = \mathbf{u}_x K_2 \tag{8-16}$$

where the superscripts denote regions (1) and (2). The boundary conditions (other than tangential **E** zero at the conductors, which we have already satisfied) are

$$D_x^{(1)} = D_x^{(2)} \qquad \text{at } x = d_1$$

$$V_1 - V_2 = \int_0^{d_1} E_x\,dx + \int_{d_1}^{d_1+d_2} E_x\,dx$$

Evaluating these equations, using Eqs. (8-16), we obtain

$$\epsilon_1 K_1 = \epsilon_2 K_2 \qquad V_1 - V_2 = K_1 d_1 + K_2 d_2$$

This pair of equations may now be solved for K_1 and K_2, giving

$$K_1 = \frac{\epsilon_2(V_1 - V_2)}{\epsilon_1 d_2 + \epsilon_2 d_1} \qquad K_2 = \frac{\epsilon_1(V_1 - V_2)}{\epsilon_1 d_2 + \epsilon_2 d_1} \tag{8-17}$$

Substitution of these constants into Eqs. (8-16) gives the desired answer. The evaluation of the scalar potential in the dielectrics requires the evaluation of two more constants.

D. Conducting sphere and point charge. In Sec. 5-5 we saw that the equipotential surface $V = 0$ due to two point charges of opposite sign

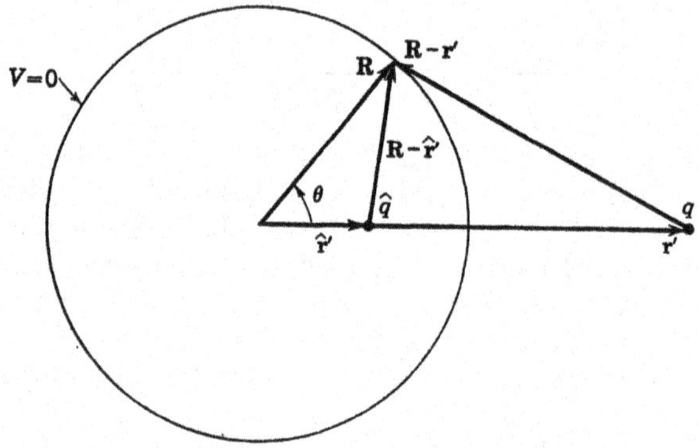

FIG. 8-4. Grounded conducting sphere and point charge.

was a sphere surrounding the smaller charge (see Fig. 5-9*c*). This solution is therefore also the solution for a *grounded* ($V = 0$) conducting sphere. We can start from a knowledge of the two point charges and obtain the necessary size and location of the equivalent conducting sphere. However, let us now look at the problem with the aim of finding the equivalent (or image) point charge, given the sphere and the external point charge. The sphere of radius R and the external point charge q are represented in Fig. 8-4. We know that it is possible to find a charge $\hat{q}$ equivalent to the conducting sphere. The potential from q and $\hat{q}$ is given by

$$V = \frac{q}{4\pi\epsilon|\mathbf{r} - \mathbf{r}'|} + \frac{\hat{q}}{4\pi\epsilon|\mathbf{r} - \hat{\mathbf{r}}'|} \tag{8-18}$$

where $\hat{q}$ and $\hat{r}'$ are unknown. The boundary condition is $V = 0$ at $\mathbf{r} = \mathbf{R}$, for which Eq. (8-18) reduces to

$$\frac{q}{|\mathbf{R} - \mathbf{r}'|} = \frac{-\hat{q}}{|\mathbf{R} - \hat{\mathbf{r}}'|}$$

Referring to Fig. 8-4, we see that

$$|\mathbf{R} - \mathbf{r}'|^2 = R^2 + r'^2 - 2Rr' \cos \theta$$
$$|\mathbf{R} - \hat{\mathbf{r}}'|^2 = R^2 + \hat{r}'^2 - 2R\hat{r}' \cos \theta$$

which, coupled with the preceding equation, leads to

$$\frac{q^2}{\hat{q}^2} = \frac{r'}{\hat{r}'}\left[\frac{(R^2/r') + r' - 2R \cos \theta}{(R^2/\hat{r}') + \hat{r}' - 2R \cos \theta}\right]$$

It is apparent that this equation is satisfied for all values of θ if

$$\frac{r'}{\hat{r}'} = \frac{q^2}{\hat{q}^2} \qquad r'\hat{r}' = R^2$$

Therefore, the equivalent or image charge $\hat{q}$ is situated at

$$\hat{r}' = \frac{R^2}{r'} \tag{8-19}$$

which is called the *inverse point to r′ in the sphere R*. The magnitude of the image charge is

$$\hat{q} = \frac{-qR}{r'} \tag{8-20}$$

the minus sign being chosen because we already know that the image charge must be opposite in sign to the external charge. Note that $\hat{q}$ must be the total charge on the conducting sphere, for $\oiint \mathbf{D} \cdot d\mathbf{s}$ over the surface of the sphere gives $\hat{q}$. Note also that this charge is *not* uniformly distributed over the surface of the sphere, as it was in the case of the isolated charged sphere. Incidentally, this problem is one for which we have a source, q, in the region of interest. The first term of Eq. (8-18) is the particular solution to the field equations, and the second term of Eq. (8-18) is the complementary solution. Note that Fig. 8-4, satisfying Eqs. (8-19) and (8-20), represents also the solution to the problem of a point charge $\hat{q}$ inside a conducting sphere $r = R$ kept at zero potential. In this case, q becomes the image charge, producing the complementary solution.

In the problem that we have just solved we find that there is a net charge $\hat{q}$ on the conducting sphere. There arises the question of the origin of this charge. Remember that we specified that the sphere was grounded, which implies that it is able to draw from some reservoir of charge any charge needed to keep its surface at zero potential. The earth acts as such a reservoir of charge, so we can imagine that the sphere is (or was) connected to the earth by a thin wire.

If the metal sphere were insulated instead of grounded, no net charge could appear on the sphere, and its surface would not be at zero potential. We can amend our previous solution by adding still another equivalent, or image, point charge at the center of the sphere, since this would not alter the equipotential characteristics of the sphere. Furthermore, this additional image charge should be equal to the negative of the previous image charge; so the net charge on the sphere is zero. This is illustrated by Fig. 8-5. The solution to the problem is therefore

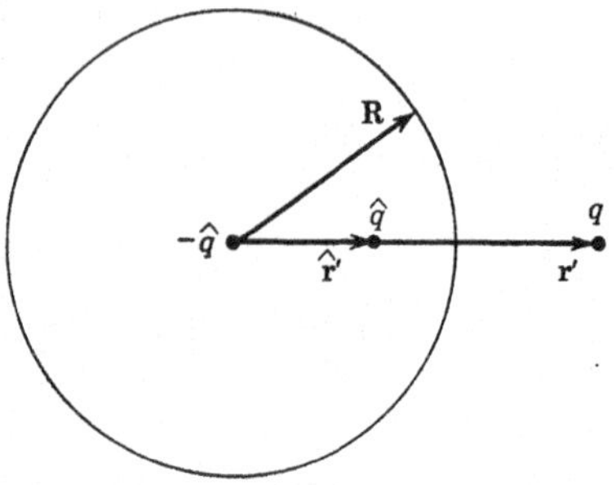

FIG. 8-5. Insulated conducting sphere and point charge.

$$V = \frac{q}{4\pi\epsilon|\mathbf{r} - \mathbf{r}'|} + \frac{\hat{q}}{4\pi\epsilon|\mathbf{r} - \hat{\mathbf{r}}'|} - \frac{\hat{q}}{4\pi\epsilon r} \tag{8-21}$$

where $\hat{q}$ is given by Eq. (8-20) and $\hat{r}'$ by Eq. (8-19). The first two terms of Eq. (8-21) give zero potential on the sphere; so the potential of the sphere is

$$V_0 = \frac{-\hat{q}}{4\pi\epsilon R} = \frac{q}{4\pi\epsilon r'} \tag{8-22}$$

where the last equality follows from Eq. (8-20). Thus, *the potential of the sphere is the same as the potential which would exist at its center if the sphere were not present.*

8-4. Steady Current. The electrostatic field requires that there be no electric current; so such a field must always be zero inside a conducting medium. If we now remove this restriction on current and admit the possibility of a steady current, we can have a time-invariant electric intensity internal to a conductor. This steady current must, however, be divergenceless; otherwise electric charge would collect and the electric field would no longer be time-invariant. All the basic equations of electrostatics will apply to the steady-current problem, but now we have the two additional equations

$$\nabla \cdot \mathbf{J} = 0 \qquad \mathbf{J} = \sigma\mathbf{E} \tag{8-23}$$

The formal mathematics of solution remains the same for the steady-current problem as for the electrostatic problem, except that the additional Eqs. (8-23) result in new types of boundary values. Note that since we are now admitting a steady current, there will also be present a steady magnetic field. However, it is the electric field that determines the current distribution, which in turn acts as a source for the magnetic field. As we pointed out previously, there is no interaction between the time-invariant electric and magnetic fields; so we do not have to consider the magnetic fields in determining the current.

The general behavior of **E**, **D**, and V at a boundary must remain the same as for electrostatics, since we have introduced no new types of sources. However, we now have to consider the additional vector quantity **J**. The current has no flow sources; so *the normal component of* **J** *must be continuous across a boundary.* There will be a discontinuity in the tangential components of **J** if there is a discontinuity in σ, for $\mathbf{J} = \sigma\mathbf{E}$ and the tangential components of **E** are continuous. The general boundary conditions for steady-current problems are summarized in Col. I of Table 8-2. These conditions must be satisfied at any conductor-to-conductor interface. Note that there exist two relationships involving normal components of vector quantities, $J_n^{(1)} = J_n^{(2)}$ and $D_n^{(1)} - D_n^{(2)} = q_s$. Since $\mathbf{D} = \epsilon\mathbf{E} = (\epsilon/\sigma)\mathbf{J}$, we find that the surface charge and the current crossing a boundary must be related by

$$q_s = \left(\frac{\epsilon_1}{\sigma_1} - \frac{\epsilon_2}{\sigma_2}\right) J_n^{(1)} \tag{8-24}$$

Thus, *if there is a steady current across a material boundary, there must be a surface charge present unless the ratio* ϵ/σ *is the same in both media.*

When one of the media is a dielectric, say medium (2), we obtain a special case of the general boundary conditions. In the dielectric, **J** is identically zero, for σ is zero. Setting $\mathbf{J}^{(2)}$ and σ_2 equal to zero in the general relationships of Table 8-2, we obtain the relationships for a conductor-to-dielectric interface. These are listed in Col. II of Table 8-2. Note that *the normal components of* **J**, **E**, *and* **D** *in the conductor are zero at the conductor-to-dielectric boundary.* The surface-charge density on such a boundary is equal to the normal component of **D** in the dielectric.

For purposes of analysis, it is often convenient to approximate a good conductor by a *perfect conductor*, which has infinite conductivity. No **E** field can exist internal to a perfect conductor; else $\mathbf{J} = \sigma\mathbf{E}$ becomes infinite. The boundary conditions at a perfect conductor in a steady-current problem are the same as for an ordinary conductor in an electrostatic problem, except that there is now a current density. Note that it is conceivable for a current to exist in a perfect conductor even though **E** is zero, for σ is infinite. The boundary conditions at a perfect conductor are obtained from the general relationships by setting **E** and **D** equal to zero within the perfect conductor. These conditions are summarized in Col. III of Table 8-2.

TABLE 8-2. BOUNDARY CONDITIONS FOR STEADY-CURRENT PROBLEMS
The direction of the normal is into region (1).

Field quantity	(I) General (two conductors)	(II) (1) conductor, (2) dielectric	(III) (1) conductor, (2) perfect conductor
J	$J_n^{(1)} = J_n^{(2)}$ $\frac{\epsilon_1}{\sigma_1} J_n^{(1)} - \frac{\epsilon_2}{\sigma_2} J_n^{(2)} = q_s$ $\frac{J_t^{(1)}}{\sigma_1} = \frac{J_t^{(2)}}{\sigma_2}$	$J_n^{(1)} = 0$ $\mathbf{J}^{(2)} = 0$	$J_n^{(1)} = J_n^{(2)}$ $\frac{\epsilon_1}{\sigma_1} J_n^{(1)} = q_s$ $J_t^{(1)} = 0$
E	$E_t^{(1)} = E_t^{(2)}$ $\sigma_1 E_n^{(1)} = \sigma_2 E_n^{(2)}$ $\epsilon_1 E_n^{(1)} - \epsilon_2 E_n^{(2)} = q_s$	$E_t^{(1)} = E_t^{(2)}$ $E_n^{(1)} = 0$ $\epsilon_2 E_n^{(2)} = -q_s$	$E_t^{(1)} = 0$ $\sigma_1 E_n^{(1)} = J_n^{(2)}$ $\epsilon_1 E_n^{(1)} = q_s$
D	$D_n^{(1)} - D_n^{(2)} = q_s$ $\frac{\sigma_1}{\epsilon_1} D_n^{(1)} = \frac{\sigma_2}{\epsilon_2} D_n^{(2)}$ $\frac{D_t^{(1)}}{\epsilon_1} = \frac{D_t^{(2)}}{\epsilon_2}$	$D_n^{(2)} = -q_s$ $D_n^{(1)} = 0$ $\frac{D_t^{(1)}}{\epsilon_1} = \frac{D_t^{(2)}}{\epsilon_2}$	$D_n^{(1)} = q_s$ $\frac{\sigma_1}{\epsilon_1} D_n^{(1)} = J_n^{(2)}$ $D_t^{(1)} = 0$
V	$V^{(1)} = V^{(2)}$ $\sigma_1 \frac{\partial V^{(1)}}{\partial n} = \sigma_2 \frac{\partial V^{(2)}}{\partial n}$ $\epsilon_1 \frac{\partial V^{(1)}}{\partial n} - \epsilon_2 \frac{\partial V^{(2)}}{\partial n} = -q_s$	$V^{(1)} = V^{(2)}$ $\frac{\partial V^{(1)}}{\partial n} = 0$ $\epsilon_2 \frac{\partial V^{(2)}}{\partial n} = q_s$	$V^{(1)} = \text{constant}$ $\sigma_1 \frac{\partial V^{(1)}}{\partial n} = -J_n^{(2)}$ $\epsilon_1 \frac{\partial V^{(1)}}{\partial n} = -q_s$

A problem of practical interest is that of determining the current distribution in a wire. Suppose that we approximate a portion of the wire by a cylinder of finite length and that we assume that the ends of the cylinder are equipotential surfaces. The boundary conditions are then (*a*) $J_n = 0$ over the side of the cylinder and (*b*) $J_t = 0$ over the ends of the cylinder. A solution to the field equations satisfying these boundary conditions is $\mathbf{J} = \mathbf{u}_z J_0$, where the z axis is the cylinder axis and J_0 is a constant. Thus, *the steady current is uniformly distributed across the cross section of a cylindrical conductor, if its cross sections are equipotential surfaces.* This analysis is independent of the shape of the cross section of the wire. It does not apply if the current varies with time.

Let us consider another example to illustrate the steady-current problem. Consider two perfectly conducting concentric spheres, between which there are two conducting media, as shown in Fig. 8-6. There is an insulated wire connecting the inner and outer spheres, in series with a battery, causing a steady current. It is assumed that the wire has negligible effect on the current distribution between the inner and outer spheres. This is a spherically symmetric problem, so the current (and also **E** and **D**) will be of the form

$$\mathbf{J}^{(1)} = \frac{\mathbf{u}_r K_1}{r^2} \qquad \mathbf{J}^{(2)} = \frac{\mathbf{u}_r K_2}{r^2} \tag{8-25}$$

where the superscripts (1) and (2) denote media (1) and (2), respectively. Across the boundary of the two media ($r = R_1$) we must have the normal

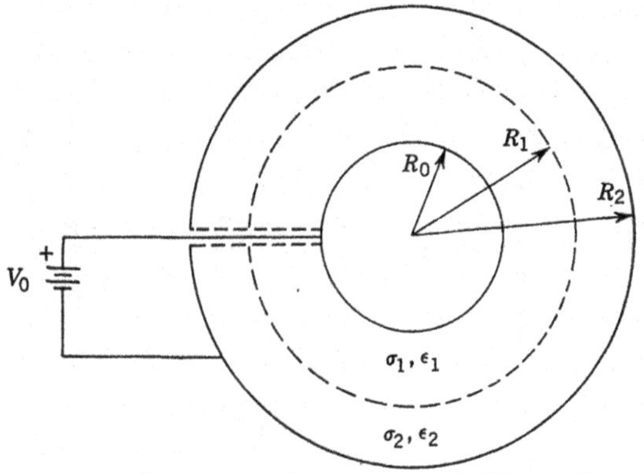

FIG. 8-6. Concentric spheres, two conducting media.

component of **J** continuous, giving

$$K_1 = K_2 \tag{8-26}$$

The electric intensity in each region is $\mathbf{E} = \mathbf{J}/\sigma$. The voltage applied between the two perfectly conducting spheres is V_0; so along a radial path

$$\begin{aligned} V_0 = \int_{R_0}^{R_2} E_r\, dr &= \int_{R_0}^{R_1} \frac{K_1}{\sigma_1 r^2}\, dr + \int_{R_1}^{R_2} \frac{K_1}{\sigma_2 r^2}\, dr \\ &= K_1 \left(\frac{R_1 - R_0}{\sigma_1 R_0 R_1} + \frac{R_2 - R_1}{\sigma_2 R_1 R_2} \right) \end{aligned} \tag{8-27}$$

This determines the constant $K_1 = K_2$. Note that there must be a surface density of charge on the sphere $r = R_1$, according to Eq. (8-24).

8-5. Magnetostatics. The fundamental equations of the magnetostatic field are

$$\nabla \times \mathbf{H} = \mathbf{J} \qquad \nabla \cdot \mathbf{B} = 0 \qquad \mathbf{B} = \mu \mathbf{H} \tag{8-28}$$

Note that **B** is a field having no flow sources. If the region is homogeneous, then **H** also has no flow sources. In view of the divergenceless character of **B**, there always exists a vector potential **A** such that $\mathbf{B} = \nabla \times \mathbf{A}$. However, we have no occasion to apply any boundary conditions to the vector potential; so we shall not consider it here. In regions where **J** is zero, **H** is the gradient of a scalar magnetic potential. It is sometimes convenient to solve boundary-value problems in terms of the magnetic scalar potential. This would be analogous to the use of the scalar potential in electrostatics.

There are no flow sources of **B** in nature; so *the normal component of* **B** *must be continuous across any material boundary.* The equivalent boundary condition for **H** would be $\mu_1 H_n^{(1)} = \mu_2 H_n^{(2)}$. Current is the vortex source of **H**, so in general *the tangential components of* **H** *may be discontinuous if a surface distribution of current exists.* The precise nature of this discontinuity is given by Eq. (4-22). The general behavior of **B** and **H** at a boundary are summarized by Col. I of Table 8-3.

TABLE 8-3. BOUNDARY CONDITIONS FOR THE MAGNETOSTATIC FIELD
The direction of the normal is into region (1).

Field quantity	(I) General	(II) No surface current
B	$B_n^{(1)} = B_n^{(2)}$ $\mathbf{u}_n \times \left(\frac{\mathbf{B}^{(1)}}{\mu_1} - \frac{\mathbf{B}^{(2)}}{\mu_2}\right) = \mathbf{J}_s$	$B_n^{(1)} = B_n^{(2)}$ $\frac{B_t^{(1)}}{\mu_1} = \frac{B_t^{(2)}}{\mu_2}$
H	$\mathbf{u}_n \times (\mathbf{H}^{(1)} - \mathbf{H}^{(2)}) = \mathbf{J}_s$ $\mu_1 H_n^{(1)} = \mu_2 H_n^{(2)}$	$H_t^{(1)} = H_t^{(2)}$ $\mu_1 H_n^{(1)} = \mu_2 H_n^{(2)}$

At any ordinary material boundary no surface current exists, for this would require an infinite conductivity. Thus, the tangential components of **H** are continuous in most cases, as shown in Col. II of Table 8-3. Sometimes the concept of a perfect electric conductor is used as a boundary in magnetostatics. Since the "building up" of a magnetic field in a perfect conductor requires a time rate of change of an electric field, no magnetic field can exist internal to a perfect conductor. For this idealized case, it is possible to have a surface density of current. As we have already seen, a closely spaced coil of current-carrying wire can often be approximated by a surface current. Another approximation to a surface current would be the current in a thin conducting sheet. Unless such a special situation exists, a material boundary in a magnetostatic field is free from surface current.

An example of a magnetostatic boundary-value problem is a straight wire encased in a circular magnetic sheath, as shown in Fig. 8-7. We have rotational symmetry, the only vortex source of **H** being the filament of current; so the **H** field must be the same as if the wire were in free space. Thus, as shown in Sec. 4-6,

$$\mathbf{H} = \frac{i}{2\pi\rho}\mathbf{u}_\phi \tag{8-29}$$

The **B** field is given by $\mathbf{B} = \mu\mathbf{H}$ in each region, that is,

$$\begin{aligned} \mathbf{B} &= \frac{\mu i}{2\pi\rho}\mathbf{u}_\phi \qquad \rho < a \\ \mathbf{B} &= \frac{\mu_0 i}{2\pi\rho}\mathbf{u}_\phi \qquad \rho > a \end{aligned} \tag{8-30}$$

Note that this solution satisfies the boundary conditions listed in Table 8-3.

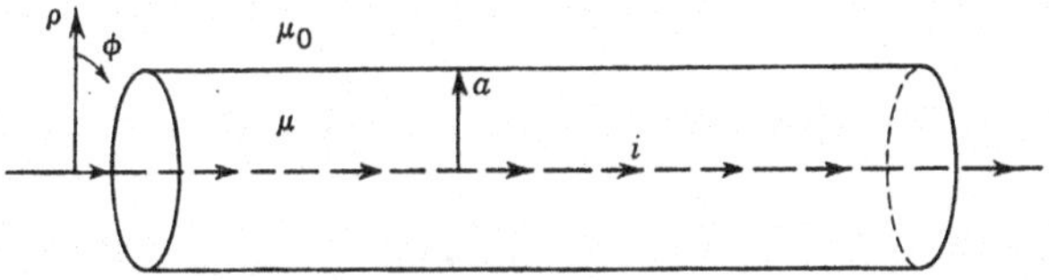

FIG. 8-7. A current-carrying wire encased in a magnetic sheath.

8-6. Image Theory. The solution to a boundary-value problem consists of a particular solution, due to sources within the region of interest, and a complementary solution, due to sources outside the region of interest. According to the equivalence principle, an infinite number of different distributions of sources outside the region of interest may produce the desired complementary solution within the region. In certain special cases, corresponding to each source element within the region of interest we can find an *image* source element outside the region which gives rise to the correct complementary solution. In this section we shall consider some of these cases.

To keep the discussion general, we shall use the word *conductor* to denote a surface over which the tangential components of some field vector vanish. The only true conducting surface in nature is the dielectric-to-conductor boundary of electrostatics, which is a conductor with respect to **E** and **D**. A conductor in electrostatic problems can also be approximated by the boundary of a medium for which $\epsilon \to \infty$, for **E** internal to this boundary must be zero, else $\mathbf{D} = \epsilon\mathbf{E}$ would be infinite and give rise to an infinite electric energy density. In steady-current problems, the conducting surface (as we are now using the term) must be a perfect conductor; else $\mathbf{E} = \mathbf{J}/\sigma$ could exist along its surface. In magnetostatic problems, a medium for which $\mu \to \infty$ approximates a conductor with respect to **B** and **H**, for **H** internal to it must be zero; else $\mathbf{B} = \mu\mathbf{H}$ would be infinite and would give rise to an infinite magnetic energy density. To indicate which field vectors have vanishing tangential components on a surface, we shall use the term *electric conductor* to denote a conductor with respect to **E** and **D** and the term *magnetic conductor* to denote a conductor with respect to **B** and **H**. Mathematically speaking, these electric and magnetic conductors are simply boundary conditions.

We shall use the term *anticonductor* to designate a surface over which the normal component of some field vector vanishes. The only true anticonducting surface in nature is the conductor-to-dielectric interface in the steady-current problem, at which the normal components of **J**, **E**, and **D** are zero in the conductor. The perfect electric conductor can be

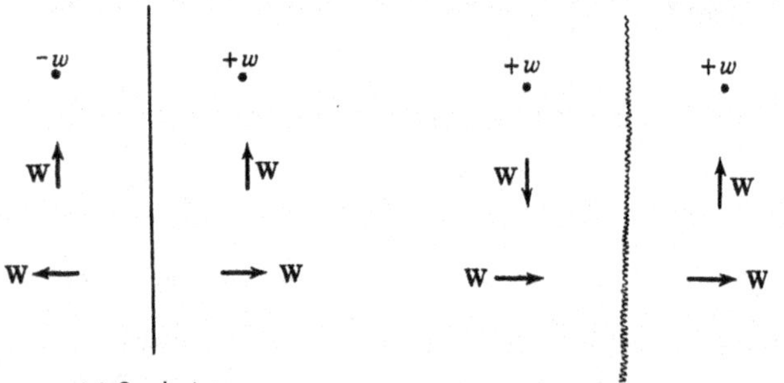

(a) Conductor (b) Anti-conductor

Fig. 8-8. Images for plane boundaries.

considered to be a magnetic anticonductor, for no magnetic field can build up internal to it. (This is not true of actual conductors in magnetostatic problems.) Similarly, the perfect magnetic conductor can be considered to be an electric anticonductor.

The most common application of image theory is to plane boundaries. Two elements of flow source, equal in magnitude but opposite in sign, produce zero tangential field over the plane bisecting the line joining them. Therefore, the image of an element of flow source in front of a conducting plane is an element of flow source of equal magnitude and opposite sign located at the *image point*. This is shown in Fig. 8-8*a*. As shown in Sec. 6-3, the field from an element of vortex source encircles the axis of the element according to the right-hand rule. Two elements of vortex source parallel to each other and perpendicular to the line joining them produce zero tangential field over the plane bisecting the line joining them. Therefore, the image of an element of vortex source parallel to the plane boundary is a vortex source of equal magnitude and in the same direction located at the image point. This also is shown in Fig. 8-8*a*. An element of vortex source perpendicular to a conducting plane requires an image vortex source of equal magnitude but in the opposite direction located at the image point, as shown in Fig. 8-8*a*.

For any arbitrary orientation of the element of vortex source, we merely decompose the element into its parallel and perpendicular components, and image each component separately.

If the plane boundary is an anticonductor, the images must be formed as shown in Fig. 8-8*b*. The boundary condition is zero normal component of field over the plane. This is met for flow sources by making the image source the same sign as the actual source. For elements of vortex source parallel to the boundary, the image must now be in the direction opposite to the direction of the original source. For elements

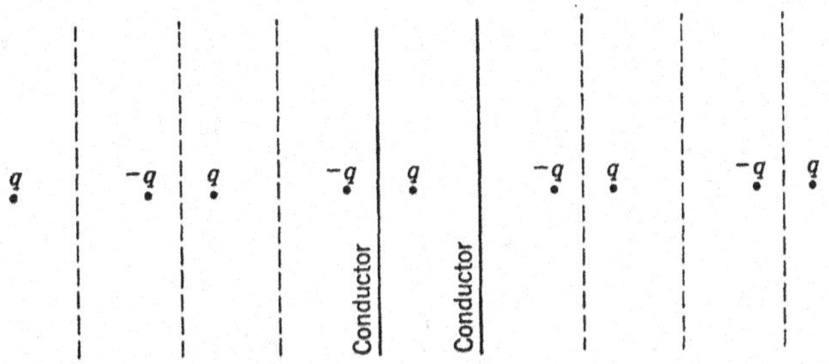

FIG. 8-9. Images for a point charge between parallel conducting plates.

of vortex source perpendicular to the boundary, the image has the same direction as the original source.

Image theory can be applied to surfaces of other shapes, for the equivalent sources producing the complementary solution can always be thought of as image sources. However, the term "image" is usually restricted to those cases for which each element of the equivalent sources corresponds to a single element of the actual source. This happens for conducting spheres, as is shown in Part *D* of Sec. 8-3. Another example is the conducting cylinder, for which each *line* element of charge has an image (see Fig. 5-15). There is, however, no simple image for a point-flow source in a conducting cylinder. A flow source has a single image in an anticonducting hyperboloid of revolution, as can be seen from a consideration of Fig. 5-9*b*. We can also use multiple images, arranged to give the appropriate conducting or anticonducting surfaces. The solutions to problems involving plane discontinuities in dielectric or magnetic media involve functions that can be identified with simple equivalent sources and that can also be considered as a part of image theory. We shall give some examples illustrating the use of image theory in its general sense.

A. Charge between conducting plates. In Fig. 8-9, the two solid lines represent parallel conducting plates with a charge q between them. The infinite system of images shown produces zero tangential **E** field over the two conducting surfaces and therefore must produce the required field between the two conducting plates. Note that this system of images can be built up by imaging first the given charge in one of the conducting plates, then imaging both these charges in the other plate, then imaging the two new charges in the first plate, and so on. The contribution of the images to the field varies inversely as the distance squared. In calculating the field, we need consider only enough of these images to obtain the desired accuracy.

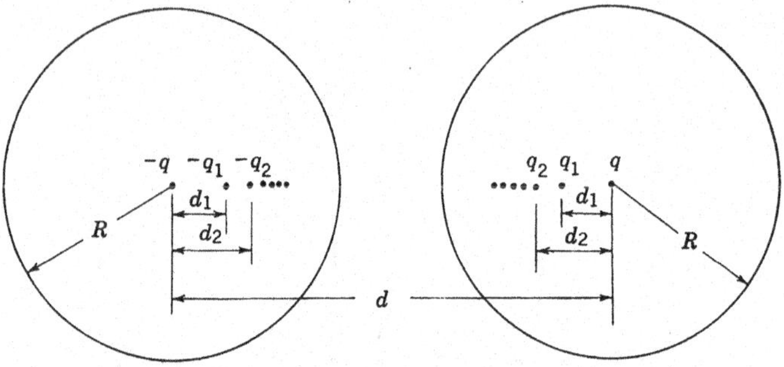

FIG. 8-10. Images for two charged conducting spheres.

B. Charged conducting spheres. The system of point charges of Fig. 8-10, where

$$\begin{aligned} d_1 &= \frac{R^2}{d} & q_1 &= \frac{qR}{d} \\ d_2 &= \frac{R^2}{d - d_1} & q_2 &= \frac{q_1 R}{d - d_1} \\ d_3 &= \frac{R^2}{d - d_2} & q_3 &= \frac{q_2 R}{d - d_2} \\ &\text{etc.} & &\text{etc.} \end{aligned} \tag{8-31}$$

produces zero tangential **E** over the two spheres of radius R. This is therefore the solution to the problem of two oppositely charged spheres having net charge

$$q_{\text{total}} = q + q_1 + q_2 + q_3 + \cdots \tag{8-32}$$

Note that this system of charges is built up by taking the two charges $+q$ and $-q$ and imaging them progressively in the two spheres according to Eqs. (8-19) and (8-20). In this problem, the image charges become progressively smaller in magnitude, in contrast with the preceding problem where the images become progressively more distant. Both situations give convergent solutions.

C. Charge near a dielectric interface. This problem consists of a plane interface between two dielectrics, with a point charge q in region

(1), as shown in Fig. 8-11*a*. We must find solutions for regions (1) and (2) that satisfy the boundary conditions

$$E_t^{(1)} = E_t^{(2)} \qquad D_n^{(1)} = D_n^{(2)} \tag{8-33}$$

over the dielectric interface. If we consider the field in region (1) to be produced by the sources of Fig. 8-11*b* and the field in region (2) to be produced by the source of Fig. 8-11*c*, then E_t and D_n in both regions

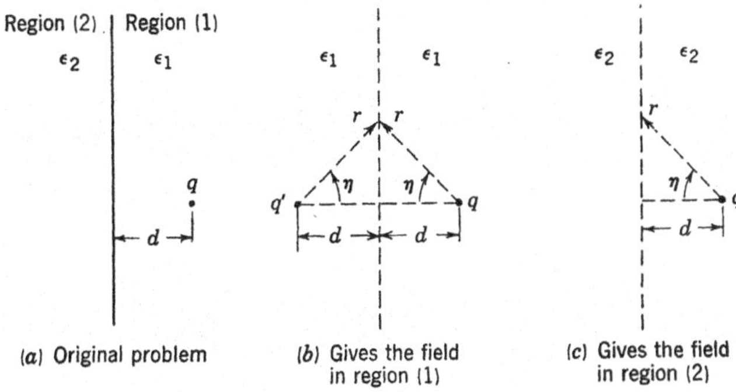

FIG. 8-11. Point charge adjacent to a plane dielectric interface.

will have the same functional form over the interface. We need only adjust the magnitudes of the image charges q' and q'' to satisfy Eqs. (8-33). From a consideration of Fig. 8-11, it follows that

$$E_t^{(1)} = \frac{\sin \eta}{4\pi\epsilon_1 r^2}(q + q')$$

$$E_t^{(2)} = \frac{\sin \eta}{4\pi\epsilon_2 r^2} q''$$

$$D_n^{(1)} = \frac{-\cos \eta}{4\pi r^2}(q - q')$$

$$D_n^{(2)} = \frac{-\cos \eta}{4\pi r^2} q''$$

Both of Eqs. (8-33) are satisfied if

$$\frac{q + q'}{\epsilon_1} = \frac{q''}{\epsilon_2} \qquad q - q' = q''$$

Solving for the image charges, we have

$$q' = \frac{\epsilon_1 - \epsilon_2}{\epsilon_1 + \epsilon_2} q \qquad q'' = \frac{2\epsilon_2}{\epsilon_1 + \epsilon_2} q \tag{8-34}$$

Note that, as $\epsilon_2 \to \infty$, $q' \to -q$ and the interface becomes a conducting boundary as seen from region (1). As $\epsilon_1 \to \infty$, $q' \to q$ and the interface becomes an anticonducting boundary as seen from region (1). Note also that as $\epsilon_2 \to \infty$, the field in region (2) becomes double the field for the same charge in a homogeneous medium.

D. Current near a magnetic interface. This problem is to determine the field due to a straight current-carrying wire parallel to a plane interface between two magnetic materials, as illustrated by Fig. 8-12*a*. It is similar to the preceding problem except that the source is now a

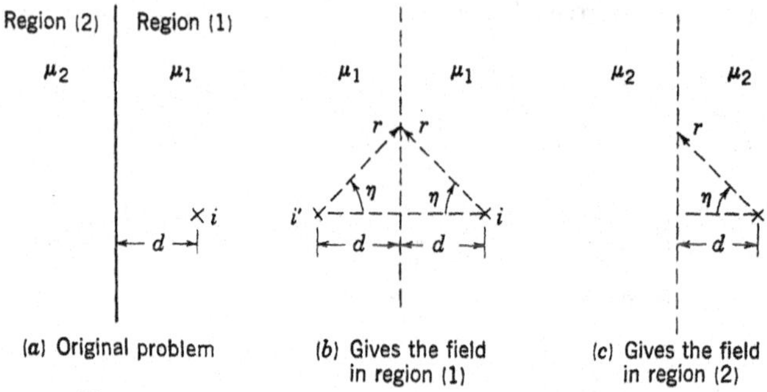

FIG. 8-12. Line current parallel to a magnetic interface.

vortex source. The boundary conditions to be satisfied at the interface are

$$H_t^{(1)} = H_t^{(2)} \qquad B_n^{(1)} = B_n^{(2)} \tag{8-35}$$

In each region, solutions of the same functional form over the interface can be obtained by introducing images as shown in Figs. 8-12*b* and *c*. From a consideration of the geometry of Fig. 8-12, it follows that

$$H_t^{(1)} = \frac{\cos \eta}{2\pi r}(i - i')$$

$$H_t^{(2)} = \frac{\cos \eta}{2\pi r} i''$$

$$B_n^{(1)} = \frac{\mu_1 \sin \eta}{2\pi r}(i + i')$$

$$B_n^{(2)} = \frac{\mu_2 \sin \eta}{2\pi r} i''$$

Both of Eqs. (8-35) are satisfied if

$$i - i' = i'' \qquad \mu_1(i + i') = \mu_2 i''$$

Solving for the image currents, we have

$$i' = \frac{\mu_2 - \mu_1}{\mu_2 + \mu_1} i \qquad i'' = \frac{2\mu_1}{\mu_1 + \mu_2} i \tag{8-36}$$

Note that, as $\mu_2 \to \infty$, $i' \to i$ and the interface becomes a (magnetically) conducting boundary with respect to region (1). As $\mu_1 \to \infty$, $i' \to -i$ and the interface acts as an anticonducting boundary. Also, as $\mu_1 \to \infty$, the field in region (2) becomes double the field for the same current in a homogeneous medium. Note the difference between Eqs. (8-36) and (8-34), which illustrates the difference in behavior between vortex sources and flow sources near such an interface.

8-7. Problems with Plane Boundaries. Using a combination of the concepts already presented, we can treat a large number of boundary-value problems involving plane boundaries. The manner in which this is done is shown in this section.

Consider the problem of determining a field in the half-space $z > 0$ in terms of its values over the boundary $z = 0$. To simplify the discussion, we shall assume that all sources are in the region $z < 0$. (If there are sources in the region $z > 0$, we need merely add the appropriate particular solution due to these sources.) This "original problem" is shown in Fig. 8-13*a*. We can, of course, determine the field in the region $z > 0$ by applying the equivalence principle, obtaining the equivalent sources

$$w_s = \mathbf{u}_n \cdot \mathbf{C} \qquad \mathbf{W}_s = \mathbf{u}_n \times \mathbf{C} \tag{8-37}$$

over the plane $z = 0$. This is illustrated by Fig. 8-13*b*. But, according to the uniqueness theorems, we do not need to know both the normal and the tangential components of $\mathbf{C}$; so let us set up another equivalent problem as follows. Over the $z = 0$ plane we place a conductor, and just in front of the conductor we place the surface distribution of vortex sources

$$\mathbf{W}_s = \mathbf{u}_n \times \mathbf{C} \tag{8-38}$$

This is shown in Fig. 8-13*c*. We have now "forced" the tangential components of $\mathbf{C}$ over the plane $z = 0$ to be the same as those of the original problem. (At the conductor the tangential components are zero, and across $\mathbf{W}_s$ the discontinuity in the tangential components is just that discontinuity required to bring them up to their values in the original problem.) Therefore, because of uniqueness, the field $z > 0$ must be the same as the field of the original problem. Still another equivalent problem can be set up as follows. Over the $z = 0$ plane we place an anticonductor, and just in front of the anticonductor we place the surface distribution of flow sources

$$w_s = \mathbf{u}_n \cdot \mathbf{C} = C_n \tag{8-39}$$

This is shown in Fig. 8-13*d*. This time we have forced the normal component of $\mathbf{C}$ over the plane $z = 0$ to be the same as the normal component in the original problem. (At the anticonductor the normal component is zero, and across w_s the discontinuity in the normal component is just that discontinuity required to bring the component up to its value in the original problem.) Again, according to uniqueness, the field $z > 0$ must be the same as the field in the original problem.

We can now apply image theory to the configurations involving a plane conductor or a plane anticonductor. Referring to Fig. 8-8, we see that

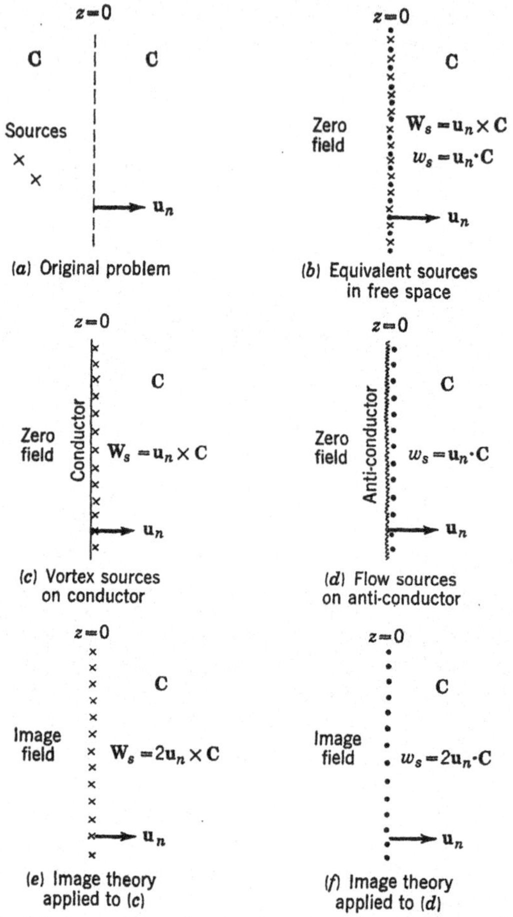

FIG. 8-13. All of these configurations produce the same field in the region $z > 0$.

the image of a vortex source parallel to a conductor is a vortex source in the same direction, located at the image point. Therefore, the images of the vortex sources of Fig. 8-13c are vortex sources in the same direction as their originals. But these sources are only an incremental distance in front of the conductor, so the images will be only an incremental distance behind the conductor. Thus the sources and their images for all practical purposes coincide. We arrive at the conclusion that

$$\mathbf{W}_s = 2\mathbf{u}_n \times \mathbf{C} \tag{8-40}$$

acting without the conductor, produces the same field $z > 0$. This is illustrated by Fig. 8-13*e*. Note that these sources produce an "image field" in the region $z < 0$, characteristic of image theory, but that this field is of no interest to us. The sources of Fig. 8-13*e* are equivalent to the original problem for the region $z > 0$, but they differ from those of Fig. 8-13*b* in that they do not produce zero field $z < 0$. Analogously, we can apply image theory to the configuration of Fig. 8-13*d*, and we arrive at the conclusion that

$$w_s = 2\mathbf{u}_n \cdot \mathbf{C} = 2C_n \tag{8-41}$$

acting without the anticonductor, produces the same field $z > 0$. This is shown by Fig. 8-13*f*. These sources are equivalent to the original problem for $z > 0$ and produce an "image field" in the region $z < 0$.

Reasoning other than that used above can be followed to establish the equivalences of Fig. 8-13. For example, instead of "constructing" the conductor and anticonductor models, as we have done, we can arrive at them starting from the original equivalence principle. Considering Fig. 8-13*b*, we note that there is zero field in the region $z < 0$; so we may place any material in this region without affecting the field $z > 0$. Thus, we may back the sources with a conductor. The images of the flow sources in the conductor are equal but of opposite sign; so the flow sources plus their images produce no field. Thus, they need not be considered, and we have Fig. 8-13*c*. Next, we may back the sources of Fig. 8-13*b* with an anticonductor. This time the images of the vortex sources just cancel the original vortex sources, producing no field. They therefore need not be considered, and we have Fig. 8-13*d*.

We might inquire how it is possible for Figs. 8-13*b*, *c*, and *d* to produce the same field everywhere. This can be explained as follows. The vortex sources of Fig. 8-13*c* induce on the conductor the same flow-source distribution as is shown in Fig. 8-13*b*. The flow sources of Fig. 8-13*d* induce on the anticonductor the same vortex-source distribution as in Fig. 8-13*b*.

Finally, we might have justified the representations of Figs. 8-13*e* and *f* in one bold step, using considerations of symmetry. The fields on either side of the plane of vortex sources in Fig. 8-13*e* must be equal in magnitude, with tangential components oppositely directed. Therefore, the discontinuity in $\mathbf{u}_n \times \mathbf{C}$ across the $z = 0$ plane is just twice what it would be if $\mathbf{C} = 0$ for $z < 0$. A similar argument justifies Fig. 8-13*f*.

In electrostatics, instead of a knowledge of the tangential or normal components of the field, we are often given the scalar potential over the boundary. If V is continuous, the tangential component of $\mathbf{E}$ can immediately be found according to $\mathbf{E} = -\nabla V$ along the boundary, and the preceding theory applies. However, a discontinuity in potential is sometimes encountered, for example, across a small gap between two conductors. This is illustrated by Fig. 8-14*a*. The electric intensity in the gap becomes very large, for $\mathbf{E} \cdot \Delta\mathbf{l}$ must equal the potential difference. The equivalent magnetic current also becomes very large, for $\mathbf{K}_s = \mathbf{E} \times \mathbf{u}_n$. This is a large current confined to a small transverse

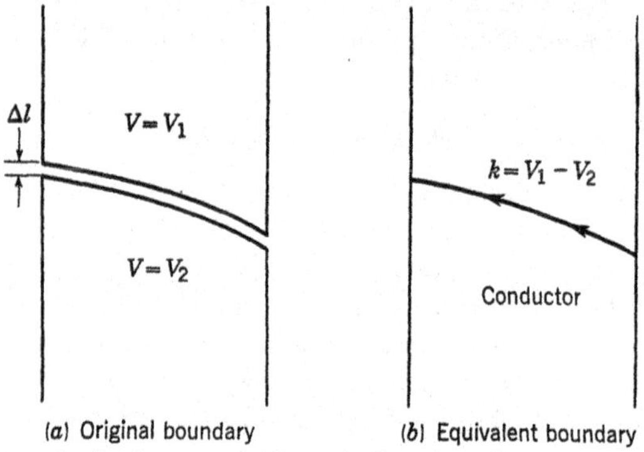

FIG. 8-14. A discontinuity in potential is equivalent to a filament of magnetic current on a conductor.

cross section, giving effectively a filament of magnetic current k. This is shown in Fig. 8-14*b*. The strength of the current must be

$$k = V_1 - V_2 \tag{8-42}$$

which we show as follows. Across the gap (top to bottom) of the original problem $\int \mathbf{E} \cdot d\mathbf{l}$ is $V_1 - V_2$. In the equivalent problem, $\int \mathbf{E} \cdot d\mathbf{l}$ across the current is equal to k, since we can close the loop along the conductor where tangential $\mathbf{E}$ is zero. Therefore we have Eq. (8-42). Note that we should pick the reference direction of k according to the left-hand rule. When the boundary is a plane, we can apply image theory. This gives an image equal to and coincident with the current filament, so a current of *twice* the value of Eq. (8-42) acting without the conductor is needed.

The procedure of "building up" the desired field in a given region by first placing a conductor or anticonductor over the boundary and then placing surface vortex or flow sources over the surface, is applicable to boundaries of any shape. However, we can apply image theory only if these boundaries are planes. In other words, we cannot solve for the field using the methods of this section unless the boundaries are planes. The extension of the methods of this section to regions bounded by more than one plane involves the use of multiple images, and will be illustrated by one of the following examples.

A problem having a single infinite-plane boundary is shown in Fig. 8-15*a*. It consists of a conductor at zero potential covering the $z = 0$ plane, except for a disk of radius a raised to a potential V_0. Thus, tangential **E** is zero over the $z = 0$ plane except at the boundary of the disk, at which we have a discontinuity in potential. The original problem is therefore equivalent to a loop of ϕ-directed magnetic current $k = -V_0$ on a conductor covering the entire $z = 0$ plane. To this representation we can apply image theory, obtaining a loop of magnetic

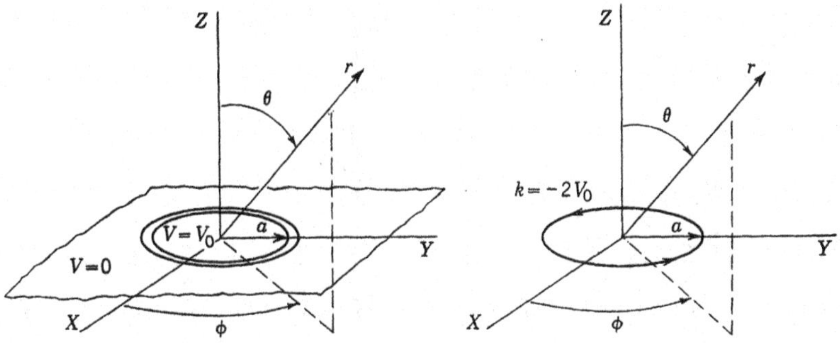

(*a*) Original problem (*b*) Equivalent problem

FIG. 8-15. A disk of constant potential in a plane of zero potential.

current $k = -2V_0$ acting in free space. This gives the same field $z > 0$ as does the original problem, and is illustrated by Fig. 8-15*b*. Thus, the problem is exactly dual to the problem for a loop of electric current, Fig. 6-4. Consulting our table of duals, Table 7-2, we see that we merely replace

$$\mathbf{H} \text{ by } \mathbf{E} \qquad i \text{ by } -k = 2V_0$$
$$\mathbf{B} \text{ by } \mathbf{D} \qquad \mu \text{ by } \epsilon$$

in the solution to the current loop and we have the solution to the problem of Fig. 8-15*a*. For example, the field along the z axis is given by the dual to Eq. (6-27), that is

$$D_z = \epsilon E_z = \frac{\epsilon V_0 a^2}{(a^2 + z^2)^{3/2}} \tag{8-43}$$

If the field point is distant from the disk, that is, if $r \gg a$, then the field reduces to the dipole field. This is the dual to Eq. (6-58), so if $r \gg a$,

$$\mathbf{E} = \frac{V_0 a^2}{2r^3} (\mathbf{u}_r 2 \cos\theta + \mathbf{u}_\theta \sin\theta) \tag{8-44}$$

Note that this field is exactly the same as the field from a charge dipole of strength $ql = 2V_0 \pi a^2 \epsilon$, as we can see by comparing Eq. (8-44) with Eq. (5-80).

For our second example, consider a pair of parallel plates at zero potential, closed at one end by a plate of potential V_0, as shown in Fig. 8-16a. Construction of an equivalent problem is accomplished by taking a conducting trough the same shape as the original boundary and placing two filaments of magnetic current along the two corners. These should be of magnitude V_0 and should flow in opposite directions. We now image them in the $x = 0$ plane, obtaining two currents of magnitude $2V_0$ between two parallel conductors infinite in both the $+x$ and $-x$ directions. Finally, we image these two currents successively in the top

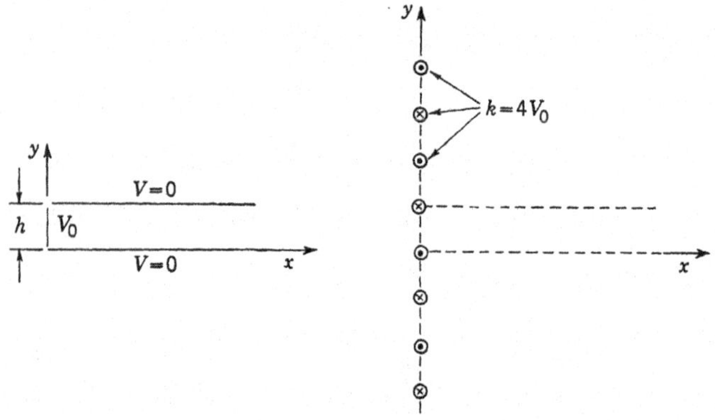

(a) Original problem (b) Equivalent problem

FIG. 8-16. Two parallel plates of zero potential closed at one end by a plate of potential V_0.

and bottom walls, obtaining currents of magnitude $4V_0$, acting in homogeneous space. This final equivalence is shown in Fig. 8-16b.

For practice, let us obtain the explicit mathematical solution for this problem. The line currents of Fig. 8-16b we label k_n such that $n = 0$ indicates the current at $y = 0$, $n = 1$ indicates the current at $y = h$, $n = -1$ indicates the current at $y = -h$, etc. The field from each filament of current is given by Eq. (6-46) with $\mathbf{H}$ replaced by $\mathbf{E}$ and with i replaced by $-k_n = -(-1)^n k$. The field from all currents is then given by the superposition

$$\mathbf{E}(\boldsymbol{\rho}) = \frac{-1}{2\pi} \sum_{n=-\infty}^{\infty} \frac{(-1)^n k \mathbf{u}_z \times (\boldsymbol{\rho} - \boldsymbol{\rho}'_n)}{|\boldsymbol{\rho} - \boldsymbol{\rho}'_n|^2} \tag{8-45}$$

where

$$\boldsymbol{\rho} = \mathbf{u}_x x + \mathbf{u}_y y$$
$$\boldsymbol{\rho}'_n = \mathbf{u}_y nh$$

Performing the cross product, we have

$$\mathbf{u}_z \times (\boldsymbol{\rho} - \boldsymbol{\rho}'_n) = -\mathbf{u}_x(y - nh) + \mathbf{u}_y x$$

Finally, substituting this into Eq. (8-45), letting $k = 4V_0$, and separating rectangular components, we have

$$E_x(x,y) = \frac{2V_0}{\pi} \sum_{n=-\infty}^{\infty} \frac{(-1)^n(y - nh)}{x^2 + (y - nh)^2}$$
$$E_y(x,y) = \frac{-2V_0}{\pi} \sum_{n=-\infty}^{\infty} \frac{(-1)^n x}{x^2 + (y - nh)^2} \tag{8-46}$$

For calculation purposes, we need take only a sufficient number of terms of the summations to give the desired accuracy. This solution converges rapidly for x small, but slowly for x large. By a method called "separation of variables" (which we do not consider in this text) it is possible to obtain a solution which converges rapidly for x large, but slowly for x small.

8-8. Other Methods of Solution. Although many problems can be solved by the techniques given in this chapter, we have only sampled the theory of boundary-value problems. So that the reader may be aware of other methods, we shall list some of the more common ones and indicate the types of problems for which they might be useful. These methods are treated in many of the books listed in the Bibliography.

A. Separation of variables. This is a method by which numerous "possible solutions" to Laplace's equation can be obtained. In fact, for certain types of regions the set of solutions obtained is "complete"; that is, any well-behaved field within such a region can be expressed in terms of these solutions. Use of the method of separation of variables is, however, restricted to certain types of coordinate systems. The rectangular, cylindrical, and spherical coordinate systems are among those for which this method is applicable.

B. Complex function theory. In regions of regularity, the real and imaginary parts of a function of a complex variable are solutions to the two-dimensional Laplace equation. They are therefore possible solutions to boundary-value problems. A problem involving complicated boundaries can often be "transformed" by conformal transformation into a problem for which the solution is known. A wide variety of problems can be treated in this manner. However, the procedure is restricted to two-dimensional problems.

C. Inversion. This is a method whereby a known solution involving plane or spherical conducting boundaries can be "inverted" to give the solution to another problem involving plane or spherical conducting boundaries. However, only a relatively small number of problems can be treated by this method. Many of the problems solved by inversion can be solved directly by image theory (see Prob. 8-15).

D. Green's-function method. An extension of the divergence theorem leads to an integral relationship known as "Green's theorem." The use of certain functions in Green's theorem leads to solutions of boundary-value problems. This Green's-function method is fundamentally the same as use of the equivalence principle, and all problems solved by the former can be solved by the latter. The Green's-function method tends to obscure the physical interpretation that we have given to the equivalence principle.

E. Integral equations. An integral equation is an equation which contains the unknown function under an integral sign. For example, Eq. (8-1) is an integral equation for the unknown **C**. It is sometimes possible to solve such equations, but there are no general methods. It is usually easier to solve a given problem by other methods than by an integral-equation method.

F. Variational method. When only a single "characteristic parameter" of a problem is desired, such as capacitance or resistance, it is often possible to obtain a close approximation to this parameter without solving the boundary-value problem. This is done by manipulating the integral equation into a "stationary" formula for the desired parameter. The parameter as given by a stationary formula is relatively insensitive to variations of the unknown about its true value. Substitution of a good guess at the unknown into the stationary formula then yields a close approximation to the desired parameter.

G. Difference equations. When it is impractical to obtain an analytic solution to Laplace's equation, it is sometimes practical to obtain approximate solutions as follows. A lattice of a discrete number of points is chosen within the region of interest, and the difference equation corresponding to Laplace's equation is written. Solution of this difference equation gives the desired approximate solution. It is sometimes possible to allow the lattice spacing to become vanishingly small and to obtain an analytic solution. A numerical procedure for solving the difference equation is known as the "relaxation method."

H. Field mapping. Approximate solutions to two-dimensional fields involving closed conducting boundaries can be obtained by a graphic method known as "field mapping." This method makes use of the fact that the field lines and equipotential lines are everywhere perpendicular to each other. A plot of both field lines and equipotential lines must therefore form a system of "curvilinear squares," which can be constructed reasonably accurately by visual inspection. This method is useful for evaluating resistance or capacitance in two-dimensional problems that cannot be solved analytically.

I. Experimental models. It is, of course, possible to construct a model of the electrostatic or magnetostatic system under consideration and to make desired measurements. However, it is also possible to take other physical systems having the same equations and to use the concept of duality. For example, Laplace's equation is also encountered in the theory of incompressible fluid flow. Thus, the flow lines of a fluid-flow model correspond to the field lines of the dual electrostatic or magnetostatic problem. A good presentation of such modeling techniques is found in Rogers, "Introduction to Electric Fields."*

QUESTIONS FOR DISCUSSION

8-1. In reference to Fig. 8-1 and Eq. (8-1), what restriction (in addition to being source-free) must be placed on the behavior of $\mathbf{C}^{(1)}$?

8-2. In view of the concepts of Sec. 7-3, show that the scalar electrostatic potential in any physical problem is always well-behaved in a region of empty space. Therefore, uniqueness based on Eq. (8-4) always applies to the problem of determining the electrostatic field in empty space.

8-3. In view of the concepts of Sec. 7-4, show that the magnetostatic vector potential in any physical problem is always well-behaved in a region of empty space. Therefore, uniqueness based on Eq. (8-5) always applies to the problem of determining the magnetostatic field in empty space.

8-4. Discuss the philosophy of the uniqueness proofs of Sec. 8-2.

8-5. Discuss qualitatively the "transient behavior" of the field, current, and charge, as we bring a point charge into the vicinity of a conductor.

8-6. Show that the boundary condition $E_t^{(1)} = E_t^{(2)}$ follows from the boundary condition $V^{(1)} = V^{(2)}$.

8-7. Why does the solution to a boundary-value problem for V involve the determination of more constants than does the solution for $\mathbf{E}$?

8-8. List the "known field solutions" that have been used in the text thus far, and indicate the regions for which these solutions are source-free.

8-9. What are the characteristics of a "grounded conductor"?

8-10. What is the origin of the surface charge of Eq. (8-24)?

8-11. Discuss the differences between the boundary conditions of Table 8-2 and those of Table 8-1.

8-12. Add to Table 8-3 the behavior of the magnetic vector potential $\mathbf{A}$ at a material boundary.

8-13. Discuss the concepts of "conductor" and "anticonductor" surfaces.

8-14. Justify qualitatively the image theory represented by Fig. 8-8.

8-15. Do the image sources of a problem actually exist?

8-16. What is the actual source of the image fields for the problem of Fig. 8-11?

8-17. What is the actual source of the image fields for the problem of Fig. 8-12?

8-18. Discuss the arguments used to establish the equivalences of Fig. 8-13.

8-19. What modification of the theory of Sec. 8-7 would be required if there were sources in the region $z > 0$ in Fig. 8-13a?

8-20. Obtain Eq. (8-42) from the equivalence principle, using a limiting procedure. (Let the discontinuity in V be approximated by a rapid, but continuous, variation in V.)

* W. E. Rogers, "Introduction to Electric Fields," McGraw-Hill Book Company, Inc., New York, 1954.

PROBLEMS

8-1. Consider the region $a < \rho < b$, $0 < z < c$ (ring-shaped region). Let the field within this region be $\mathbf{C} = \mathbf{u}_\phi/\rho$ (the line-current field). Show that a scalar potential of $\mathbf{C}$ is $g = -\phi$. Note that $g(\phi = 0) = 0$ and that $g(\phi = 2\pi) = -2\pi$, giving a discontinuity at the $\phi = 0$ plane. Thus, g is not well-behaved, and uniqueness based on Eq. (8-4) is not justified. Find a second source-free field $\hat{\mathbf{C}}$ which has the same normal component over the given boundary as does the given field $\mathbf{C}$.

8-2. Modify the region of Prob. 8-1 by introducing a "barrier" over the $\phi = 0$ plane. The boundary is now considered to be the surface of the region defined in Prob. 8-1 plus the cross section $\phi = 0$. The scalar potential $g = -\phi$, $0 < \phi < 2\pi$, is well-behaved in this new region, and uniqueness based on Eq. (8-4) applies. Show that the fields $\mathbf{C}$ and $\hat{\mathbf{C}}$ of Prob. 8-1 now have different normal components over this new boundary. (The region of Prob. 8-2 is "simply connected" while the region of Prob. 8-1 is "multiply connected." A region is simply connected if all possible closed contours within the region can be continuously contracted to a point without passing out of the region.)

8-3. Consider the region defined by $a < r < b$ (spherical-shell region). Let the field within this region be $\mathbf{C} = \mathbf{u}_r/r^2$ (point-charge field). Show that this field is source-free in the given region. Also show that a vector potential of field $\mathbf{C}$ is $\mathbf{G} = -\mathbf{u}_\phi\ (r \tan\theta)^{-1}$. Show that this vector potential is not well-behaved, and uniqueness based on Eq. (8-5) is not justified. Find a second source-free field $\hat{\mathbf{C}}$ having the same tangential components over the given boundary as does the given field $\mathbf{C}$.

8-4. Modify the region of Prob. 8-3 by introducing two "corridors" between the outer and inner spheres along the z axis. The boundary is now considered to be the surface of the original spheres plus the tubes connecting them. Show that the vector potential $\mathbf{G} = -\mathbf{u}_\phi(r \tan\theta)^{-1}$ is well-behaved in this new region. Therefore, uniqueness based on Eq. (8-5) now applies. Show that the fields $\mathbf{C}$ and $\hat{\mathbf{C}}$ of Prob. 8-3 now have different tangential components over this new boundary. (The region of Prob. 8-4 is "simply bounded," while the region of Prob. 8-3 is "multiply bounded." A region is simply bounded if it is possible to travel continuously through all points on the bounding surface without leaving the surface.)

8-5. Give a qualitative discussion of the results of Probs. 8-1 to 8-4, using known properties of flow sources and vortex sources.

8-6. Consider a pair of conductors covering the planes $x = 0$ and $x = d$. Let the potentials be $V = 0$ at $x = 0$ and $V = V_0$ at $x = d$. The region $y < 0$ contains vacuum, and the region $y > 0$ contains a dielectric of capacitivity ϵ. Determine V, $\mathbf{E}$, and $\mathbf{D}$ in the region between the two plates, and q_s on each plate.

8-7. Consider the conducting cylinders $\rho = a$ and $\rho = b$, $b > a$, between which a potential difference V_0 exists. Let the region $a < \rho < c$ contain a dielectric of capacitivity ϵ_1 and let the region $c < \rho < b$ contain a dielectric of capacitivity ϵ_2. Determine $\mathbf{E}$ and $\mathbf{D}$ in the region $a < \rho < b$ and determine q_s on each conductor.

8-8. Consider a pair of conducting spheres covering $r = a$ and $r = b$, $b > a$, between which a potential difference V_0 exists. Let the region $0 < \theta < \pi/2$ contain a dielectric ϵ_1 and let the region $\pi/2 < \theta < \pi$ contain a dielectric ϵ_2. Determine $\mathbf{E}$ and $\mathbf{D}$ in the region between the conducting spheres and determine q_s on the spheres.

8-9. Consider a conducting sphere of radius R and a dipole ql located at a distance d from the center of the sphere, $d > R$. If the dipole points radially outward from the sphere, determine the appropriate image dipole for calculating the field $r > R$. If the dipole points perpendicular to the radial direction, determine the appropriate image dipole. Show that a dipole of arbitrary orientation can always be considered as the superposition of one dipole pointing in the radial direction and one pointing perpendicular to the radial direction.

8-10. Take the solution to the problem of Fig. 8-5, and let $r' \to \infty$ and $q \to \infty$ so that the field of q in the vicinity of the sphere $r = R$ remains finite and becomes uniform. Denote this uniform field by E_0, that is

$$E_0 = \frac{q}{4\pi\epsilon(r')^2}$$

Show that as $r' \to \infty$, $\hat{r}' \to 0$ and the image charges $\hat{q}$ form a dipole

$$ql = \hat{q}\hat{r}'$$

Thus, show that a conducting sphere of radius R placed in a uniform field E_0 produces the same field $r > R$ as a point dipole

$$ql = 4\pi\epsilon E_0 R^3$$

at the origin. The total field external to the sphere is, of course, the sum of E_0 plus the field of the dipole.

8-11. Consider a conducting cylinder $\rho = R$ and a parallel filament of charge q_l = constant at ρ', where $\rho' > R$. Using the results of the example in Sec. 5-7, determine the image of the line charge in the conducting cylinder. If the potential $V \to 0$ as $\rho \to \infty$, determine the potential of the conducting cylinder. Is it possible to add an additional line charge at the cylinder axis making $V = 0$ both at the conducting cylinder and at infinity?

8-12. Using the results of Prob. 8-11, obtain the field produced by a conducting cylinder placed perpendicular to a uniform field E_0. Use two line charges on opposite sides of the conducting cylinder, and let these line charges recede to infinity.

8-13. Consider two parallel conducting cylinders, supporting equal but opposite charges per unit length. Let the radius of one cylinder be a, that of the other b, and the separation of their axes d, where $d > a + b$. Using the solution to the example in Sec. 5-7, determine V, **E**, and **D** external to the cylinders and determine q_s on each cylinder.

8-14. Using the results of the analysis of the problem of Fig. 5-11, determine the surface-charge density on an insulated conducting ellipsoid maintained at a potential V_0 with respect to infinity.

8-15. Consider the two intersecting spheres and the three point charges shown in Fig. 8-17. Show that $\hat{q}$ is at the image point to q in both spheres if the spheres intersect orthogonally. What must the magnitude of $\hat{q}$ be for $\hat{q}$ to be the image of q in both spheres? Justify the statement that the outside surface of the intersecting spheres is an equipotential surface, and calculate this potential. Hence, the field external to a pair of charged intersecting spheres is that of the three point charges. Give several other problems involving conductors and point charges for which Fig. 8-17 represents the solution.

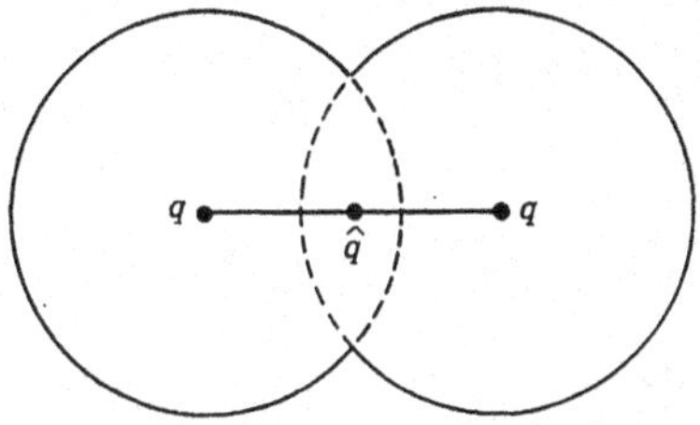

FIG. 8-17. Spheres and point charges for Prob. 8-15.

8-16. In the problem of Fig. 8-3, suppose that the two materials, $0 < x < d_1$ and $d_1 < x < (d_1 + d_2)$, have conductivities σ_1 and σ_2. Solve for **J**, **E**, and **D** between the conducting plates. Determine q_s on the three boundaries $x = 0$, $x = d_1$, and $x = d_1 + d_2$. Show that the solution for **J** is the same as the solution for **D** in Sec. 8-3*C*, if ϵ is replaced by σ.

8-17. Consider the perfectly conducting cylinders $\rho = a$ and $\rho = b$, $b > a$, between which a potential difference V_0 exists. Let the region $0 < \phi < \pi$ be filled with material characterized by σ_1, ϵ_1, and let the region $\pi < \phi < 2\pi$ contain vacuum. Determine **J**, **E**, and **D** everywhere in the region $a < \rho < b$ and determine q_s on all boundaries.

8-18. Consider the perfectly conducting spheres $r = a$ and $r = b$, $b > a$, between which a potential difference V_0 exists. Let the region $0 < \theta < \theta_0$ be filled with a material characterized by σ_1, ϵ_1, and let there be vacuum elsewhere. Determine **J**, **E**, and **D** everywhere in the region $a < r < b$ and determine q_s on all boundaries.

8-19. Consider the two parallel, perfectly conducting cylinders half buried in earth, shown in Fig. 8-18. Determine **J** in the earth and q_s on the cylinders. (*Hint:* Use the field of parallel line charges.)

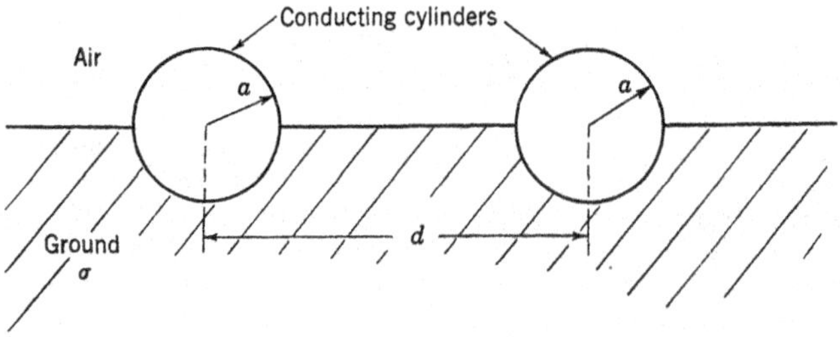

FIG. 8-18. Resistance system for Prob. 8-19.

8-20. Given that a filament of current i flows along the z axis, that the region $y < 0$ is filled with a material having inductivity μ, and that vacuum exists in the region $y > 0$, determine **B** and **H** everywhere.

8-21. Suppose that the two conductors of Fig. 8-3 support surface currents $\mathbf{J}_s = \mathbf{u}_y J_0$ at $x = 0$, and $\mathbf{J}_s = -\mathbf{u}_y J_0$ at $x = d_1 + d_2$, with J_0 a constant. Let the two media be characterized by μ_1 and μ_2. Determine **B** and **H** between the two conductors, assuming no field $x < 0$ and $x > d_1 + d_2$.

8-22. External to a sphere $r = R$, whose large permeability is approximated by $\mu = \infty$, there exists a small loop of current. Considering the loop to be a magnetic dipole of moment iS, pointing radially outward from the sphere, determine the magnetic field. (*Hint:* Replace the loop by its equivalent magnetic charge dipole, and use results dual to those of Prob. 8-9.)

8-23. A cylinder $\rho = a$ of permeability μ is placed in a uniform field $\mathbf{H} = \mathbf{u}_x H_0$. Show that the field produced by the magnetization of the cylinder is that of a line dipole $\rho > a$ and that it is uniform $\rho < a$. The total field is the sum of the "impressed" uniform field plus the "magnetization" field of the cylinder. Determine **M** and the total **H** and **B** everywhere. Let $\mu \to \infty$, and compare the result with that of Prob. 8-12. State the dual electrostatic problem, and give its answer.

8-24. A sphere $r = a$ of permeability μ is placed in a uniform field $\mathbf{H} = \mathbf{u}_z H_0$. Show that the field produced by the magnetization of the sphere is that of a point dipole $r > a$ and that it is uniform $r < a$. Determine **M** and the total **H** and **B** everywhere. Let $\mu \to \infty$, and compare the result with that of Prob. 8-10. State the dual electrostatic problem, and give its answer.

8-25. Consider an arbitrary vector field **C**, its possible scalar potential g, and its possible vector potential **G**. Admit the possibility of surface-flow sources w_s and surface-vortex sources W_s. Make up a table of boundary conditions, showing the behavior of **C**, g, and **G** at a boundary.

8-26. Given conductors covering the planes $y = 0$ and $z = 0$, and a point charge q

at $(0,a,b)$, determine the system of image charges correctly giving the field in the quarter-space $y > 0$, $z > 0$. Determine q_s on the conductors.

8-27. The half-space $x < 0$ is filled with a dielectric of capacitivity ϵ, and there is vacuum $x > 0$. A dipole $\mathbf{q}l = \mathbf{u}_x ql$ is located at $x = d$. Determine the field everywhere. State the dual magnetic problem.

8-28. A circular loop of wire carrying current i and of radius R lies on the plane surface of an infinitely large block of magnetic matter of inductivity μ. Find **B** and **H** along the axis of the loop in both regions. What is the bound pole surface density at the center of the loop?

8-29. Two small conducting spheres are buried in the ground and connected by insulated wires to a battery. This is shown in Fig. 8-19. Treating the ground as

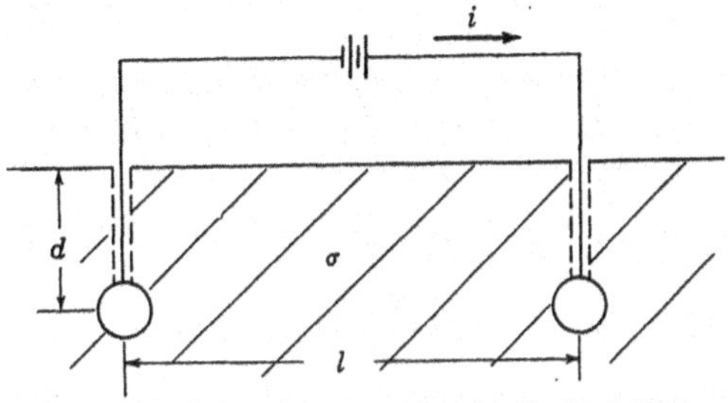

FIG. 8-19. Grounding system for Prob. 8-29.

a homogeneous conductor and the spheres as point sources, set up the image system from which the current distribution in the ground may be obtained.

8-30. Consider a pair of concentric conducting spheres of radii a and b, and a point charge q at a radius c, where $a < c < b$. Determine the system of images (magnitude and location) from which the field in the region between the spheres may be obtained.

8-31. The half-plane $y = 0$, $x > 0$ is covered by a conductor maintained at a potential V_0, and the half-plane $y = 0$, $x < 0$ is covered by a conductor maintained at a potential $-V_0$. Obtain (*a*) the magnetic current in vacuum equivalent in the region $y > 0$; (*b*) the electric charge in vacuum equivalent in the region $y > 0$; (*c*) the magnetic current equivalent in the region $y < 0$; and (*d*) the electric charge equivalent in the region $y < 0$. Sketch the **E** lines for the original problem.

8-32. Consider the quarter-space $x > 0$, $y > 0$, shown in Fig. 8-20. Along the boundary planes, $V = 0$ except over the portion $x < a$, $y < a$, as shown. Determine the magnetic current in vacuum equivalent in the original quarter-space. Solve for the **E** field in the quarter-space.

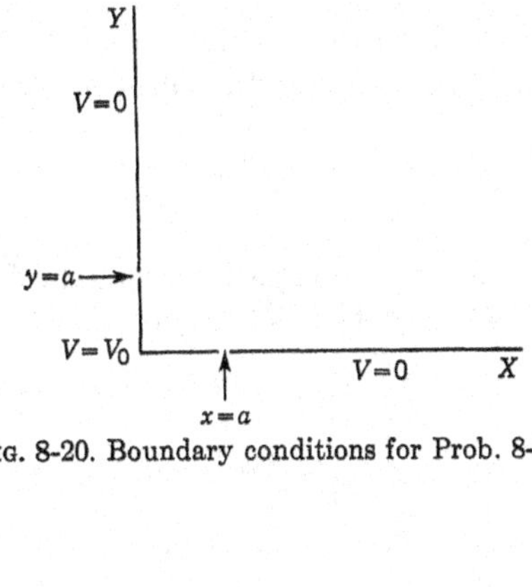

FIG. 8-20. Boundary conditions for Prob. 8-32.

CHAPTER 9

CIRCUIT ELEMENTS

9-1. Introduction. A circuit element can be viewed as a configuration of matter, two or more points of which are designated "terminals." Suppose that a d-c source has been applied to a two-terminal element and that steady-state conditions exist. The *d-c resistance* R is defined as the ratio of the terminal voltage v to the terminal current i, that is

$$R = \frac{v}{i} \tag{9-1}$$

The rate at which energy is transferred to the element is $P = vi$. In linear matter, this energy is dissipated as heat, and an alternative definition of d-c resistance is

$$R = \frac{P_d}{i^2} = \frac{v^2}{P_d} \tag{9-2}$$

In addition to the dissipation of energy, there may also be a storage of energy. To account for the electric energy W_e, we say that the element has a *d-c capacitance* C given by

$$C = \frac{2W_e}{v^2} \tag{9-3}$$

When there is no conduction of current through the element, we have the alternative definition

$$C = \frac{q}{v} \tag{9-4}$$

where q is the net charge on the positive terminal. To account for the magnetic energy W_m, we say that the element has a *d-c inductance* L given by

$$L = \frac{2W_m}{i^2} \tag{9-5}$$

When there is a perfect conduction path through the element (no terminal voltage), we have the alternative definition

$$L = \frac{\psi}{i} \tag{9-6}$$

where ψ is the net magnetic flux linking the current. The d-c equivalent circuit of a simple two-terminal element is, therefore, as given by Fig. 9-1. This circuit does not necessarily remain a valid representation of the element when v and i vary with respect to time.

When the stored energy is small compared to the rate of energy dissipation, we say that the element is a *resistor*. When the magnetic energy and the rate of energy dissipation are small compared to the electric energy, we say that the element is a *capacitor*. When the electric energy and the rate of energy dissipation are small compared to the magnetic energy, we say that the element is an *inductor*. For example, consider

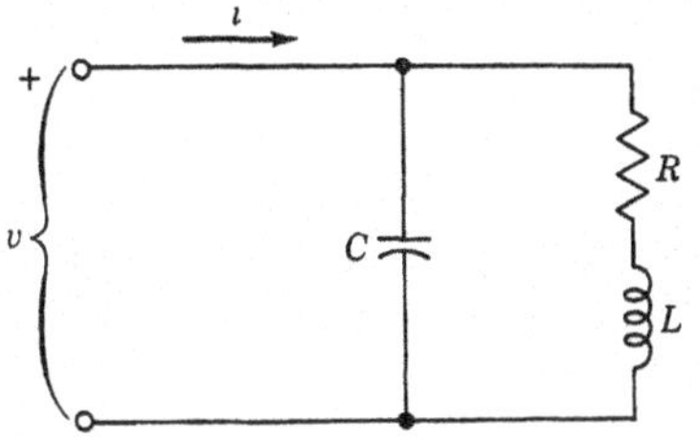

FIG. 9-1. D-c equivalent circuit for a two-terminal element.

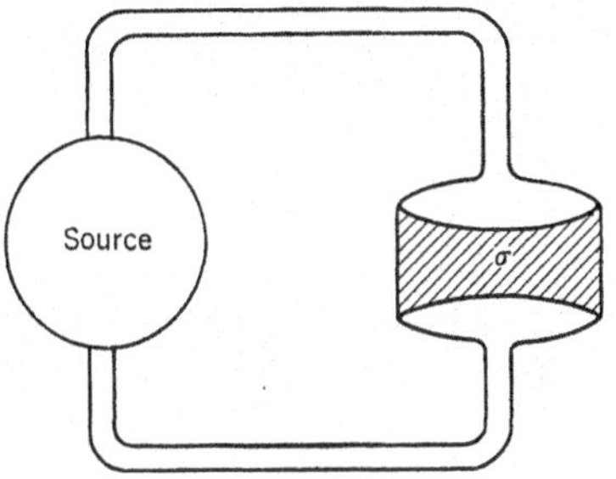

FIG. 9-2. A circuit element prototype.

the prototypic element of Fig. 9-2. We take the wire and plates to be a perfect conductor and the surrounding medium to be a perfect insulator. When the block of material has an intermediate value of σ, the element is a resistor. As $\sigma \to 0$, the element becomes a capacitor. As $\sigma \to \infty$, the element becomes an inductor.

9-2. Resistance. Basically, the determination of resistance involves obtaining the solution to a boundary-value problem, from which v and i may then be determined. An actual resistor consists of a conducting medium with good conductors covering part of its surface and good insulators covering the rest. We shall idealize the problem, and assume perfect conductors and perfect insulators bounding a conducting medium. The boundary conditions for the problem are (*a*) that the tangential components of **E** and **J** are zero over the perfect conductor surface, and (*b*) that the normal components of **E** and **J** are zero over the perfect insulator surface. Furthermore, we shall consider only a homogeneous conducting medium; so **E** and **J** have no sources (flow or vortex) within the conducting medium. The problem is, therefore, to find solutions to the source-free field equations, satisfying the above boundary conditions. For a linear medium, R is a function only of the geometry and of the conductivity of the medium.

The most common geometry encountered in practice is that of the uniform cylinder, shown in Fig. 9-3. We take the ends to be equipotential surfaces (perfectly conducting surfaces) and the sides to be perfectly insulated. A source-free field satisfying the appropriate boundary con-

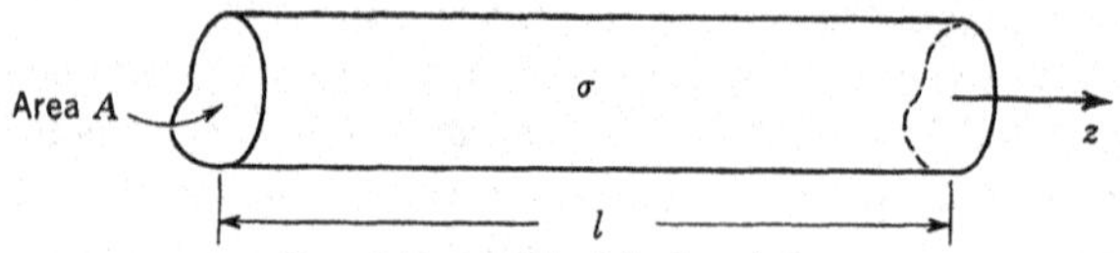

FIG. 9-3. A cylindrical resistor.

ditions is the one-dimensional field

$$\mathbf{E} = \mathbf{u}_z K$$

where K is a constant and z is measured in the direction of the cylinder axis. The voltage between the two ends of the cylinder is

$$v = \int_0^l E_z\, dz = K \int_0^l dz = Kl$$

where l is the length of the cylinder. The current through the cylinder is

$$i = \iint \sigma E_z\, ds = \sigma K \iint ds = \sigma K A$$

where A is the cross-sectional area of the cylinder. Therefore, the resistance according to Eq. (9-1) is

$$R = \frac{1}{\sigma}\frac{l}{A} \tag{9-7}$$

The reciprocal of conductivity is commonly called the *resistivity* of the medium. Note that Eq. (9-7) is the same result that we obtained in Sec. 3-3, showing that our previous analysis was exact.

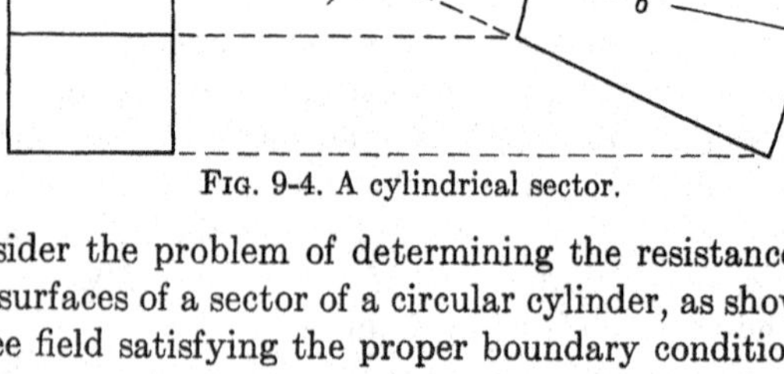

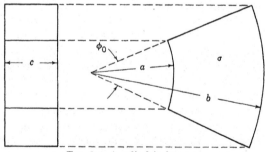

FIG. 9-4. A cylindrical sector.

Now consider the problem of determining the resistance between the two curved surfaces of a sector of a circular cylinder, as shown in Fig. 9-4. A source-free field satisfying the proper boundary conditions is the line-charge field

$$\mathbf{E} = \frac{\mathbf{u}_\rho K}{\rho}$$

where ρ is the radial distance from the cylinder axis. The voltage between the two surfaces $\rho = a$ and $\rho = b$ is

$$v = \int_a^b E_\rho \, d\rho = K \int_a^b \frac{d\rho}{\rho} = K \log \frac{b}{a}$$

The current through the sector is

$$i = \iint \sigma E_\rho \, ds = \sigma K \int_0^{\phi_0} d\phi \int_0^c dz = \sigma K \phi_0 c$$

where ϕ_0 is the angle subtended, in radians. The resistance according to Eq. (9-1) is therefore

$$R = \frac{1}{\sigma} \frac{\log (b/a)}{c\phi_0} \tag{9-8}$$

In the limit as $\phi_0 \to 0$, $a \to b$, Eq. (9-8) reduces to Eq. (9-7).

We can also determine the resistance between the two inclined flat surfaces of the sector of Fig. 9-4. For this case, the appropriate source-free field is that of the line-vortex source along the $\rho = 0$ axis. Proceeding in the usual manner, we obtain

$$R = \frac{1}{\sigma} \frac{\phi_0}{c \log (b/a)} \tag{9-9}$$

Derivation of this is left to the reader. The resistance between the two flat parallel sides of the sector of Fig. 9-4 would be given by Eq. (9-7) with $A = (b^2 - a^2)\phi_0/2$ and $l = c$.

As a final example, let us take a case for which we have not yet encountered the appropriate source-free field but for which this field may be readily determined. Consider the spherical sector defined by $r = a$, $r = b$, $\theta = \theta_0$, $\theta = \pi - \theta_0$, as shown in Fig. 9-5. We wish to determine the resistance between the two flat surfaces. Note that the required boundary conditions can be satisfied if the **E** field has only a θ component. Also, in view of the symmetry of the problem, there must be no variation of **E** with respect to ϕ (the angle about the ring). Thus we let

$$\mathbf{E} = \mathbf{u}_\theta E_\theta(r,\theta)$$

which, within the sector, must satisfy

$$\nabla \cdot \mathbf{E} = \frac{1}{r \sin \theta} \frac{\partial}{\partial \theta} (E_\theta \sin \theta) = 0$$

$$\nabla \times \mathbf{E} = \mathbf{u}_\phi \frac{1}{r} \frac{\partial}{\partial r} (rE_\theta) = 0$$

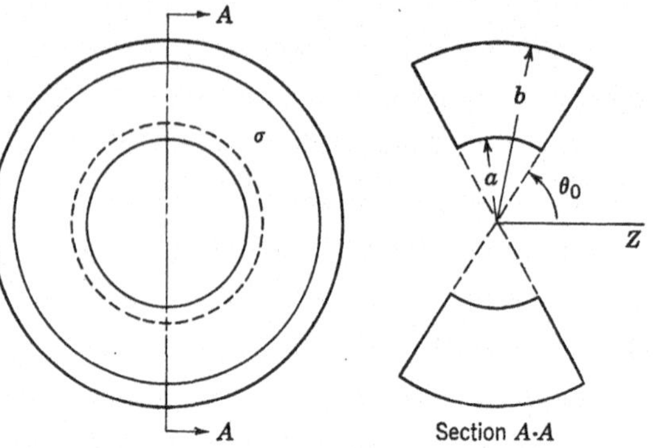

FIG. 9-5. A spherical sector.

The first of these equations requires that $E_\theta \sim 1/\sin\theta$ and the second that $E_\theta \sim 1/r$. Therefore, a possible solution is

$$E_\theta = \frac{K}{r \sin\theta}$$

with K a constant. We now obtain the voltage between the two flat faces according to

$$v = \int_{\theta_0}^{\pi-\theta_0} E_\theta r\, d\theta = K \int_{\theta_0}^{\pi-\theta_0} \frac{d\theta}{\sin\theta} = 2K \log \cot \frac{\theta_0}{2}$$

The current flowing through the sector is

$$i = \iint \sigma E_\theta\, ds = \sigma K \int_0^{2\pi} d\phi \int_a^b dr = 2\pi\sigma K(b-a)$$

Therefore, the resistance according to Eq. (9-1) is given by

$$R = \frac{\log \cot(\theta_0/2)}{\sigma\pi(b-a)} \tag{9-10}$$

To obtain the resistance between the two curved surfaces of Fig. 9-5, we would use the point-charge field. This results in a resistance of

$$R = \frac{b-a}{\sigma 4\pi ab \cos\theta_0} \tag{9-11}$$

We could make the problem more general by considering the sector formed by terminating the ring of Fig. 9-5 at two ϕ = constant planes.

9-3. An Approximation. The geometry of a resistor is often such that an exact solution to the boundary-value problem cannot readily be obtained. In this case we can resort to approximate methods of evaluating the resistance. The following is a procedure whereby upper and lower bounds to the correct value of resistance may be established. We first prove the following theorem: *If the conductivity of any part of a resistor is increased, the total resistance is either decreased or unchanged. If the conductivity of any part of a resistor is decreased, the total resistance is either increased or unchanged.*

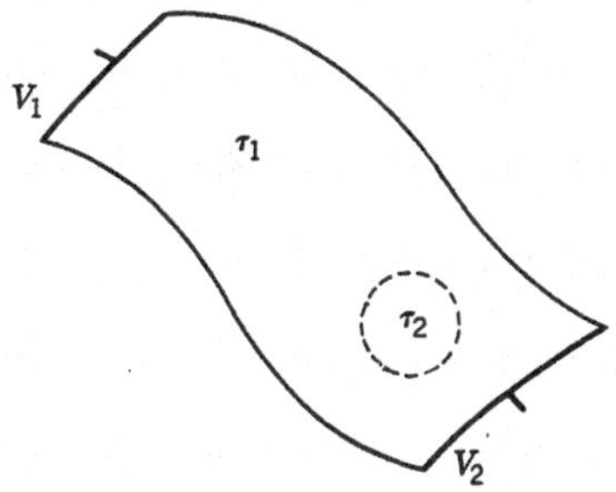

FIG. 9-6. The conductivity within τ_2 is changed from σ to $\hat{\sigma}$.

Consider the resistor of Fig. 9-6. Originally it consists of a homogeneous material of conductivity σ, in which the fields **E**, **J**, and V exist. Suppose that the conductivity within the small region τ_2 is changed to $\hat{\sigma}$ and that the terminal potentials V_1 and V_2 are kept constant. The new fields we denote by $\hat{\mathbf{E}}$, $\hat{\mathbf{J}}$, and $\hat{V}$. The rates of energy dissipation in the resistor before and after the change are

$$P = \iiint \mathbf{E} \cdot \mathbf{J}\, d\tau \qquad \hat{P} = \iiint \hat{\mathbf{E}} \cdot \hat{\mathbf{J}}\, d\tau$$

respectively. The change in power is therefore

$$\begin{aligned} \delta P = \hat{P} - P &= \iiint (\hat{\mathbf{J}} \cdot \hat{\mathbf{E}} - \mathbf{J} \cdot \mathbf{E})\, d\tau \\ &= \iiint [(\hat{\mathbf{J}} - \mathbf{J}) \cdot \hat{\mathbf{E}} + \mathbf{J} \cdot (\hat{\mathbf{E}} - \mathbf{E})]\, d\tau \\ &= \iiint (\delta \mathbf{J} \cdot \hat{\mathbf{E}} + \delta \mathbf{E} \cdot \mathbf{J})\, d\tau \end{aligned}$$

But $V = \hat{V}$ over the ends of the resistor, $J_n = 0$ over the sides, and $\nabla \cdot \mathbf{J} = 0$ throughout; so

$$\begin{aligned} \iiint \delta \mathbf{E} \cdot \mathbf{J}\, d\tau &= -\iiint \nabla\, \delta V \cdot \mathbf{J}\, d\tau = -\iiint \nabla \cdot (\delta V\, \mathbf{J})\, d\tau \\ &= -\oint\!\!\oint \delta V\, \mathbf{J} \cdot d\mathbf{s} = 0 \end{aligned}$$

Thus, the change in power reduces to

$$\delta P = \iiint \delta \mathbf{J} \cdot \hat{\mathbf{E}}\, d\tau \tag{9-12}$$

The proof of $\iiint \delta \mathbf{E} \cdot \mathbf{J}\, d\tau = 0$ is also valid for **J** replaced by $\hat{\mathbf{J}}$, which we can write in the form

$$\iiint_{\tau_1} \delta \mathbf{E} \cdot \hat{\mathbf{J}}\, d\tau = -\iiint_{\tau_2} \delta \mathbf{E} \cdot \hat{\mathbf{J}}\, d\tau$$

But in τ_1 the conductivity remains unchanged; so

$$\delta \mathbf{E} \cdot \hat{\mathbf{J}} = \frac{1}{\sigma}\, \delta \mathbf{J} \cdot \hat{\mathbf{J}} = \delta \mathbf{J} \cdot \hat{\mathbf{E}}$$

in τ_1, and the above equation can be written as

$$\iiint_{\tau_1} \delta \mathbf{J} \cdot \hat{\mathbf{E}}\, d\tau = -\iiint_{\tau_2} \delta \mathbf{E} \cdot \hat{\mathbf{J}}\, d\tau$$

Eq. (9-12) can now be written as

$$\begin{aligned}\delta P &= \iiint_{\tau_2} \delta\mathbf{J} \cdot \hat{\mathbf{E}}\, d\tau + \iiint_{\tau_1} \delta\mathbf{J} \cdot \hat{\mathbf{E}}\, d\tau \\ &= \iiint_{\tau_2} (\delta\mathbf{J} \cdot \hat{\mathbf{E}} - \delta\mathbf{E} \cdot \hat{\mathbf{J}})\, d\tau\end{aligned}$$

Expanding out the integrand, we have

$$\begin{aligned}\delta\mathbf{J} \cdot \hat{\mathbf{E}} - \delta\mathbf{E} \cdot \hat{\mathbf{J}} &= (\hat{\sigma}\hat{\mathbf{E}} - \sigma\mathbf{E}) \cdot \hat{\mathbf{E}} - (\hat{\mathbf{E}} - \mathbf{E}) \cdot \hat{\sigma}\hat{\mathbf{E}} \\ &= (\hat{\sigma} - \sigma)\mathbf{E} \cdot \hat{\mathbf{E}} = \delta\sigma\, \mathbf{E} \cdot \hat{\mathbf{E}}\end{aligned}$$

so the final result is

$$\delta P = \iiint_{\tau_2} \delta\sigma\, \mathbf{E} \cdot \hat{\mathbf{E}}\, d\tau \tag{9-13}$$

For small changes in σ, $\hat{\mathbf{E}} \approx \mathbf{E}$ in τ_2, and δP is positive or negative according as $\delta\sigma$ is positive or negative. Since $R = v^2/P$ and v has been kept constant, the theorem of the first paragraph follows.

Let us now see how we may make use of this theorem. If we take thin sheets of a perfect conductor and place them roughly perpendicular to the current lines, we are in effect forcing these surfaces to be equipotentials. By the above theorem, this procedure can only decrease the total resistance, giving us a lower bound to the correct R. If we take thin sheets of a perfect insulator and place them roughly parallel to the current lines, we are in effect forcing the current to follow a specific path. By the above theorem, this procedure can only increase the total resistance, giving us an upper bound to the correct R.

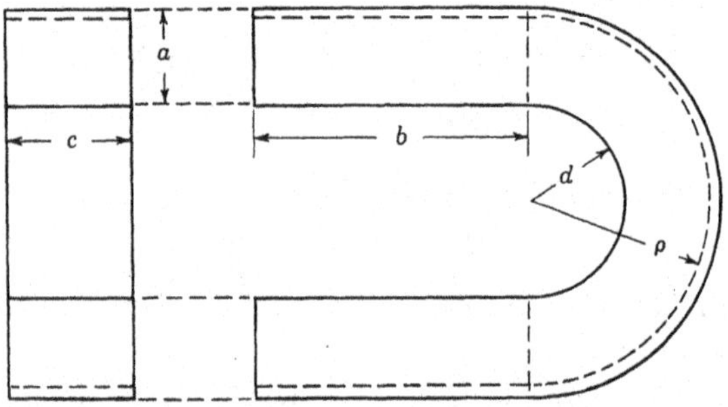

FIG. 9-7. A horseshoe-shaped resistor.

As an example, consider the problem of finding the resistance between the two end faces of the horseshoe-shaped resistor of Fig. 9-7. If we place thin sheets of a perfect conductor across the cross sections at which the curvature begins, we divide the resistor into three parts, each of which we have already analyzed. According to Eq. (9-7), the resistance of each leg is

$$R_{\text{leg}} = \frac{1}{\sigma}\frac{b}{ac}$$

and according to Eq. (9-9), the resistance of the cylindrical sector is

$$R_{\text{sector}} = \frac{1}{\sigma}\frac{\pi}{c \log (1 + a/d)}$$

The total resistance is the sum of these three parts; so a lower bound to the correct resistance is

$$R_{\min} = \frac{1}{\sigma c}\left[\frac{2b}{a} + \frac{\pi}{\log (1 + a/d)}\right] \tag{9-14}$$

If we now place a system of thin sheets of a perfect insulator parallel to the outer edge of the horseshoe, we form a system of resistances in parallel. The conductance of each strip, according to Eq. (9-7), is given by

$$\frac{1}{R_{\text{strip}}} = \frac{\sigma c \, d\rho}{2b + \pi\rho}$$

where $d\rho$ is the thickness of the strip. The total conductance of the system of strips is the reciprocal of an upper bound to the correct resistance. This is found by summing over all strips, that is

$$\frac{1}{R_{\max}} = \sigma c \int_d^{d+a} \frac{d\rho}{2b + \pi\rho}$$

giving

$$R_{\max} = \frac{\pi}{\sigma c} \Big/ \log \left[\frac{\pi(d + a) + 2b}{\pi d + 2b}\right] \tag{9-15}$$

The resistance of the original resistor must lie between the resistances of Eqs. (9-14) and (9-15). To illustrate, suppose that

$$a = b = c = d = 1 \text{ meter}$$

We calculate

$$R_{\min} = \frac{6.54}{\sigma} \qquad R_{\max} = \frac{6.60}{\sigma}$$

so a value $R = 6.57/\sigma$ we know to be correct to within ± 0.5 per cent.

9-4. Capacitance. The capacitance problem, like the resistance problem, is basically one of determining the solution to a boundary-value problem, from which the charge q and the voltage v may in turn be determined. In a resistor problem, we have

$$\frac{1}{R} = \frac{i}{v} \qquad \mathbf{J} = \sigma\mathbf{E} \qquad i = \iint \mathbf{J} \cdot d\mathbf{s}$$

while in a capacitor problem, we have

$$C = \frac{q}{v} \qquad \mathbf{D} = \epsilon\mathbf{E} \qquad q = \iint \mathbf{D} \cdot d\mathbf{s}$$

In both cases, **E** is a source-free field satisfying appropriate boundary conditions. Thus, C, **D**, and ϵ in a capacitor problem are duals of $1/R$, **J**, and σ in a resistance problem. There is, however, one important difference between the two cases. In resistance problems, at a conductor (σ finite) to insulator (σ zero) boundary, $J_n = E_n = 0$. That is, we have an anticonducting boundary. In capacitance problems, there is no material for which ϵ is zero; so there are no anticonducting boundaries. Therefore, for a resistance problem to have a physically meaningful dual capacitance problem, σ must be nonzero everywhere that **E** exists. Such problems are generally more difficult than those for which σ exists only in a given region. Using the above duality relationship, if we have the solution to a resistor problem, we have only to replace $1/R$ by C and σ by ϵ to obtain the solution to the dual capacitor problem, and vice versa. Thus, for conductors in a homogeneous medium of infinite extent, the equation

$$CR = \frac{\epsilon}{\sigma} \tag{9-16}$$

holds if R and C are for dual problems. Unfortunately, those resistance problems that we have solved in the preceding section do not have useful capacitance duals except in special cases.

The simplest capacitor problem for which an exact solution is possible is that of two concentric conducting spheres, of radii a and b, $a < b$. We already know that the field in this case is that of a point charge,

$$V(r) = \frac{q}{4\pi\epsilon r} + K$$

where q is the charge on the inner sphere. The potential difference between the two spheres is

$$v = V(a) - V(b) = \frac{q}{4\pi\epsilon}\left(\frac{1}{a} - \frac{1}{b}\right)$$

Therefore, the capacitance according to Eq. (9-4) is

$$C = 4\pi\epsilon\left(\frac{ab}{b-a}\right) \tag{9-17}$$

Note that the solution to the dual resistor problem is Eq. (9-11) with $\theta_0 = 0$ and that Eq. (9-16) is satisfied. Other capacitance problems that can be readily solved with our present knowledge are concentric ellipsoids (using the solution to Fig. 5-11) and nonconcentric spheres (using the method of Sec. 8-6B).

The idea of capacitance can be extended to two-dimensional problems, that is, to problems for which there is no variation of field quantities with respect to one spatial direction. The simplest of these is the case of two concentric conducting cylinders, Fig. 8-2, to which the solution is known. The total charge on each infinitely long cylinder is infinite; so the total capacitance would be infinite. However, we can determine the capacitance per unit length of the concentric cylinders by interpreting the q of Eq. (9-4) as the charge per unit length. The field is that of the line charge,

$$V(\rho) = \frac{-q_l}{2\pi\epsilon} \log \rho + K$$

where q_l is the charge per unit length on the inner cylinder. The potential difference between the two cylinders is therefore

$$v = V(a) - V(b) = \frac{-q_l}{2\pi\epsilon} (\log a - \log b)$$

where a and b are the radii of the inner and outer cylinders, respectively. The charge per unit length is $q = q_l$; so the capacitance per unit length is

$$\frac{C}{l} = \frac{2\pi\epsilon}{\log (b/a)} \tag{9-18}$$

The solution to the dual resistor problem is Eq. (9-8) with $\phi_0 = 2\pi$. The capacitance of parallel but nonconcentric cylinders can also be found, using the field of two line charges, which is illustrated by Fig. 5-15.

The basic one-dimensional capacitance problem is that of two parallel conducting plates of infinite extent. Again the total capacitance becomes infinite, but we may determine a capacitance per unit area by interpreting the q of Eq. (9-4) as the charge per unit area. For this problem we use the one-dimensional field and in the usual manner we obtain the capacitance per unit area as

$$\frac{C}{A} = \frac{\epsilon}{d} \tag{9-19}$$

where d is the separation of the plates. Note that this is the same solution that we postulated for the parallel-plate capacitor of Sec. 4-4. It is, of course, approximate in that case because of fringing effects at the edge of the plates. The dual resistor problem is the problem of the cylindrical resistor, Eq. (9-7).

It would be helpful to know the exact effect of the fringing field on the capacitance of a finite parallel-plate capacitor or a finite concentric-cylinder capacitor. Unfortunately, such problems have never been solved exactly, and even approximate solutions require mathematical methods beyond the scope of this book. We can, however, determine whether the exact capacitance (including fringing effects) will be higher or lower than the approximate capacitance (neglecting fringing effects). We know the effect of changing σ in the dual resistance problem; so we can state the following dual theorem for a capacitance problem. *If the capacitivity of any part of a capacitor is increased, the total capacitance is either increased or unchanged. If the capacitivity of any part of a capacitor is decreased, the total capacitance is either decreased or unchanged.* When we neglect fringing, we are solving exactly the problem for which $\epsilon = 0$ in the external medium. In other words, the capacitivity of the external medium has been decreased from ϵ_0 to zero. Therefore, *the value of capacitance obtained by neglecting fringing is always less than the true capacitance.*

9-5. Capacitance Systems. The two-terminal capacitor can be viewed as a special case of a system of conductors immersed in a perfect dielectric. Suppose we have N conducting bodies, all initially uncharged, as suggested by Fig. 9-8. If a unit charge is placed on the jth conductor, the ith conductor assumes a potential $V_i = p_{i,j}$. The field equations are linear; so if we place q_j coulombs of charge on the jth conductor, the potential of the ith conductor becomes $V_i = p_{i,j}q_j$. Now if charges q_j are placed on the respective conductors $j = 1, 2, \ldots, N$, the potential of the ith conductor will be the superposition

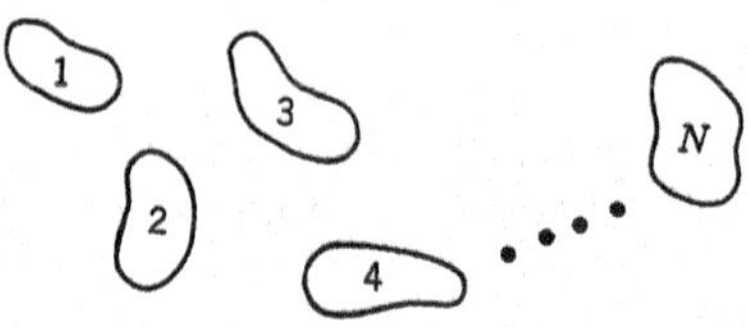

FIG. 9-8. A system of conductors.

$$V_i = \sum_{j=1}^{N} p_{i,j}q_j$$

This relationship holds for each conductor, so we have the set of equations

$$\begin{aligned} V_1 &= p_{1,1}q_1 + p_{1,2}q_2 + \cdots + p_{1,N}q_N \\ V_2 &= p_{2,1}q_1 + p_{2,2}q_2 + \cdots + p_{2,N}q_N \\ &\cdots\cdots\cdots\cdots\cdots\cdots \\ V_N &= p_{N,1}q_1 + p_{N,2}q_2 + \cdots + p_{N,N}q_N \end{aligned} \tag{9-20}$$

The constants $p_{i,j}$ have been given the name *coefficients of potential.* They depend only upon the geometry of the system and upon the capacitivity of the dielectric.

Reciprocity assumes the following special form for systems of conductors. Let V_i^a be the conductor potentials produced by net charges q_i^a and let V_i^b be those produced by q_i^b. Then

$$\sum_{i=1}^{N} V_i^a q_i^b = \sum_{i=1}^{N} V_i^b q_i^a \tag{9-21}$$

which is called *Green's reciprocation theorem.* It is proved as follows. All charges are distributed in surface layers on the conductors, as shown in Sec. 8-3. The reciprocity integral, Eq. (5-82), therefore assumes the form

$$\iint V^b q_s{}^a \, ds = \iint V^a q_s{}^b \, ds$$

where the integration is carried out over all conductors. The potential on each conductor is a constant; so the above equation reduces to

$$\sum_i V_i{}^b \oiint q_{s,i}{}^a \, ds = \sum_i V_i{}^a \oiint q_{s,i}{}^b \, ds$$

Since these integrals give the net charge on each conductor, Eq. (9-21) is proved.

Some special properties of the coefficients of potential can readily be obtained. Suppose that the a field is produced by a charge $q_j{}^a$ on the jth conductor, that the b field is produced by a charge $q_i{}^b$ on the ith conductor, and that all other conductors are uncharged. Equation (9-21) then reduces to

$$V_i{}^a q_i{}^b = V_j{}^b q_j{}^a$$

but in this special case $V_i{}^a = p_{i,j} q_j{}^a$ and $V_j{}^b = p_{j,i} q_i{}^b$; so we have

$$p_{i,j} q_j{}^a q_i{}^b = p_{j,i} q_i{}^b q_j{}^a$$

from which it follows that

$$p_{i,j} = p_{j,i} \tag{9-22}$$

Thus, only $N(N+1)/2$ coefficients of potential are independent. Physically, Eq. (9-22) means that if a charge q on the jth conductor raises the potential of the ith conductor to V, then the same charge q on the ith conductor raises the potential of the jth conductor to V also. Other properties of the coefficients follow from the basic definition. Since the $p_{i,j}$ are the potentials due to a unit positive charge on the jth conductor, all $p_{i,j}$ must be positive. Also, the conductor having the charge must be the most positive, so we have

$$p_{i,i} \geq p_{i,j} \geq 0 \tag{9-23}$$

We can solve Eqs. (9-20) for the charges in terms of the potentials, obtaining

$$\begin{aligned} q_1 &= c_{1,1}V_1 + c_{1,2}V_2 + \cdots + c_{1,N}V_N \\ q_2 &= c_{2,1}V_1 + c_{2,2}V_2 + \cdots + c_{2,N}V_N \\ &\cdots\cdots\cdots\cdots\cdots\cdots \\ q_N &= c_{N,1}V_1 + c_{N,2}V_2 + \cdots + c_{N,N}V_N \end{aligned} \tag{9-24}$$

The $c_{i,i}$ are called *coefficients of capacitance* and the $c_{i,j}$, $i \neq j$, are called *coefficients of electrostatic induction.* The logic of these names can be seen as follows. When all $V_i = 0$ except V_j, Eqs. (9-24) become

$$q_i = c_{i,j}V_j$$

For $i = j$, $q_i = q_j$ is the charge on the jth conductor, whose potential is V_j. The ratio $c_{j,j} = q_j/V_j$ is then the capacitance between the jth conductor and all others grounded (kept at zero potential); hence the name "coefficient of capacitance." For $i \neq j$, q_i is the charge "induced" on the ith conductor (drawn from the ground reservoir of charge) by raising the potential of the jth conductor to V_j; hence the name "coefficient of induction." When all conductors are grounded except the jth conductor and this is kept at 1 volt, then the $c_{i,j}$ are equal to the charges q_i on the various conductors. It follows from reciprocity, Eq. (9-21), that

$$c_{i,j} = c_{j,i} \tag{9-25}$$

the proof being similar to that for Eq. (9-22). To maintain the jth conductor positive at 1 volt and the other conductors at zero potential, q_j must be positive and all other charges negative. Therefore

$$\begin{aligned} c_{j,j} &\geq 0 \\ c_{i,j} &\leq 0 \qquad i \neq j \end{aligned} \tag{9-26}$$

Also, since the total charge of the system cannot be negative under these conditions, we have

$$\sum_{i=1}^{N} c_{i,j} \geq 0 \tag{9-27}$$

Although the $c_{i,j}$ have the dimensions of capacitance, they do not correspond directly to the usual circuit quantity of capacitance. We can, however, rearrange Eqs. (9-24) in terms of the potential differences

$$v_{i,j} = V_i - V_j$$

obtaining

$$\begin{aligned} q_1 &= C_{1,1}V_1 + C_{1,2}v_{1,2} + \cdots + C_{1,N}v_{1,N} \\ q_2 &= C_{2,1}v_{2,1} + C_{2,2}V_2 + \cdots + C_{2,N}v_{2,N} \\ &\cdots\cdots\cdots\cdots\cdots\cdots\cdots\cdots \\ q_N &= C_{N,1}v_{N,1} + C_{N,2}v_{N,2} + \cdots + C_{N,N}V_N \end{aligned} \tag{9-28}$$

The $C_{i,j}$ are called the *direct capacitances* and are, evidently, the usual circuit capacitances when $i \neq j$. The V_i are potential differences with respect to zero potential; so the $C_{i,i}$ can be interpreted as the capacitances to zero potential, or ground. The circuit equivalent of an N-body system requires N terminals plus a ground terminal. This is shown for a three-body system in Fig. 9-9. Comparing Eqs. (9-28) with Eqs. (9-24), it is seen that

$$\begin{aligned} c_{i,i} &= \sum_{j=1}^{N} C_{i,j} \\ c_{i,j} &= -C_{i,j} \qquad i \neq j \end{aligned} \tag{9-29}$$

Solving these for the C's, we have

$$C_{i,i} = \sum_{j=1}^{N} c_{i,j} \qquad \text{(9-30)}$$
$$C_{i,j} = -c_{i,j} \qquad i \neq j$$

We have, of course, $C_{i,j} = C_{j,i}$ from Eq. (9-25). It also follows from the properties of the $c_{i,j}$ that all $C_{i,j}$ are positive or zero.

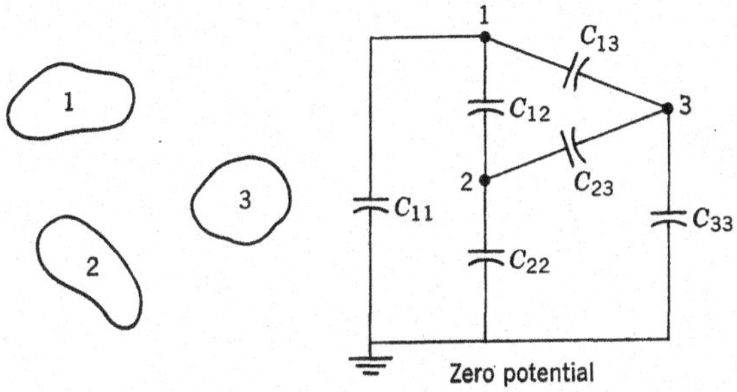

(a) Three body system (b) Equivalent circuit

FIG. 9-9. A system of conductors represented by a network of capacitors.

If we go to a one-body "system," Eqs. (9-28) reduce to

$$q_1 = C_{1,1}V_1 \qquad \text{(9-31)}$$

where $C_{1,1}$ is called the *self-capacitance* of the body. It can be interpreted as the capacitance between the body and a sphere of infinite radius. A charge of q coulombs on a sphere of radius a raises its potential with respect to infinity to $V = q/4\pi\epsilon a$, so the self-capacitance of a sphere is

$$C_{1,1} = 4\pi\epsilon a \qquad \text{(9-32)}$$

Note that this is the limiting value of Eq. (9-17) as $b \to \infty$, that is, as the outer sphere becomes infinitely large. If the finite sphere is one element of a system of conductors, the $C_{1,1}$ of Eqs. (9-28) is *not* the self-capacitance of the isolated sphere but the capacitance obtained when the sphere is in the presence of all other bodies kept at the same potential.

Since three capacitive parameters are required to represent a two-body system, we might inquire about the validity of representing circuit capacitors by a single capacitance. We now know that the three-capacitor network of Fig. 9-10*b* correctly represents a two-body system. When we determined the parameter C in Sec. 9-3, we used the total charge on each conductor. Therefore we were evaluating the total capacitance between the two bodies, or

$$C = C_{1,2} + \frac{C_{1,1}C_{2,2}}{C_{1,1} + C_{2,2}} \qquad \text{(9-33)}$$

However, for the commonly used capacitors of circuit theory, $C_{1,1}$ and $C_{2,2}$ are very small compared with $C_{1,2}$; so $C \approx C_{1,2}$. Let us illustrate this for the capacitor of Fig. 9-10*a*, constructed of two conducting hemispheres separated by a narrow cut. The self-capacitance of each hemisphere is found by maintaining both halves at the same potential, say by joining them with a wire across the gap. Each hemisphere will carry

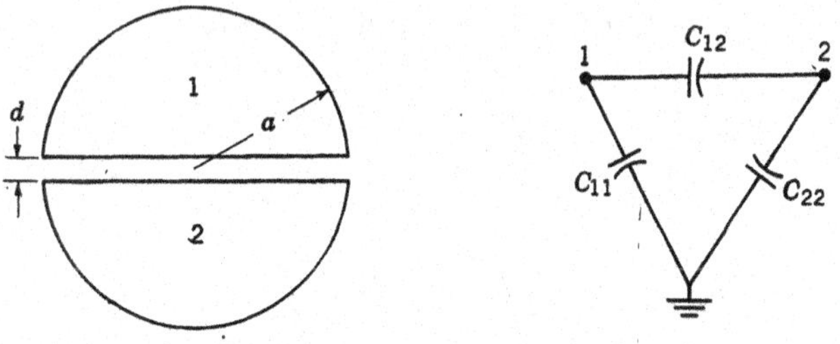

(*a*) Capacitor (*b*) Equivalent circuit

FIG. 9-10. A hemispherical capacitor.

half of the total charge; so the self-capacitance of each will be one-half that of a complete sphere, Eq. (9-32), or

$$C_{1,1} = C_{2,2} = 2\pi\epsilon a \tag{9-34}$$

The mutual-capacitance is found by maintaining one hemisphere at zero potential and placing a charge on the other. This results in a parallel-plate capacitor for all practical purposes, so long as the separation d is small. We therefore have

$$C \approx C_{1,2} = \frac{\epsilon\pi a^2}{d} \tag{9-35}$$

The ratio of the self-capacitance to the mutual-capacitance is thus

$$\frac{C_{1,1}}{C_{1,2}} = \frac{2d}{a} \tag{9-36}$$

which can be made as small as desired by making d small. In circuit theory, the self-capacitances are usually neglected. Incidentally, the total self-capacitance of any capacitor is always less than (or equal to) that of a sphere enclosing it. This follows from the theorem in the last paragraph of Sec. 9-4 if we note that a dielectric with $\epsilon \to \infty$ is equivalent to a conducting medium (see Sec. 8-6).

The capacitance which is effective in changing the potential of one conductor when the potential of another conductor is changed we call the "effective" capacitance. In a two-body system, the effective capacitance would be $C_{1,2}$. In a three-body system, the effective capacitance is $C_{1,2}$ in parallel with the series combination of $C_{1,3}$ and $C_{2,3}$ (see Fig.

9-9b). If conductor 3 were grounded, the effective capacitance between 1 and 2 would be reduced to $C_{1,2}$, although the capacitance to ground is increased. Conductors 1 and 2 are said to be partially shielded in this case, which is illustrated by Fig. 9-11a. We obtain total shielding if conductor 3 completely encloses one of the other two conductors and is grounded. This is illustrated by Fig. 9-11b. For the situation shown, $C_{1,1} = C_{1,2} = 0$; so when conductor 3 of Fig. 9-9b is grounded, there remains no effective capacitance between terminals 1 and 2.

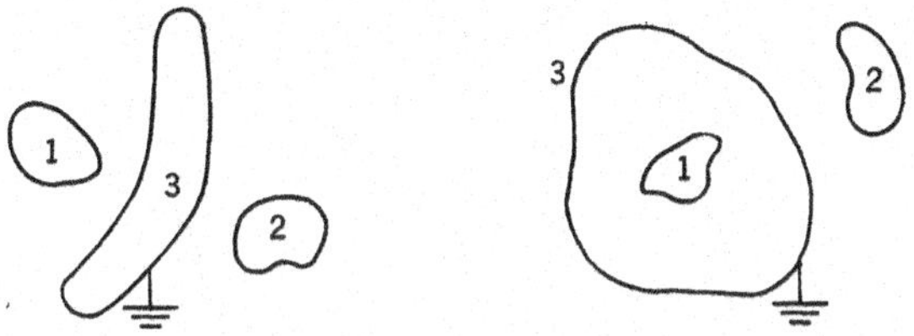

(a) Partial shielding (b) Total shielding

FIG. 9-11. Electrostatic shielding.

9-6. Resistance Systems. As we have noted earlier, for every capacitance problem involving conductors and perfect dielectrics, there exists a dual resistance problem involving perfect conductors and conducting media. The general resistance system is that of Fig. 9-8 if we consider the bodies to be perfectly conducting and the surrounding medium to be imperfectly conducting. The current leaving each body in the resistance problem is the dual of the charge maintained on each body in the capacitance problem. Since charge cannot be created in the physical world, we must visualize each perfectly conducting body in the resistance problem as being supplied by electric current via an insulated wire. It is assumed that these insulated wires are small enough so that their effect on the current distribution in the conducting medium is negligible. When we talk of the current leaving a body, we mean the current leaving through the surrounding medium, exclusive of the current through the insulated wire.

In a manner dual to that used to establish Eqs. (9-20), we have the set of equations

$$\begin{aligned} V_1 &= r_{1,1}i_1 + r_{1,2}i_2 + \cdots + r_{1,N}i_N \\ V_2 &= r_{2,1}i_1 + r_{2,2}i_2 + \cdots + r_{2,N}i_N \\ &\cdots\cdots\cdots\cdots\cdots\cdots\cdots \\ V_N &= r_{N,1}i_1 + r_{N,2}i_2 + \cdots + r_{N,N}i_N \end{aligned} \tag{9-37}$$

The $r_{i,j}$ are called the *coefficients of resistance*, although, by analogy to the dual electrostatic problem, they could also be called "resistive coefficients of potential." Equations (9-37) can be solved for the i's giving the set

$$\begin{aligned} i_1 &= g_{1,1}V_1 + g_{1,2}V_2 + \cdots + g_{1,N}V_N \\ i_2 &= g_{2,1}V_1 + g_{2,2}V_2 + \cdots + g_{2,N}V_N \\ &\cdots\cdots\cdots\cdots\cdots\cdots \\ i_N &= g_{N,1}V_1 + g_{N,2}V_2 + \cdots + g_{N,N}V_N \end{aligned} \tag{9-38}$$

dual to Eqs. (9-24). We shall call the $g_{i,j}$ *coefficients of conductance.* By procedure analogous to what was done in the capacitance system, we can now rearrange Eqs. (9-38) in terms of the potential differences $v_{i,j} = V_i - V_j$, and obtain

$$\begin{aligned} i_1 &= G_{1,1}V_1 + G_{1,2}v_{1,2} + \cdots + G_{1,N}v_{1,N} \\ i_2 &= G_{2,1}v_{2,1} + G_{2,2}V_2 + \cdots + G_{2,N}v_{2,N} \\ &\cdots\cdots\cdots\cdots\cdots\cdots \\ i_N &= G_{N,1}v_{N,1} + G_{N,2}v_{N,2} + \cdots + G_{N,N}V_N \end{aligned} \tag{9-39}$$

These equations are dual to Eqs. (9-28). The $G_{i,j}$ are called the *direct conductances,* and are the usual circuit conductances, $G_{i,i}$ being interpreted as the conductance to ground (zero potential). A picture of the equivalent circuit of Eqs. (9-39) would be Fig. 9-9 with the capacitors replaced by resistors. All of the other equations of Sec. 9-5 also apply, in the dual sense, to the resistance problem. Thus, we have a reciprocity theorem

$$\sum_{j=1}^{N} V_j^a i_j^b = \sum_{j=1}^{N} V_j^b i_j^a \tag{9-40}$$

and restrictions on the coefficients as follows. The coefficients of resistance satisfy

$$r_{i,j} = r_{j,i} \qquad r_{i,i} \geq r_{i,j} \geq 0 \tag{9-41}$$

The coefficients of conductance and mutual-conductance satisfy

$$\begin{aligned} &g_{i,j} = g_{j,i} \qquad g_{j,j} \geq 0 \\ &\sum_{i=1}^{N} g_{i,j} \geq 0 \qquad g_{i,j} \leq 0 \qquad i \neq j \end{aligned} \tag{9-42}$$

and the direct conductances satisfy

$$G_{i,i} = \sum_{j=1}^{N} g_{i,j} \qquad \begin{aligned} G_{i,j} &= -g_{i,j} \qquad i \neq j \\ G_{i,j} &\geq 0 \end{aligned} \tag{9-43}$$

Because of the duality of the capacitance system to the resistance system, an analysis of the former is equivalent to an analysis of the latter, and vice versa, provided the geometries are the same. Given the capacitance-system parameters, the parameters of the resistance-system analogue are

$$r_{i,j} = p_{i,j}\frac{\epsilon}{\sigma} \qquad g_{i,j} = c_{i,j}\frac{\sigma}{\epsilon} \qquad G_{i,j} = C_{i,j}\frac{\sigma}{\epsilon} \tag{9-44}$$

These relationships are established in the same manner as Eq. (9-16).

In a one-body system, Eqs. (9–39) reduce to

$$i_1 = G_{1,1}V_1 \tag{9-45}$$

The "self-conductance" $G_{1,1}$ is called the *grounding conductance*, and its reciprocal

$$R_g = \frac{1}{G_{1,1}} \tag{9-46}$$

is called the *grounding resistance*. This can be thought of as the resistance between a perfectly conducting body and a sphere of infinite radius. We shall see below that the direct resistance between two *distant* objects is the sum of the two grounding resistances. Thus, the parameter R_g is useful in the analysis of systems grounded in the sense of burying a metal object in the ground. In an actual grounding system, the objects are near the surface of the ground. For more precise results, we can analyze a system having one half of space filled with a conductor, the other half an insulator. This modification can be fairly simply accomplished through the use of image theory.

The grounding resistance of certain elementary shapes of conductors in a medium of infinite extent can be readily obtained. A spherical conductor of radius a has a grounding resistance of

$$R_g = \frac{1}{4\pi\sigma a} \tag{9-47}$$

obtained directly from Eq. (9-32), using duality. To obtain the grounding resistance of a long thin wire, we can approximate it by a long thin ellipsoid and use analysis of Fig. 5-11. In the dual capacitance problem, the potential of an ellipsoid of minor diameter d and interfocal distance l is given by substituting $z = y = 0$, $x = d/2$ in Eq. (5-57), that is

$$V = \frac{q_l}{4\pi\epsilon}\log\left[\frac{l/2 + \sqrt{(d/2)^2 + (l/2)^2}}{-l/2 + \sqrt{(d/2)^2 + (l/2)^2}}\right]$$

When $d \ll l$, we can approximate the square root by

$$\sqrt{\left(\frac{d}{2}\right)^2 + \left(\frac{l}{2}\right)^2} \approx \frac{l}{2} + \frac{d^2}{4l}$$

and the potential reduces to

$$V \approx \frac{q_l}{4\pi\epsilon} \log \frac{4l}{d} \qquad d \ll l$$

The total charge on the ellipsoid is

$$q = q_l l$$

so the self-capacitance of the ellipsoid according to Eq. (9-31) is

$$C_{1,1} = \frac{2\pi\epsilon l}{\log (2l/d)} \qquad d \ll l \tag{9-48}$$

Approximating a wire by an ellipsoid, we have by duality the grounding resistance of a wire of length l and diameter d given by

$$R_g \approx \frac{\log (4l/d)}{4\pi\sigma l} \qquad d \ll l \tag{9-49}$$

Another geometry of interest is that of a buried plate. The thin circular disk is a problem that can be solved exactly using spheroidal coordinates,* the result being

$$R_g = \frac{1}{8\sigma a} \tag{9-50}$$

where a is the radius of the disk. An approximate analysis can be made by using the field of a uniform disk of surface charge (see Prob. 9-16).

A grounding system is formed by two buried objects, fed equal but opposite currents as shown in Fig. 9-12. If the objects are distant from each other, their mutual presence has a negligible effect on the current distribution leaving each object. We can then superimpose the single-body fields to obtain the total field. Distant from any object, the field of that object vanishes as the point-source field. Therefore, at object a, the total field is essentially the single-body field of a, denoted V^a; and similarly for object b. The resistance between the two distant objects is thus

FIG. 9-12. Two perfectly conducting bodies immersed in a conducting medium.

$$R = \frac{V(a) - V(b)}{i} \approx \frac{V^a + V^b}{i} = R_g{}^a + R_g{}^b \tag{9-51}$$

where $R_g{}^a$ and $R_g{}^b$ are the grounding resistances of each object. Note that $V(b) \approx -V^b$, for the current reference is reversed (see Fig. 9-12).

* Ernst Weber, "Electromagnetic Fields: Theory and Application," John Wiley & Sons, Inc., New York, 1950, vol. I, p. 452.

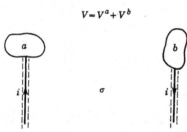

Therefore, *the total resistance of a ground system is the sum of the individual grounding resistances, providing the objects are widely separated.* If the objects are not buried deep enough to approximate the ground by a medium of infinite extent, we can modify the individual grounding resistances as follows.

Fig. 9-13*a* shows a conductor buried close to a conductor-dielectric interface. The boundary is anticonducting with respect to **E**; so image theory according to the concepts of Sec. 8-6 applies. The images in an anticonducting surface are of the same sign, giving Fig. 9-13*b* as the image equivalent to Fig. 9-13*a* in the conducting medium. The potential at the conducting body is the same in both Figs. 9-13*a* and *b*. The current leaving the body in Fig. 9-13*a* is one-half that leaving both bodies in

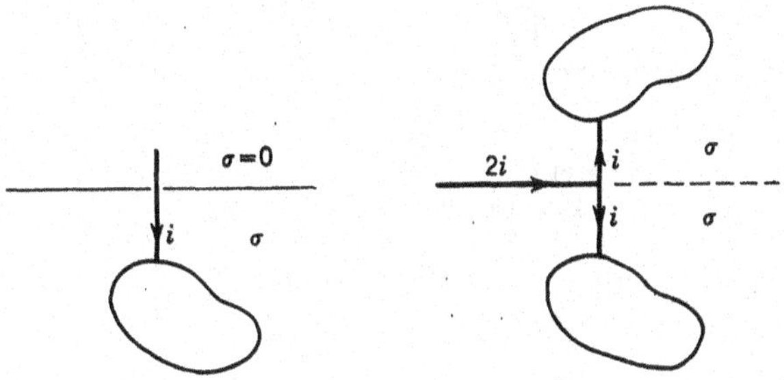

(*a*) Original problem (*b*) Image problem

FIG. 9-13. A perfect conductor buried near a conductor-dielectric interface.

Fig. 9-13*b*. Therefore, *the grounding resistance of the original problem is twice the total grounding resistance of the image problem.* We also can use the converse of this procedure. Any plane of symmetry (both in geometry and in feed) is an anticonducting surface. Therefore, *a half-buried object symmetrical about the conductor-dielectric interface has twice the grounding resistance of the object buried in the conducting medium of infinite extent.* For example, a sphere half-buried has a grounding resistance twice that of Eq. (9-47). A long thin ellipsoid half buried along either its major or its minor axis has a grounding resistance twice that of Eq. (9-49). The same is true of a wire, which may be approximated by an ellipse. A disk laid on the surface of the ground has a grounding resistance twice that of Eq. (9-50).

9-7. Inductance. The calculation of inductance involves determination of the magnetic field from knowledge of the current. The current distribution is determined by the electric field, and is usually assumed to be known. From the field solution, the inductance is found either by determining the magnetic energy and using Eq. (9-5) or by finding the flux linkage and using Eq. (9-6). It is not always necessary to determine the magnetic field explicitly, for the flux linkage can be found in terms of the magnetic vector potential according to Eq. (6-7).

The simplest type of problem is that for which the magnetic flux is more or less confined to a given magnetic circuit. This can be accomplished either by providing a very high permeability path or by spacing a winding so that it forms a current sheet which confines the field to a specific path. Suppose we have the situation illustrated by Fig. 9-14*a*. The flux is confined largely to the path of high μ; that is, $B_n \approx 0$ at the

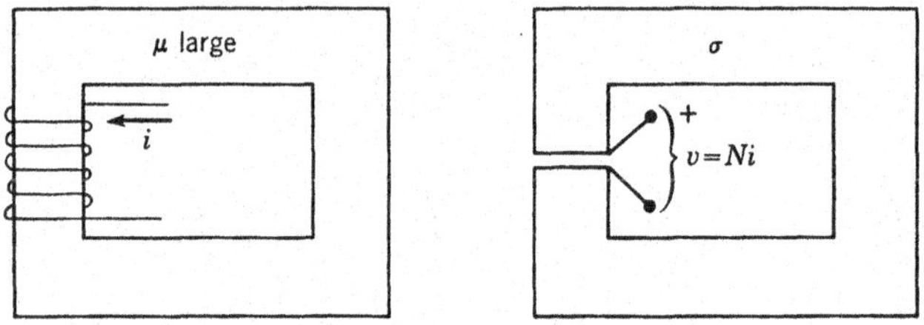

(*a*) Magnetic problem (*b*) Equivalent resistance problem

FIG. 9-14. The magnetic circuit.

surface of the magnetic material. The magnetomotive force, Eq. (4-17), once around the circuit gives

$$u = \oint \mathbf{H} \cdot d\mathbf{l} = Ni$$

where N is the number of turns carrying the current i. The problem is thus the dual of the resistance problem of Fig. 9-14*b*, with the quantities $\mathbf{B}$, $\mathbf{H}$, μ, u, ψ dual to $\mathbf{J}$, $\mathbf{E}$, σ, v, i, respectively. The *reluctance* $\mathcal{R}$ in the magnetic problem is defined as the dual of the resistance in the resistance problem, that is

$$\mathcal{R} = \frac{u}{\psi} = \frac{Ni}{\psi} \tag{9-52}$$

To obtain the reluctance from the dual resistance problem, we have only to replace R by $\mathcal{R}$ and σ by μ. Thus, for problems involving only one homogeneous medium, the equation

$$\mu\mathcal{R} = \sigma R \tag{9-53}$$

holds when $\mathcal{R}$ and R are for dual problems. Once the reluctance of the magnetic circuit is known, we can determine the inductance from Eq. (9-52) and Eq. (9-6). We must remember, however, that the ψ of Eq. (9-6) is the flux through a surface bounded by the wire, that is, the flux linkage, whereas the ψ of Eq. (9-52) is the flux through a simple cross section of the magnetic circuit. Therefore, the ψ of Eq. (9-6) is N times the ψ of Eq. (9-52), and the inductance is given by

$$L = \frac{N^2}{\mathcal{R}} \tag{9-54}$$

The reciprocal of reluctance has been given the name *permeance.*

As an example, suppose that the magnetic circuit is a circular torus of rectangular cross section. The dual resistance problem is that of Fig. 9-4 with $\phi_0 = 2\pi$. The reluctance of the magnetic circuit, obtained from Eqs. (9-9) and (9-53), is therefore

$$\mathcal{R} = \frac{2\pi}{\mu c \log (b/a)}$$

where a and b are the inner and outer radii of the torus and c is its thickness. The inductance according to Eq. (9-54) is then

$$L = \frac{\mu}{2\pi} N^2 c \log \frac{b}{a} \tag{9-55}$$

This formula is quite precise if the winding is uniformly distributed around the torus, so that it approximates a sheet of current. If the dimensions of the cross section are small compared with the radius of the torus, we have the approximation

$$L \approx \frac{\mu N^2 A}{2\pi\rho} \tag{9-56}$$

where A is the cross-sectional area and ρ is the average radius of the torus. This follows from the corresponding approximation in the resistance problem.

The reluctance concept actually holds for any inductance problem, although its primary usefulness is for the magnetic-circuit-type problem. We know that the resistance of the dual resistor problem is increased by a decrease in conductivity at any point. Therefore, the reluctance is increased by a decrease in inductivity at any point, and the inductance is increased. We thus have the theorem: *If the inductivity of any part of an inductor is increased (decreased), the total inductance is either increased (decreased) or unchanged.* When we use the approximation that the flux is confined to the magnetic circuit, we are solving exactly the problem for which $\mu = 0$ in the external medium. Therefore, *the value of inductance obtained by neglecting leakage flux is always less than the true inductance.*

When the current exists more or less in isolated wires in a magnetically homogeneous medium, the problem becomes more difficult. Suppose we approximate a portion of the circuit by a long straight wire. Close to the wire, the field is given by

$$\mathbf{H} = \mathbf{u}_\phi \frac{i}{2\pi\rho}$$

and the energy density by

$$w_m = \tfrac{1}{2}\mu H^2 = \frac{\mu i^2}{8\pi^2\rho^2}$$

If the wire is taken to be a true filament (zero radius), the energy in a volume element containing $\rho = 0$ becomes infinite. We conclude that the magnetic energy associated with a circuit of infinitely thin wires is infinite, and thus has an infinite inductance. The actual radius of the wire must therefore be considered in inductance calculations. A procedure for doing this is as follows. The flux linking all current, that is, the flux through a surface bounded by the inner edge of the wire, is calculated and called the *external flux* ψ_{ext}. The parameter

$$L_{\text{ext}} = \frac{\psi_{\text{ext}}}{i} \tag{9-57}$$

is called the *external inductance*. Rather than calculate the partial flux linkages over the wire itself, we determine the rest of the inductance from energy considerations. The *internal inductance* is defined by

$$L_{\text{int}} = \frac{2W_{\text{int}}}{i^2} \tag{9-58}$$

where W_{int} is the magnetic energy internal to the wire. The total inductance of the circuit is then taken to be

$$L = L_{\text{ext}} + L_{\text{int}} \tag{9-59}$$

Calculation of an external inductance from flux linkage and of an internal inductance from energy is exact only when **B** is everywhere tangential to the conductor surface. This approximation is quite good so long as the wires are thin.

The internal inductance of a wire is usually approximated by that for a long straight wire. The current in the wire is taken to be uniformly distributed across its cross section; so the current density is $J = i/(\pi a^2)$ where a is the radius of the wire. Taking a closed circular path concentric with the axis of the wire, the magnetomotive force must equal the current enclosed. From symmetry, **H** is rotationally symmetric; so

$$\int_0^{2\pi} H_\phi \, \rho \, d\phi = \int_0^{2\pi} \int_0^{\rho} \frac{i}{\pi a^2} \rho \, d\rho \, d\phi$$

giving

$$H_\phi = \frac{i\rho}{2\pi a^2} \tag{9-60}$$

so long as $\rho \leq a$. The energy density within the wire is therefore

$$w_m = \tfrac{1}{2}\mu H^2 = \frac{\mu i^2 \rho^2}{8\pi^2 a^4}$$

Taking a section of length l, the internal magnetic energy is

$$W_{\text{int}} = \int_0^l dz \int_0^a d\rho \int_0^{2\pi} \rho \, d\phi \left(\frac{\mu i^2 \rho^2}{8\pi^2 a^4}\right) = \frac{\mu i^2 l}{16\pi}$$

From Eq. (9-58), the internal inductance per unit length is found to be

$$\frac{L_{\text{int}}}{l} = \frac{\mu}{8\pi} \tag{9-61}$$

where μ is the permeability of the wire. Equation (9-61) is used to approximate the internal inductance of any thin wire. In most problems, the internal inductance is small compared to the external inductance, and more precise values are not necessary.

It is also difficult to calculate the external inductance exactly, for the current must then be treated as a volume distribution. The field external to an infinitely long straight wire is the same as it would be if the total current were concentrated along its axis. To simplify calculations, this is usually used as an approximation for other configurations of wire. We assume that the field external to a thin wire is approximately the same as the field from a filament of current along the axis of the wire.

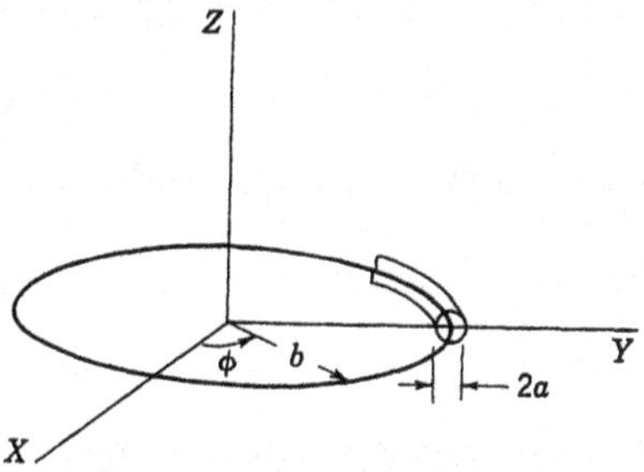

FIG. 9-15. The circular loop.

To illustrate the above concepts, consider a single-turn circular loop, as shown in Fig. 9-15. To obtain the external flux, we apply Eq. (6-7) to the inner boundary of the wire. This gives

$$\psi_{\text{ext}} = (b - a) \int_0^{2\pi} A_\phi(b - a, \phi, 0) \, d\phi$$

where a is the radius of the wire and b is the radius of the loop. From symmetry, the vector potential must be independent of ϕ; so

$$\psi_{\text{ext}} = 2\pi(b - a) A_\phi(b - a, \phi, 0) \tag{9-62}$$

We now assume that the external field is approximately the same as it would be if the total current were concentrated along the axis of the wire. The solution to this problem is given by Eq. (6-31). For $z = 0$ and $\rho = b - a$ with $a \ll b$, the α parameter is nearly unity. The elliptic integrals can then be approximated by

$$\left.\begin{aligned} K(\alpha) &\approx \log\left[\frac{4}{1-\alpha^2}\right] \\ E(\alpha) &\approx 1 \end{aligned}\right\} \alpha \approx 1$$

For our case

$$1 - \alpha^2 = 1 - \left(\frac{4b(b-a)}{2b-a}\right)^2 \approx \left(\frac{a}{2b}\right)^2$$

so Eq. (6-31) reduces to

$$A_\phi(b-a, \phi, 0) \approx \frac{\mu}{2\pi}\left(\log\frac{8b}{a} - 2\right)$$

Therefore, according to Eqs. (9-62) and (9-57), the external inductance is given by

$$L_{\text{ext}} \approx \mu b\left(\log\frac{8b}{a} - 2\right) \tag{9-63}$$

where $(b - a)$ has been replaced by b. The internal inductance is approximated by Eq. (9-61) multiplied by the length of the wire $2\pi b$. Thus the total inductance of the loop is given by

$$L \approx \mu b\left(\log\frac{8b}{a} - \frac{7}{4}\right) \tag{9-64}$$

if the wire has the same permeability as the external medium.

9-8. Mutual-inductance. Given two or more circuits, a current in one circuit may produce a magnetic flux linking another circuit. The *mutual-inductance* $M_{1,2}$ between circuit 1 and 2 is defined by

$$M_{1,2} = \frac{\psi_{1,2}}{i_1} \tag{9-65}$$

where $\psi_{1,2}$ is the magnetic flux produced by i_1 linking circuit 2. Since the circuits are physically separated, we can treat isolated wires as true filaments for the calculation of mutual-inductance. In this respect, the problem is simpler than that of calculating self-inductance.

When the problem can be treated as a magnetic circuit, the mutual-inductance can be found quite readily. We need only determine the proportion of the total flux produced by one winding which links the other winding, and multiply by the number of turns in the second winding. For the special case in which the total flux links both windings, the mutual-inductance is simply

$$M_{1,2} = \frac{N_1 N_2}{\mathcal{R}} \tag{9-66}$$

where N_1 and N_2 are the number of turns in coils 1 and 2, respectively, and $\mathcal{R}$ is the reluctance of the entire magnetic circuit. This would be the case if a second winding were placed on the core of Fig. 9-14*a*. Such an analysis neglects the leakage flux, of course, and gives a somewhat higher value of $M_{1,2}$ than the true value.

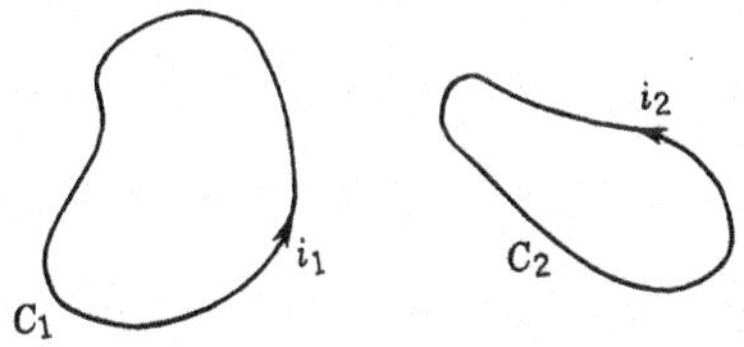

FIG. 9-16. A pair of circuits.

Suppose we now consider a pair of circuits in a homogeneous medium, as illustrated by Fig. 9-16. The magnetic field produced by i_1 can be expressed in terms of the vector potential

$$\mathbf{A}_1 = \frac{\mu i_1}{4\pi} \oint_{C_1} \frac{d\mathbf{l}_1}{|\mathbf{r} - \mathbf{r}_1|}$$

The magnetic flux due to i_1 linking circuit 2 is found according to

$$\psi_{1,2} = \oint_{C_2} \mathbf{A}_1 \cdot d\mathbf{l}_2$$

Combining the above two equations with Eq. (9-65), we have the mutual-inductance between the two circuits given by

$$M_{1,2} = \frac{\mu}{4\pi} \oint_{C_2} \oint_{C_1} \frac{d\mathbf{l}_1 \cdot d\mathbf{l}_2}{|\mathbf{r}_1 - \mathbf{r}_2|} \tag{9-67}$$

Note that this expression is symmetrical in the parameters of circuits 1 and 2, proving the reciprocity relationship

$$M_{1,2} = M_{2,1} \tag{9-68}$$

Equation (9-68) is also valid for nonhomogeneous media, although we have not proved it.

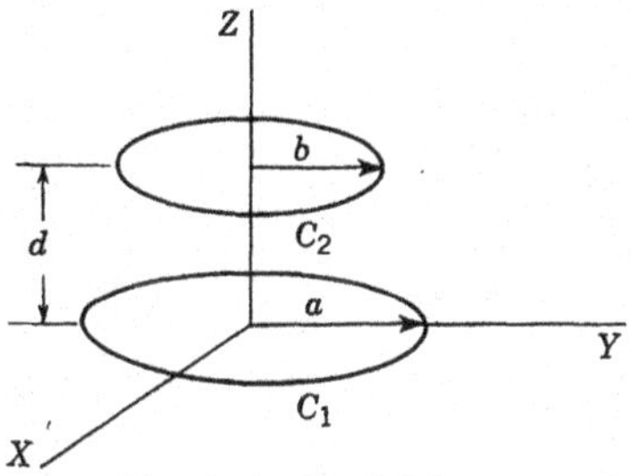

FIG. 9-17. Coaxial loops.

The classic example of calculating mutual-inductance is the problem of coaxial loops, Fig. 9-17. Considering loop 1 as carrying a current i, the vector potential is given by Eq. (6-31). The flux-linking loop 2 is

$$\psi_{1,2} = \int_0^{2\pi} A_\phi^{(1)}(b,\phi,d)\, b\, d\phi = 2\pi b A_\phi^{(1)}(b,\phi,d)$$

Substituting for $A_\phi^{(1)}$ from Eq. (6-31) and using Eq. (9-65), we have

$$M_{1,2} = \frac{2\mu}{\alpha} \sqrt{ab} \left[\left(1 - \frac{\alpha^2}{2}\right) K(\alpha) - E(\alpha) \right] \tag{9-69}$$

where
$$\alpha = 2 \sqrt{\frac{ab}{(a+b)^2 + d^2}}$$

Note that when we calculated the inductance of a loop, the external-inductance calculation was the same as the calculation for the mutual-inductance between C_1 along the wire axis and C_2 along the inner edge of the wire. The same procedure can be used for arbitrarily shaped loops and is often referred to as *the calculation of self-inductance by selected mutual-inductance.*

QUESTIONS FOR DISCUSSION

9-1. Why is the equivalent circuit of Fig. 9-1 not necessarily valid when the voltage and current vary with time? In a-c circuit theory, why would you expect this equivalent circuit to be more nearly correct at low frequencies than at high frequencies?

9-2. The classification of physical two-terminal elements into resistors, capacitors, and inductors is one of degree, not of kind. Discuss this statement.

9-3. Suppose we attempt to verify Eq. (9-7) experimentally. What are some of the difficulties that might arise?

9-4. Justify the theorem of the first paragraph of Sec. 9-3 by d-c circuit concepts. Visualize the volume as a network of resistors.

9-5. For the resistor of Fig. 9-7, why is the resistance not given exactly by summing the resistances of each leg according to Eq. (9-7) and the resistance of the curved part according to Eq. (9-9)?

9-6. For every capacitor problem, there is a dual resistor problem; but the converse is not true. Why?

9-7. Why are fringing effects a problem in capacitor problems, but usually not in resistor problems?

9-8. What are the dimensions and mksc units of the capacitive coefficients of potential? Justify your answer.

9-9. Discuss the significance of Green's reciprocation theorem.

9-10. Show that the first of Eqs. (9-29) is consistent with Fig. 9-9*b*.

9-11. In a one-body system, show that $c_{i,i} = C_{i,i}$ and therefore that it is also the self-capacitance of the body.

9-12. In the theory of electrostatic shielding, when we decrease the effective capacitance between two conductors by grounding a third conductor, is the capacitance to ground always increased?

9-13. Justify and interpret Eqs. (9-37). What are the dimensions of the coefficients of resistance?

9-14. Discuss the concept of grounding resistance.

9-15. Given two distant conductors buried in the ground, which of the following determines whether or not we may approximate the ground by an infinite medium: the depth below ground level or the separation of the conductors?

9-16. What complications arise if we attempt to use the magnetic circuit concept in problems for which the flux is not confined to a given region?

9-17. Discuss the approximations involved in using Eqs. (9-57) to (9-59).

9-18. What does Eq. (9-64) become if the wire is made of a magnetic material having inductivity μ, different from the surrounding medium?

9-19. Why can current-carrying wires be treated as true filaments for the calculation of mutual-inductance, but not for the calculation of self-inductance?

9-20. Discuss the calculation of self-inductance by the method of selected mutual-inductance.

PROBLEMS

9-1. Show that Eq. (9-8) reduces to Eq. (9-7) in the limit $a \to b$, $\phi_0 \to 0$. Let the length of the sector be denoted by $l = b - a$ and the average width by $w = \phi_0(a + b)/2$.

9-2. Consider the toroidal resistor bounded by the surface generated by rotating a contour of arbitrary shape through an angle ϕ_0, as shown in Fig. 9-18. Show that the resistance between the two flat surfaces is given by

$$\frac{1}{R} = \frac{\sigma}{\phi_0} \int_a^b \frac{w(\rho)}{\rho} \, d\rho$$

where $w(\rho)$ is the width of the toroid at a radial distance ρ. Show that the above equation reduces to Eq. (9-9) if $w(\rho) = c$, a constant.

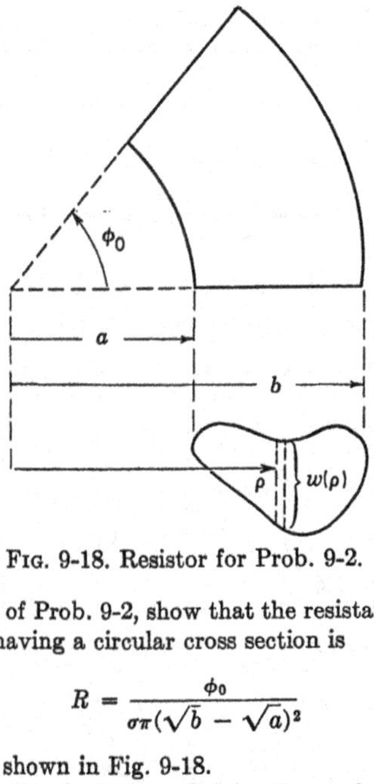

FIG. 9-18. Resistor for Prob. 9-2.

9-3. Using the formula of Prob. 9-2, show that the resistance between the flat ends of a segment of a toroid having a circular cross section is

$$R = \frac{\phi_0}{\sigma\pi(\sqrt{b} - \sqrt{a})^2}$$

where a, b, and ϕ_0 are as shown in Fig. 9-18.

9-4. Consider a conducting body bounded by the surfaces $r = a$, $r = b$, $\theta = \theta_0$, $\theta = \pi - \theta_0$, $\phi = 0$, $\phi = \phi_0$. This is a segment of the body of Fig. 9-5. Determine resistance between the two θ = constant surfaces, and show that the result reduces to Eq. (9-10) when $\phi_0 = 2\pi$.

9-5. Using the same body as in Prob. 9-4, determine the resistance between the two r = constant surfaces. Show that the result reduces to Eq. (9-11) when $\phi_0 = 2\pi$.

9-6. Prove the theorem of the first paragraph of Sec. 9-3, keeping the current constant. This is in contrast to the proof given, which keeps the potential difference constant.

9-7. The problem is to determine the resistance between the two ends of the horseshoe-shaped resistor of Fig. 9-19. By inserting perfectly conducting sheets over the junctions between the legs and the curved portion, show that a lower limit to the resistance is

$$R_{min} = \frac{1}{\sigma}\left[\frac{4b}{a^2} + \frac{\pi}{a - c \log (1 + a/c)}\right]$$

By inserting insulating strips parallel to the outer surface, show that an upper limit to the resistance is

$$R_{max} = \frac{\pi^2}{\sigma}\left[\pi a - (2b + \pi c) \log \frac{2b + \pi(a + c)}{2b + \pi c}\right]$$

Evaluate these limits for $a = c = 1$ meter, $b = 2$ meters. Within what per cent accuracy is the resistance evaluated in this case?

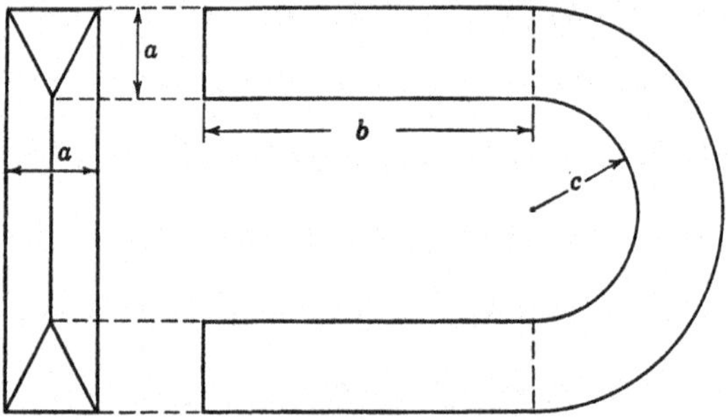

FIG. 9-19. Resistor for Prob. 9-7.

9-8. Calculate the capacitance between concentric ellipsoids (same foci) using the results of the example in Sec. 5-6. Let the ellipses be specified by the interfocal distance l and by their major axes a and b, $a > b$. The result is

$$C = \frac{4\pi\epsilon l}{\log \dfrac{(b + l)(a - l)}{(a + l)(b - l)}}$$

Show that this reduces to Eq. (9-17) when $l \to 0$ (a and b become diameters in this case).

9-9. Using the results of Prob. 8-13, show that the capacitance per unit length of two parallel wires is

$$\frac{C}{l} = \frac{2\pi\epsilon}{\cosh^{-1}\left(\dfrac{d^2 - a^2 - b^2}{2ab}\right)}$$

where a and b are the radii of the two wires and $d > a + b$ is the separation of the axes of the wires.

9-10. Determine the capacitance per unit length between the conductors in Prob. 8-7. Show that the same result can be obtained by viewing the problem as two cylindrical capacitors in series. What is the theoretical justification of this view?

9-11. Determine the capacitance between the conducting spheres of Prob. 8-8. Show that the same result can be obtained by viewing the problem as two hemispherical capacitors in parallel. What is the theoretical justification of this view?

9-12. Using the image procedure of Sec. 8-6B, show that the coefficients of capacitance and electrostatic induction for two spheres of equal radii a and separation d

between centers are

$$c_{1,1} = c_{2,2} = 4\pi\epsilon a \left(1 + \frac{1}{(d/a)^2 - 1} + \frac{1}{(d/a)^4 - 3\,(d/a)^2 + 1} + \cdots\right)$$

$$c_{1,2} = \frac{-4\pi\epsilon a^2}{d} \left(1 + \frac{1}{(d/a)^2 - 2} + \cdots\right)$$

Determine also the coefficients of potential and the direct capacitances in terms of the $c_{i,j}$.

9-13. From the results of Prob. 8-15, show that the self-capacitance of two spheres of equal radii a intersecting orthogonally is

$$C_{1,1} = 2\pi\epsilon a(4 - \sqrt{2})$$

9-14. Suppose we have the three parallel plane conductors as shown in Fig. 9-20. They extend a distance l into the paper. Evaluate the direct capacitances between

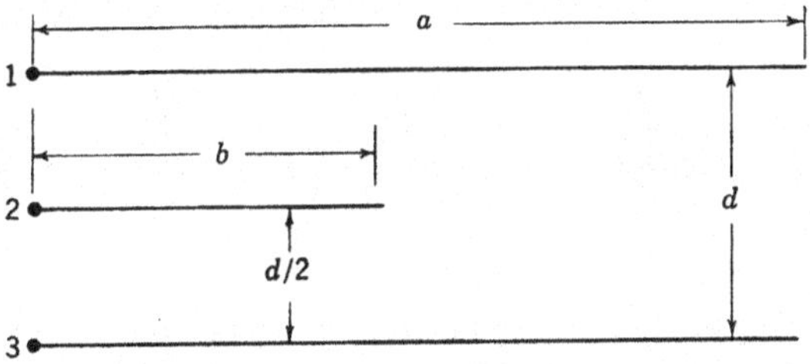

FIG. 9-20. Parallel conducting plates for Prob. 9-14.

terminals 1-2, 2-3, and 1-3, neglecting fringing. What is the effective capacitance between terminals 1 and 3 if terminal 2 is "floating"? What is the effective capacitance between terminals 1 and 3 if terminal 2 is grounded? What is the increase in capacitance from 1 to ground and from 3 to ground in the latter case?

9-15. State the resistance problems dual to Probs. 9-12 to 9-14, and give formulas for the dual resistance-system parameters.

9-16. Suppose we wished to determine the approximate grounding resistance of a circular disk and we did not know of the exact solution, Eq. (9-50). In Sec. 5-8, we saw that the equipotential lines of an infinite-plane sheet of charge were parallel to the sheet. The same should be approximately true very near a disk of uniform charge, especially over the central region. From the results of Prob. 5-16, we can evaluate the potential at the center of a disk of uniform charge. Using this to approximate the potential of a charged conducting disk, find the approximate capacitance. By duality, give the approximate grounding resistance of a buried metal plate. Show that this result is 26 per cent greater than the exact result, Eq. (9-50). What are the probable sources of error?

9-17. A telegraph circuit uses No. 10 B and S wire (0.1019 inch diameter) and 2-inch-diameter grounding rods driven 10 feet into the ground. If the conductivity of the ground is $\sigma = 10^{-3}$, calculate the grounding resistance of a single rod. Calculate the resistance of a circuit using two grounding rods and a single wire if the stations are separated l miles. Find the smallest value of l for which the single-wire system has less resistance than a two-wire system.

9-18. Two parallel cylinders of equal radii a are half buried in a ground of conductivity σ. They are separated a distance d between axes (see Fig. 8-18). Using the result of Prob. 9-9, show that the resistance between the cylinders for a length l is

$$R = \frac{\cosh^{-1}\left(\frac{d^2}{2a^2} - 1\right)}{\pi\sigma l}$$

9-19. Let d denote the distance from a plane ground surface to the center of a buried or partly buried sphere of radius a. Determine the grounding resistance for $d = 0$, $a/\sqrt{2}$, a, $2a$, $3a$, $5a$, $10a$, ∞. Plot $\sigma a R_g$ vs. d. (*Hint:* Use the results of Probs. 9-12, 9-13, and 9-15.)

9-20. Two large plane-faced copper blocks are clamped to the ends of a right circular cylinder of length l, radius a, and conductivity σ. If the blocks were perfect conductors, the resistance between blocks would be $R = l/\sigma\pi a^2$. Using Eq. (9-50) and the result of Prob. 9-16, show that the use of copper blocks introduces an additional resistance

$$\frac{0.5}{\sigma_c a} < R_{\text{additional}} < \frac{0.636}{\sigma_c a}$$

where σ_c is the conductivity of copper. A more exact treatment of Prob. 9-16 reduces the upper limit to $0.541/\sigma_c a$.*

9-21. Using the result of Prob. 9-3, show that the inductance of a closely wound toroidal coil having circular cross section is

$$L = \tfrac{1}{2}\mu N^2(\sqrt{b} - \sqrt{a})^2$$

where N is the number of turns, μ is the inductivity of the toroid, and a and b are the radii of the inner and outer edges of the toroid, respectively.

9-22. The magnetic core of Fig. 9-21 is of width a and inductivity $\mu \gg \mu_0$. Determine the approximate inductance of a coil of N turns wound on (*a*) leg A, (*b*) leg B, (*c*) leg C. State your approximations clearly.

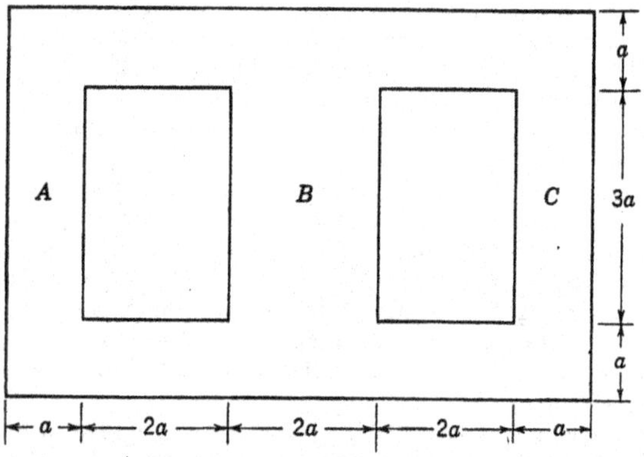

FIG. 9-21. Magnetic core for Probs. 9-22 and 9-25.

9-23. Show that a single-turn square loop, b meters on a side, fashioned of nonmagnetic wire having radius a, has an inductance

* See J. C. Maxwell, "Electricity and Magnetism," Oxford University Press, London, 1881, p. 433.

$$L \approx \mu_0 \frac{4b}{\pi} \left(\log \frac{4b}{a} - 0.776 \right)$$

if $a \ll b$. (The vector potential was asked for in Prob. 6-7.)

9-24. A pair of coaxial conductors, $\rho = a$ and $\rho = b$, $b > a$, carry equal but oppositely directed currents. Show that the inductance per unit length of the coax is given by

$$\frac{L}{l} = \frac{\mu}{2\pi} \log \frac{b}{a}$$

9-25. Using the magnetic core of Fig. 9-21 (width a, inductivity $\mu \gg \mu_o$), determine the approximate mutual-inductance between windings (*a*) N_1 on leg A and N_2 on leg B; (*b*) N_1 on leg A and N_2 on leg C. State your approximations clearly.

9-26. Consider two small loops of area S_1 and S_2 separated by a large distance r. Show that if the moment of loop 2 points radially outward from loop 1, the mutual-inductance is given by

$$M \approx \frac{\mu}{2\pi r^3} S_1 S_2 \cos \theta$$

where θ is the angle between the moment of loop 1 and the $\mathbf{r}$ direction. Show that if the moment of loop 2 is perpendicular to $\mathbf{r}$ and lies in the plane defined by the moment of loop 1 and $\mathbf{r}$, the mutual-inductance is given by

$$M \approx \frac{\mu}{4\pi r^3} S_1 S_2 \sin \theta$$

Show that, for arbitrary orientations, the mutual-inductance is a linear combination of the above two cases.

CHAPTER 10

THE ELECTROMAGNETIC FIELD

10-1. The Wave Equation. In linear homogeneous media having no conductivity, the field equations read

$$\begin{aligned} \nabla \times \mathbf{E} &= -\mu \frac{\partial \mathbf{H}}{\partial t} \qquad & \nabla \cdot \mathbf{E} = 0 \\ \nabla \times \mathbf{H} &= \epsilon \frac{\partial \mathbf{E}}{\partial t} \qquad & \nabla \cdot \mathbf{H} = 0 \end{aligned} \tag{10-1}$$

Taking the curl of the first of these equations, we have

$$\nabla \times \nabla \times \mathbf{E} = -\mu \frac{\partial}{\partial t} \nabla \times \mathbf{H}$$

Substituting for $\nabla \times \mathbf{H}$ from the second of Eqs. (10-1), we have

$$\nabla \times \nabla \times \mathbf{E} + \epsilon\mu \frac{\partial^2 \mathbf{E}}{\partial t^2} = 0$$

Finally, using the vector identity $\nabla \times \nabla \times \mathbf{E} = \nabla(\nabla \cdot \mathbf{E}) - \nabla^2 \mathbf{E}$ and the third of Eqs. (10-1), we obtain

$$\nabla^2 \mathbf{E} - \epsilon\mu \frac{\partial^2 \mathbf{E}}{\partial t^2} = 0 \tag{10-2}$$

This is called the *vector wave equation.* In rectangular coordinates and components, the wave equation is explicitly

$$\left(\frac{\partial^2}{\partial x^2} + \frac{\partial^2}{\partial y^2} + \frac{\partial^2}{\partial z^2} - \epsilon\mu \frac{\partial^2}{\partial t^2}\right) \mathbf{E} = 0 \tag{10-3}$$

which is a concise way of writing the three component equations

$$\begin{aligned} \frac{\partial^2 E_x}{\partial x^2} + \frac{\partial^2 E_x}{\partial y^2} + \frac{\partial^2 E_x}{\partial z^2} - \epsilon\mu \frac{\partial^2 E_x}{\partial t^2} &= 0 \\ \frac{\partial^2 E_y}{\partial x^2} + \frac{\partial^2 E_y}{\partial y^2} + \frac{\partial^2 E_y}{\partial z^2} - \epsilon\mu \frac{\partial^2 E_y}{\partial t^2} &= 0 \\ \frac{\partial^2 E_z}{\partial x^2} + \frac{\partial^2 E_z}{\partial y^2} + \frac{\partial^2 E_z}{\partial z^2} - \epsilon\mu \frac{\partial^2 E_z}{\partial t^2} &= 0 \end{aligned}$$

which are *scalar wave equations.* Thus, each rectangular component of **E** satisfies the scalar wave equation. In a similar manner, from Eqs.

(10-1) can be derived

$$\nabla^2 \mathbf{H} - \epsilon\mu \frac{\partial^2 \mathbf{H}}{\partial t^2} = 0 \tag{10-4}$$

Thus, **E** and **H** satisfy the same vector wave equation. This does not, of course, mean that **E** and **H** are identical, for a partial differential equation has an infinity of solutions.

So that we may begin to appreciate the significance of the wave equation, let us hypothesize a simple special case. Suppose that **E** is independent of x and y, and further that it has only an x component. Eq. (10-3) then reduces to

$$\frac{\partial^2 E_x}{\partial z^2} - \epsilon\mu \frac{\partial^2 E_x}{\partial t^2} = 0 \tag{10-5}$$

which is the one-dimensional scalar wave equation. The reader may recall that this equation is encountered in the study of lossless transmission lines. Solutions to Eq. (10-5) and their interpretation are therefore the same as solutions to the lossless-transmission-line problem. It is a mathematical consequence of Eq. (10-5) that

$$\nu = \frac{1}{\sqrt{\epsilon\mu}} \tag{10-6}$$

is a velocity of propagation of an electrical disturbance, for Eq. (10-5) has solutions of the form

$$E_x = f(z - \nu t) + g(z + \nu t) \tag{10-7}$$

where f and g are any well-behaved functions. We shall now show that $f(z - \nu t)$ represents a wave traveling in the $+z$ direction and that $g(z + \nu t)$ represents a wave traveling in the $-z$ direction.

Suppose we have $E_x = f(z - \nu t)$. At some value $t = t_1$, a plot of f vs. z is as shown in the first graph of Fig. 10-1. At a later instant of time $t = t_2$, a plot of f vs. z would be as shown in the second graph of Fig. 10-1. At a still later instant of time $t = t_3$, the plot of f vs. z would be as shown in the third graph of Fig. 10-1. Thus, the wave appears to be moving in the $+z$ direction as time progresses. To obtain the velocity at which the wave is propagated, consider some definite value of f, say $f_0 = f(k)$, where k is a constant. This is represented by a dot in Fig. 10-1. The velocity at which f_0 is moving is dz/dt where $z - \nu t = k$. Therefore $dz/dt - \nu = 0$, substantiating the statement that ν is the velocity of propagation. A similar consideration of the term $g(z + \nu t)$ would show that it represents a wave traveling in the $-z$ direction with a velocity ν.

Light is an electromagnetic wave, and the velocity at which it propagates in vacuum is called the *velocity of light* c. Thus

$$c = \frac{1}{\sqrt{\epsilon_0 \mu_0}} = 2.99790 \times 10^8 \approx 3.0 \times 10^8 \text{ meters/second} \quad (10\text{-}8)$$

is an experimentally measured constant, which shows that ϵ_0 and μ_0 are interrelated. In the mksc system of units, μ_0 is arbitrarily chosen to be $4\pi \times 10^{-7}$, which, together with Eq. (10-8), fixes the value of ϵ_0 and specifies the mksc system of units.

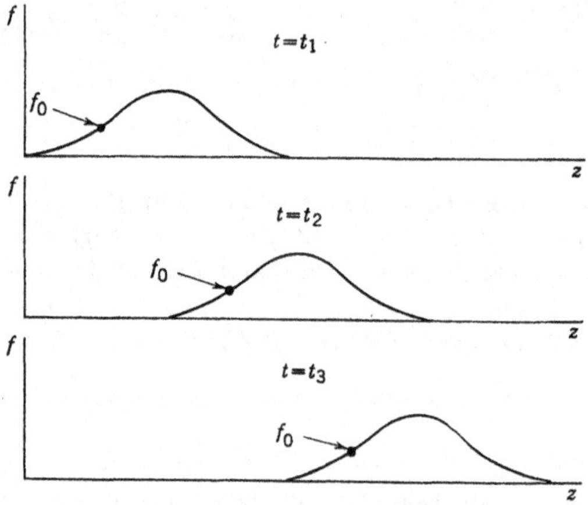

FIG. 10-1. $f(z - vt)$ represents a wave traveling in the $+z$ direction.

In a time-varying field, **E** and **H** are interrelated according to Eqs. (10-1). Given the simple traveling wave

$$E_x = f(z - vt)$$

we can obtain **H** from the first of Eqs. (10-1). Determining $\nabla \times \mathbf{E}$, we have

$$-\mu \frac{\partial \mathbf{H}}{\partial t} = \mathbf{u}_y \frac{\partial E_x}{\partial z} = \mathbf{u}_y \frac{\partial f}{\partial z}$$

Note that $$\frac{\partial}{\partial z} f(z - vt) = -v \frac{\partial}{\partial t} f(z - vt)$$

so $$\mu \frac{\partial \mathbf{H}}{\partial t} = \mathbf{u}_y v \frac{\partial f}{\partial t}$$

Integrating with respect to time, we obtain

$$\mu \mathbf{H} = \mathbf{u}_y v f + \mathbf{C}(z)$$

where **C** is independent of time and is therefore a static field. Assuming that no static field is present, we set $\mathbf{C} = 0$ and obtain

$$H_y = \frac{v}{\mu} f$$

as the only component of **H**. Substituting for ν from Eq. (10-6), we finally have

$$H_y = \sqrt{\frac{\epsilon}{\mu}}\, f = \sqrt{\frac{\epsilon}{\mu}}\, E_x \qquad (10\text{-}9)$$

Thus, **H** is functionally the same as **E** and is also a wave traveling in the $+z$ direction. Note that the **E** direction, the **H** direction, and the direction of propagation form a right-handed triad of mutually orthogonal directions, as shown in Fig. 10-2. The ratio of E_x to H_y has the dimensions of voltage per current, or of resistance, and the quantity

$$\eta = \frac{E_x}{H_y} = \sqrt{\frac{\mu}{\epsilon}} \text{ ohms} \qquad (10\text{-}10)$$

is called the *intrinsic impedance* of the medium. Note that the intrinsic impedance as defined here refers only to a *single* traveling wave in nondissipative media and that it is an example of a *wave impedance*. Using complex notation, we can extend the definition to a-c (harmonically time-varying) waves in dissipative media, much as we extended the concept of a d-c resistance to a-c impedance in circuit theory. If we have waves traveling in both the $+z$ and $-z$ directions, Eq. (10-9) no longer applies. In free space, we have

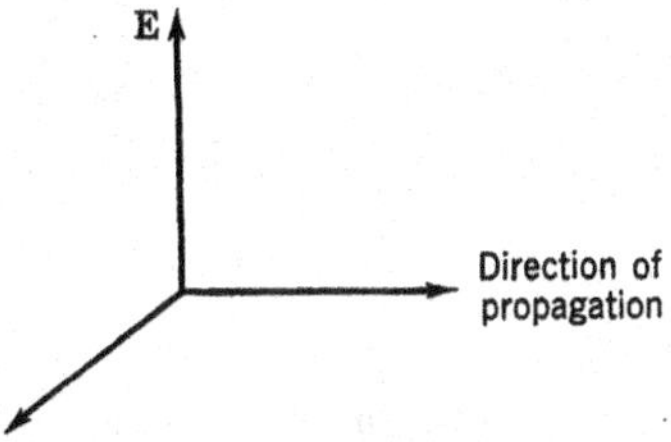

FIG. 10-2. Relationship between **E**, **H**, and the direction of propagation for a simple traveling wave.

$$\eta_0 = \sqrt{\frac{\mu_0}{\epsilon_0}} \approx 120\pi \approx 377 \text{ ohms} \qquad (10\text{-}11)$$

as the intrinsic impedance. For example, if E in a single traveling wave at some value of z and t had a value of 1 volt per meter, H would have a value of $\frac{1}{377}$ amperes per meter at that instant and that place.

It is probably in order at this time to mention something about the terminology "wave." The basic idea of a wave is that it is a phenomenon that travels through space. As indicated by the analysis of this section, an electromagnetic disturbance travels through space. We therefore use the terms "wave" and "field" interchangeably in connection with electromagnetic phenomena. When we wish to emphasize the propagating aspects of the field, we use the term "wave." Otherwise, we use the more general term "field."

10-2. Plane Waves. From an engineering viewpoint, the most important type of electromagnetic field is the so-called "a-c field," or *harmonically-time-varying field*. In this section, we shall look further into the phenomena of waves in dielectric media ($\sigma = 0$). We shall restrict consideration to single-frequency a-c fields, but the results can be extended to arbitrary time variations through the use of Fourier series and Fourier integrals. The distinguishing characteristic of a-c quantities is that all time variation appears as a sinusoid. The most general form of an a-c quantity is therefore

$$A(x,y,z) \sin [\omega t + \Phi(x,y,z)] \tag{10-12}$$

where A is the *amplitude* of the quantity, Φ is the *phase*, $\omega = 2\pi f$ is the *angular frequency* in radians per second, and f is the *frequency* in cycles per second. Note that the amplitude and phase may vary with position but that at each point the time variation is sinusoidal. Treatment of a-c phenomena can be simplified through the use of complex quantities, but we shall not need these for our elementary treatment.

As before, we start by restricting consideration to the one-dimensional problem

$$\mathbf{E} = \mathbf{u}_x E_x(z,t)$$

so our equation for $\mathbf{E}$ is Eq. (10-5), that is

$$\frac{\partial^2 E_x}{\partial z^2} - \epsilon\mu \frac{\partial^2 E_x}{\partial t^2} = 0$$

This time we specify that the field is a-c, so we assume

$$E_x = A(z) \sin [\omega t + \Phi(z)]$$

Substituting this into the preceding equation, we obtain

$$\left[\frac{d^2 A}{dz^2} - A\left(\frac{d\Phi}{dz}\right)^2 + \omega^2\epsilon\mu A\right] \sin (\omega t + \Phi) + \left[2\frac{dA}{dz}\frac{d\Phi}{dz} + A\frac{d^2\Phi}{dz^2}\right] \cos (\omega t + \Phi) = 0$$

For this equation to be valid at all instants of time, the coefficients of $\sin (\omega t + \Phi)$ and of $\cos (\omega t + \Phi)$ must be identically zero. Thus, we must satisfy the two equations

$$\begin{aligned} \frac{d^2 A}{dz^2} - A\left(\frac{d\Phi}{dz}\right)^2 + \omega^2\epsilon\mu A &= 0 \\ 2\frac{dA}{dz}\frac{d\Phi}{dz} + A\frac{d^2\Phi}{dz^2} &= 0 \end{aligned} \tag{10-13}$$

Any solution to these two equations gives the amplitude and phase of a possible a-c field.

Let us consider first the possibility that $dA/dz = 0$, that is, that the amplitude of wave is constant. The first of Eqs. (10-13) then becomes

$$\left(\frac{d\Phi}{dz}\right)^2 = \omega^2\epsilon\mu$$

Taking the square root and integrating, we have

$$\Phi = (\pm\omega\sqrt{\epsilon\mu})z + \Phi_0$$

where Φ_0 is a constant. We identify Φ_0 as the *initial phase* of the field at $z = 0$. The quantity

$$\beta = \omega\sqrt{\epsilon\mu} \tag{10-14}$$

is called the *phase constant* of the wave. Note that the second of Eqs. (10-13) is now satisfied identically, so we have the two possible solutions

$$E_x = A \sin(\omega t \pm \beta z + \Phi_0) \tag{10-15}$$

If we choose the negative sign preceding βz, we have a wave traveling in the $+z$ direction. If we choose the positive sign preceding βz, we have a

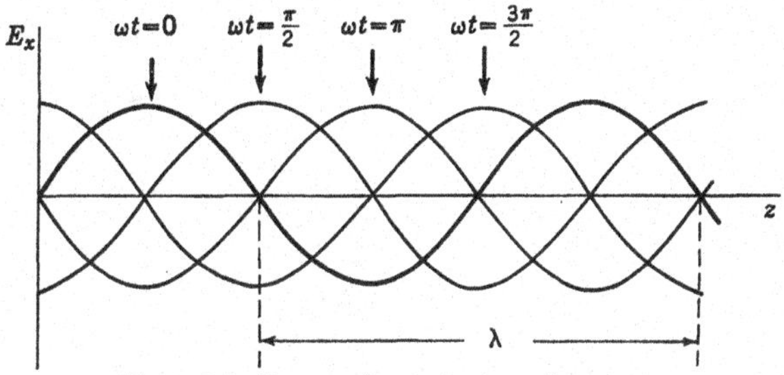

FIG. 10-3. E_x vs. z for a $+z$ traveling wave.

wave traveling in the $-z$ direction. Plots of E_x vs. z for a $+z$-traveling wave at several instants of time are given in Fig. 10-3. At a given instant of time, the distance over which the phase of the wave increases 2π is called the *wavelength* λ of the wave; so

$$\beta\lambda = 2\pi \tag{10-16}$$

The distance λ is indicated on Fig. 10-3. Eq. (10-6) is valid for any traveling wave; so the velocity of propagation is given by

$$v = \frac{1}{\sqrt{\epsilon\mu}} = \frac{\omega}{\beta} = \frac{\omega\lambda}{2\pi} = f\lambda \tag{10-17}$$

This can also be established by taking a point of constant phase and evaluating $v = dz/dt$. This velocity is called a *phase velocity* because it is the velocity at which a point of constant phase travels. The wave specified by Eq. (10-15) is called a *plane* wave because the phase Φ is constant over each plane perpendicular to the z axis. It is called a *uniform* plane wave because the amplitude is constant over each plane perpendicular to the z axis. The wave is said to be *linearly polarized* because **E** always points in the same direction in space. In particular, it is said to be linearly polarized in the x direction, for **E** points in the x direction.

To complete the picture of the traveling wave, we must also find the magnetic field. For the $+z$-traveling wave

$$E_x = A \sin (\omega t - \beta z)$$

The **H** must be given by Eqs. (10-13); so

$$H_y = \sqrt{\frac{\epsilon}{\mu}} A \sin (\omega t - \beta z)$$

We say that **E** and **H** are *in phase*, since at all points they have the same phase. At each point the spatial relationship between **E**, **H**, and the

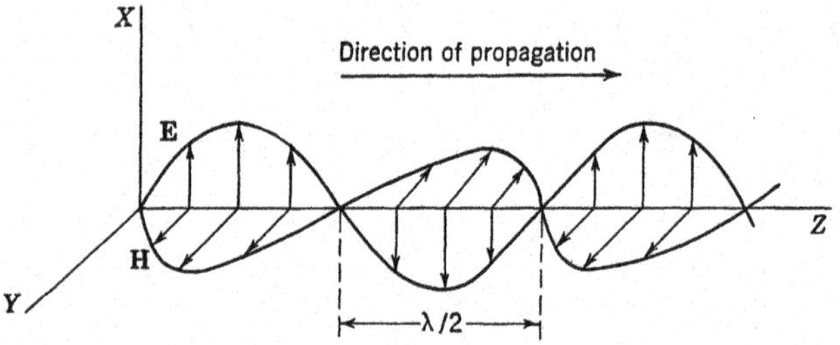

FIG. 10-4. Instantaneous picture of **E** and **H** in a traveling wave.

direction of propagation must be as shown in Fig. 10-2. The variation of H_y vs. z at any instant of time is the same as that of E_x vs. z, shown in Fig. 10-3. A vector picture of **E** and **H** in space at $t = 0$ is therefore as shown in Fig. 10-4. As time increases, the entire picture appears to move in the $+z$ direction. The electric energy density of the wave is given by Eq. (4-29), which in the present case becomes

$$w_e = \frac{\epsilon}{2} E_x^2 = \frac{\epsilon}{2} A^2 \sin^2 (\omega t - \beta z) \tag{10-18}$$

The magnetic energy density is given by Eq. (4-32), which in view of Eq. (10-9) becomes

$$w_m = \frac{\mu}{2} H_y^2 = \frac{\epsilon}{2} E_x^2 = w_e \tag{10-19}$$

Thus, *half of the energy appears as electric energy and half as magnetic energy*. The total energy density of the wave is the sum of electric and magnetic, or

$$w = w_e + w_m = \epsilon A^2 \sin^2 (\omega t - \beta z) \tag{10-20}$$

The energy in each "pulse," or each half sine wave of Fig. 10-4, travels with the wave as if it were trapped. This, however, is a special property of the type of wave being considered and is not valid for waves in general.

Returning to Eqs. (10-13), we now consider the possibility that $d\Phi/dz = 0$, that is, that the phase of the wave is constant. In this case the second of Eqs. (10-13) vanishes, and the first becomes

$$\frac{d^2A}{dz^2} + \omega^2\epsilon\mu A = 0$$

This is the well-known harmonic equation, the general solution of which is

$$A = C \sin (\beta z + \Theta)$$

where $\beta = \omega \sqrt{\epsilon\mu}$ and C and Θ are constants. The field is then given by

$$E_x = C \sin (\beta z + \Theta) \sin (\omega t + \Phi) \tag{10-21}$$

where Φ is constant. Without loss of generality we can set both Θ and Φ equal to zero, for this corresponds to a judicious choice of z and t origins.

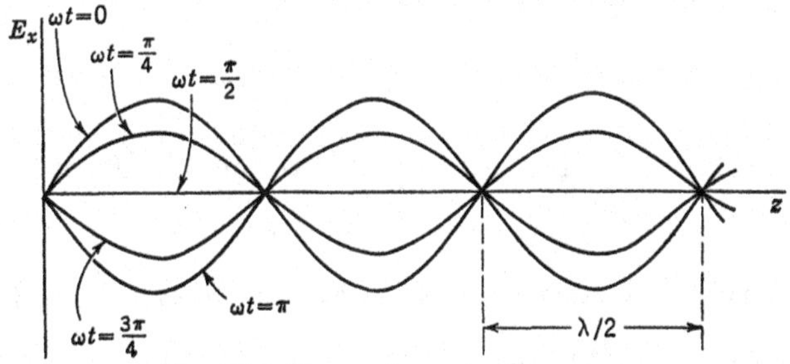

FIG. 10-5. E_x vs. z for a standing wave.

Plots of Eq. (10-21) for several instants of time are given in Fig. 10-5. Note that the wave appears stationary in space, varying only in amplitude. It is therefore called a *standing wave.* The distance between successive zeros is a half wavelength, for $\sin (\beta z)$ has two zeros in each 2π of argument. The wave is still a plane wave, for it is in phase over planes of constant z. It is still a uniform plane wave, for it has the same amplitude over planes of constant z. It is still linearly polarized in the x direction, for **E** always points in the x direction.

Using the trigonometric identity

$$\sin x \sin y = \tfrac{1}{2} \cos (x - y) - \tfrac{1}{2} \cos (x + y)$$

we can write Eq. (10-21) with $\Theta = \Phi = 0$ as

$$E_x = \frac{C}{2} \cos (\omega t - \beta z) - \frac{C}{2} \cos (\omega t + \beta z) \tag{10-22}$$

The first term is a wave traveling in the $+z$ direction, and the second term is a wave traveling in the $-z$ direction. Thus, *a standing wave is the sum of two traveling waves having equal amplitude and propagating in opposite directions.* The other possible solutions of Eqs. (10-13) correspond to oppositely directed traveling waves of unequal amplitudes. This results in a *partial standing wave*, a wave for which the amplitude varies with z but never goes completely to zero at any point.

To complete the standing wave picture, we must find the magnetic field. This can be done by direct substitution of Eq. (10-21) into the field equations. Note that Eq. (10-9) does *not* apply to the **E** of Eq. (10-21), since this is not a single traveling wave. Eq. (10-9) does, however, apply to each *traveling wave component* of the field. Let us apply Eq. (10-9) accordingly. Taking the complete field as given by Eq. (10-22), we define the $+z$-traveling component to be

$$E_x^+ = \frac{C}{2} \cos (\omega t - \beta z)$$

and the $-z$-traveling component to be

$$E_x^- = -\frac{C}{2} \cos (\omega t + \beta z)$$

The total field is the superposition of its two components, or

$$E_x = E_x^+ + E_x^- \qquad H_y = H_y^+ + H_y^-$$

The $+z$-traveling components of E_x and H_y are related by Eq. (10-9), so

$$H_y^+ = \sqrt{\frac{\epsilon}{\mu}} \frac{C}{2} \cos (\omega t - \beta z)$$

The $-z$-traveling components of E_x and H_y are related in *magnitude* in the same way, but we must preserve the direction relationship of Fig. 10-2. Therefore

$$H_y^- = \sqrt{\frac{\epsilon}{\mu}} (-E_x^-) = \sqrt{\frac{\epsilon}{\mu}} \frac{C}{2} \cos (\omega t + \beta z)$$

Summing H_z^+ and H_z^-, and using the trigonometric identity

$$\cos x \cos y = \tfrac{1}{2} \cos (x - y) + \tfrac{1}{2} \cos (x + y)$$

we obtain

$$H_y = \sqrt{\frac{\epsilon}{\mu}} C \cos (\beta z) \cos (\omega t) \tag{10-23}$$

Comparing this with the electric intensity

$$E_x = C \sin (\beta z) \sin (\omega t)$$

we note that **H** is in *phase quadrature* with **E**, since

$$\cos (\omega t) = \sin (\omega t + \pi/2)$$

We also note that **H** is displaced $\lambda/4$ in distance along the z axis with respect to **E**. A vector picture of this is given in Fig. 10-6, but it should be remembered that **E** and **H** reach the peak values at *different* times. It should also be noted that the picture remains stationary in space, there being no traveling motion. The electric energy density in the wave is given by

$$w_e = \frac{\epsilon}{2} E_x^2 = \frac{\epsilon}{2} C^2 \sin^2 (\beta z) \sin^2 (\omega t) \tag{10-24}$$

and the magnetic energy density by

$$w_m = \frac{\mu}{2} H_y^2 = \frac{\epsilon}{2} C^2 \cos^2 (\beta z) \cos^2 (\omega t) \tag{10-25}$$

Note that the *peak* values of electric and magnetic energy densities are equal, but they occur at different times and different places. The energy

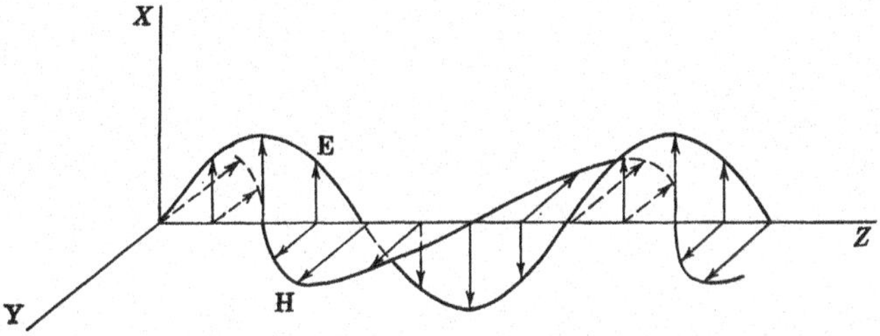

FIG. 10-6. Instantaneous picture of **E** and **H** in a standing wave.

oscillates between the electric form and the magnetic form, giving a "resonance" phenomenon.

We have spoken of electromagnetic waves as existing, but we have not yet considered their sources. We shall see in Sec. 10-6 that far from a source we have approximately a traveling plane wave. When a plane wave is normally incident on a good conductor, we have almost total reflection, obtaining a standing wave. Thus, traveling waves are associated with unobstructed "radiation," and standing waves are associated with reflection.

10-3. The Generalized Current Concept. In stationary media, we have the field equation

$$\nabla \times \mathbf{H} = \frac{\partial \mathbf{D}}{\partial t} + \mathbf{J} \tag{10-26}$$

In source-free media, we identify $\mathbf{J} = \mathbf{J}^c$ as the conduction current "caused by" the field. The rate at which energy is converted into heat, or the *power dissipation*, has the density

$$p_d = \mathbf{E} \cdot \mathbf{J}^c \tag{10-27}$$

as shown in Sec. 3-3. The term $\partial \mathbf{D}/\partial t$ has the dimensions of current, and has been given the name *displacement current density* $\mathbf{J}^d$, that is

$$\mathbf{J}^d = \frac{\partial \mathbf{D}}{\partial t} \tag{10-28}$$

Note that, in linear media,

$$\mathbf{E} \cdot \mathbf{J}^d = \epsilon \mathbf{E} \cdot \frac{\partial \mathbf{E}}{\partial t} = \frac{\epsilon}{2} \frac{\partial}{\partial t} (E^2) = \frac{\partial w_e}{\partial t} \tag{10-29}$$

so the power associated with the displacement current is the rate of change of electric energy.

To *represent* sources, we introduce the concept of an *impressed current density* $\mathbf{J}^i$. This may be such that the product $\mathbf{E} \cdot \mathbf{J}^i$ is negative, indicating a change of energy *into* electromagnetic energy. To emphasize that impressed currents are viewed as sources, we reverse our power reference condition and define the *density of power supplied by the sources* as

$$p_s = -\mathbf{E} \cdot \mathbf{J}^i \tag{10-30}$$

The representation of sources by impressed electric currents corresponds to the use of current sources in circuit theory, as we shall presently show.

We now define the *total electric current density* $\mathbf{J}^t$ as

$$\mathbf{J}^t = \mathbf{J}^d + \mathbf{J}^c + \mathbf{J}^i \tag{10-31}$$

that is, the sum of the displacement, conduction, and impressed currents. Eq. (10-26) can now be concisely written as

$$\nabla \times \mathbf{H} = \mathbf{J}^t \tag{10-32}$$

Note that the total current is divergenceless, that is

$$\nabla \cdot \mathbf{J}^t = 0 \tag{10-33}$$

since $\nabla \cdot \nabla \times \mathbf{H} = 0$ is a vector identity. Thus, we have recast Eq. (10-26) into a form like that of the equations for magnetostatics and steady current.

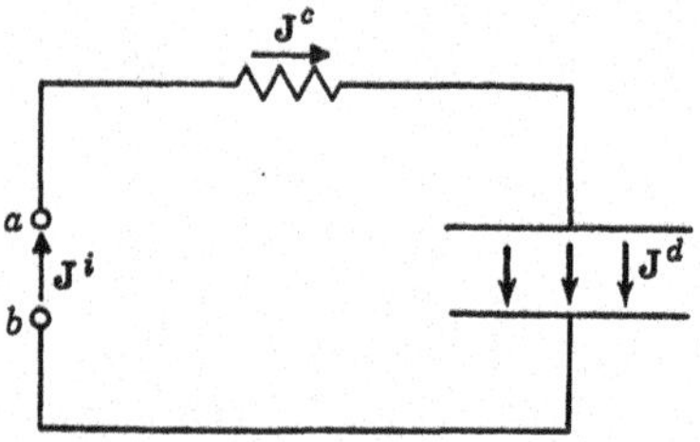

FIG. 10-7. Various types of electric current.

To illustrate the various types of electric current, consider Fig. 10-7, which represents a source connected to a resistor and a capacitor in series. The divergenceless character of total current implies that the impressed current of the source equals the conduction current in the resistor, which in turn equals the displacement current through the capacitor (we neglect leakage conduction current and stray capacitance). To illustrate how the extended concept of current reduces to the usual circuit concept of current, consider a closed surface surrounding the top plate of the capacitor. In view of Eq. (10-33), we have

$$0 = \oiint \mathbf{J}^t \cdot d\mathbf{s} = -i + \frac{d}{dt} \oiint \mathbf{D} \cdot d\mathbf{s} = -i + \frac{dq}{dt}$$

which is the circuit equation $i = dq/dt$. To illustrate that Eq. (10-30) is consistent with the usual circuit relationship for current sources, consider the impressed current to be a filament. The power supplied by the source is then

$$P_s = -\iiint \mathbf{E} \cdot \mathbf{J}^i \, d\tau = -i \int_b^a \mathbf{E} \cdot d\mathbf{l} = iv_{ab}$$

which is the circuit equation $P = vi$. Note that the current source with the impressed current removed is an open circuit, in agreement with the circuit superposition theorem.

Let us now consider the other curl equation of Maxwell, which in stationary media is

$$\nabla \times \mathbf{E} = -\frac{\partial \mathbf{B}}{\partial t} \tag{10-34}$$

In analogy to the electric case, we may define a *magnetic displacement current density* $\mathbf{K}^d$ to be

$$\mathbf{K}^d = \frac{\partial \mathbf{B}}{\partial t} \tag{10-35}$$

The total magnetic displacement current through a surface is the rate of change of magnetic flux. We can also form a power associated with magnetic-displacement current in linear media according to

$$\frac{\partial w_m}{\partial t} = \frac{\partial}{\partial t}(\mu H^2) = \mu \mathbf{H} \cdot \frac{\partial \mathbf{H}}{\partial t} = \mathbf{H} \cdot \mathbf{K}^d \tag{10-36}$$

This rate of change of magnetic energy density is analogous to the electric case, Eq. (10-29).

To *represent* sources, we can also introduce an *impressed magnetic current density* $\mathbf{K}^i$. From the symmetry of the field equations and their interpretation, it follows that the *density of power supplied by impressed magnetic currents* would be

$$p_s = -\mathbf{H} \cdot \mathbf{K}^i \tag{10-37}$$

which is analogous to Eq. (10-30). The representation of sources in terms of impressed magnetic currents corresponds to the use of voltage sources in circuit theory, as we shall presently show.

We define the *total magnetic current density* $\mathbf{K}^t$ as

$$\mathbf{K}^t = \mathbf{K}^d + \mathbf{K}^i \tag{10-38}$$

Note that there is no magnetic conduction current, a fact which tends to make magnetic currents less "real" to us. We now amend Eq. (10-34) to read

$$\nabla \times \mathbf{E} = -\mathbf{K}^t \tag{10-39}$$

in analogy to Eq. (10-34). The magnetic current is also divergenceless in character, that is

$$\nabla \cdot \mathbf{K}^t = 0 \tag{10-40}$$

for $\nabla \cdot \nabla \times \mathbf{E} = 0$ is a vector identity. We now have complete duality between electric and magnetic quantities, and we can, therefore, use dual pictures. The impressed magnetic current can be pictured as a flow of magnetic charges similar to an electric current. Remember that impressed magnetic current is used only to *represent* the effect of sources. The question of what a source "actually" is should be of no more concern to us than it is in circuit theory where we use ideal sources.

To illustrate the relationship of impressed magnetic current to the voltage source of circuit theory, consider Fig. 10-8. This represents a perfectly conducting wire encircled by a filament of magnetic current. Consider a closed path along the wire and closing across the terminals. This encloses the magnetic current, so $\oint \mathbf{E} \cdot d\mathbf{l} = k$. But $\mathbf{E} = 0$ in the wire; so $\oint \mathbf{E} \cdot d\mathbf{l} = v$, the terminal voltage. Therefore $v = k$, showing that specification of k is equivalent to specifying the terminal voltage. To illustrate that Eq. (10-37) is consistent with circuit concepts, let us apply it to the source of Fig. 10-8. The total power supplied by the source is

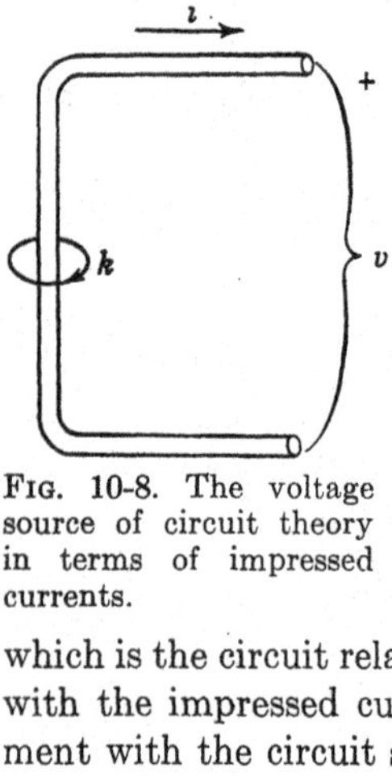

FIG. 10-8. The voltage source of circuit theory in terms of impressed currents.

$$P_s = -\iiint \mathbf{H} \cdot \mathbf{K}^i \, d\tau = k \oint \mathbf{H} \cdot d\mathbf{l} = vi$$

which is the circuit relationship. Note that the voltage source, Fig. 10-8, with the impressed current removed, becomes a short circuit, in agreement with the circuit superposition theorem.

It is possible to use the concept of an *impressed electric intensity* $\mathbf{E}^i$ in place of the concept of impressed magnetic currents. The procedure is to define a *total electric intensity* $\mathbf{E}^t$ equal to the sum of the induced (or caused) electric intensity $\mathbf{E}$ and the impressed electric intensity $\mathbf{E}^i$, that is

$$\mathbf{E}^t = \mathbf{E} + \mathbf{E}^i \tag{10-41}$$

Eq. (10-34) is then amended to read

$$\nabla \times \mathbf{E}^t = -\frac{\partial \mathbf{B}}{\partial t}$$

which, substituting from Eq. (10-41), we can express as

$$\nabla \times \mathbf{E} = -\frac{\partial \mathbf{B}}{\partial t} - \nabla \times \mathbf{E}^i \tag{10-42}$$

Comparing Eq. (10-42) with Eq. (10-39), we see that

$$\mathbf{K}^i = \nabla \times \mathbf{E}^i \tag{10-43}$$

Thus, the curl of an impressed electric intensity is equivalent to an impressed magnetic current density. We prefer to use the impressed-magnetic-current concept rather than the impressed-electric-intensity concept because: (*a*) we preserve the duality of the equations, so that the same mathematical methods of solution apply; and (*b*) it is easier to picture the magnetic current, which is dual to the electric current.

It should be remembered that, mathematically speaking, sources are those quantities that are *treated* as known quantities. Cause and effect are meaningless terms. In one problem we might treat a conduction current as an impressed current, in another problem we might treat a displacement current as an impressed current. In other words, the impressed currents are those which we *view* as sources, regardless of whether they "cause" the field or are "caused by" the field. They may, of course, be equivalent currents, obtained by extending the static equivalence principle to time-varying fields.

10-4. Poynting's Theorem. Consider the field equations written concisely as

$$\nabla \times \mathbf{H} = \mathbf{J}^t \qquad \nabla \times \mathbf{E} = -\mathbf{K}^t$$

Dot-multiplying the first equation by **E** and the second equation by **H**, we obtain equations having the dimensions of power density. Subtracting the resulting second equation from the first, we have

$$\mathbf{E} \cdot \nabla \times \mathbf{H} - \mathbf{H} \cdot \nabla \times \mathbf{E} = \mathbf{E} \cdot \mathbf{J}^t + \mathbf{H} \cdot \mathbf{K}^t$$

It is a vector identity that the left-hand side is $-\nabla \cdot (\mathbf{E} \times \mathbf{H})$; so

$$\nabla \cdot (\mathbf{E} \times \mathbf{H}) + \mathbf{E} \cdot \mathbf{J}^t + \mathbf{H} \cdot \mathbf{K}^t = 0 \tag{10-44}$$

Poynting's theorem states that

$$\mathbf{S} = \mathbf{E} \times \mathbf{H} \tag{10-45}$$

is a vector density of power flow and that

$$p_f = \nabla \cdot \mathbf{S} = \nabla \cdot (\mathbf{E} \times \mathbf{H}) \tag{10-46}$$

represents the *density of power leaving the point of consideration.* The vector $\mathbf{S} = \mathbf{E} \times \mathbf{H}$ is called the *Poynting vector* and is considered to specify both the direction and the rate of energy flow at every point in space. **E** has the dimensions voltage per length and **H** has the dimensions current per length; so $\mathbf{S} = \mathbf{E} \times \mathbf{H}$ has the dimensions power per area. The unit of **S** in mksc system is therefore the *watt per square meter.*

We have already interpreted the other terms of Eq. (10-44) in Sec.

10-3. The second term is

$$\mathbf{E} \cdot \mathbf{J}^t = \mathbf{E} \cdot \frac{\partial \mathbf{D}}{\partial t} + \mathbf{E} \cdot \mathbf{J}^c + \mathbf{E} \cdot \mathbf{J}^i$$
$$= \frac{\partial w_e}{\partial t} + p_d - p_s$$

and the last term is

$$\mathbf{H} \cdot \mathbf{K}^t = \mathbf{H} \cdot \frac{\partial \mathbf{B}}{\partial t} + \mathbf{H} \cdot \mathbf{K}^i$$
$$= \frac{\partial w_e}{\partial t} - p_s$$

Combining the two source terms according to

$$p_s = -\mathbf{E} \cdot \mathbf{J}^i - \mathbf{H} \cdot \mathbf{K}^i \tag{10-47}$$

we can now write Eq. (10-44) as

$$p_s = p_f + p_d + \frac{\partial}{\partial t}(w_e + w_m) \tag{10-48}$$

This is a statement of the *conservation of energy*. In words, it states that the density of power supplied by the sources equals the density of power leaving the point plus the density of power dissipated plus the rate of increase in electric and magnetic energy densities.

The interpretation of Eq. (10-48) is perhaps easier to visualize if we integrate the densities throughout a region. Integration of Eq. (10-48) over a region of space gives

$$\iiint p_s \, d\tau = \iiint (p_f + p_d) \, d\tau + \frac{d}{dt} \iiint (w_e + w_m) \, d\tau \tag{10-49}$$

We have already considered most of the terms in detail. The last two terms involve the total electric and magnetic energies within the region,

$$\begin{aligned} W_e &= \textstyle\iiint w_e \, d\tau = \frac{1}{2}\iiint \mathbf{E} \cdot \mathbf{D} \, d\tau \\ W_m &= \textstyle\iiint w_m \, d\tau = \frac{1}{2}\iiint \mathbf{H} \cdot \mathbf{B} \, d\tau \end{aligned} \tag{10-50}$$

The other three terms

$$\begin{aligned} P_d &= \textstyle\iiint p_d \, d\tau = \iiint \mathbf{E} \cdot \mathbf{J}^c \, d\tau \\ P_s &= \textstyle\iiint p_s \, d\tau = -\iiint (\mathbf{E} \cdot \mathbf{J}^i + \mathbf{H} \cdot \mathbf{K}^i) \, d\tau \\ P_f &= \textstyle\iiint p_f \, d\tau = \iiint \nabla \cdot \mathbf{S} \, d\tau \end{aligned} \tag{10-51}$$

are the total power dissipated within the region, the total power supplied by sources within the region, and the total power leaving the region, respectively. The last expression can be put into more convenient form by applying the divergence theorem. This gives

$$P_f = \oiint \mathbf{S} \cdot d\mathbf{s} = \oiint \mathbf{E} \times \mathbf{H} \cdot d\mathbf{s} \tag{10-52}$$

where the integration is carried out over the surface enclosing the region, $d\mathbf{s}$ pointing outward. We can now write Eq. (10-49) as

$$P_s = P_f + P_d + \frac{d}{dt}(W_e + W_m) \tag{10-53}$$

which is the statement of *conservation of energy as applied to an entire region.* In words, it states that the power supplied by sources within a region is equal to the power leaving the region plus the power dissipated within the region plus the rate of increase of electric and magnetic energy within the region.

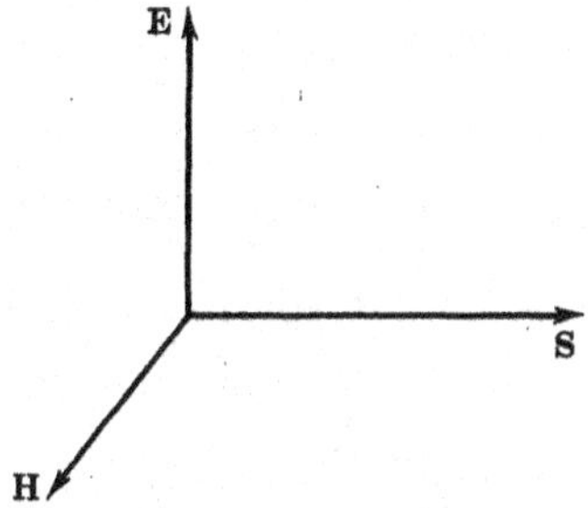

FIG. 10-9. The relationship between **E**, **H**, and **S**.

Let us now consider the Poynting vector in a little more detail. **S** is the vector product $\mathbf{E} \times \mathbf{H}$; so it is perpendicular to both **E** and **H** according to the right-hand rule, as illustrated by Fig. 10-9. Note that our equations of conservation of energy, Eqs. (10-48) and (10-53), interpret only the divergence of **S** and the closed surface integral of **S**. We could define another vector $\mathbf{S}' = \mathbf{S} + \nabla \times \mathbf{A}$, where **A** is *any* vector, and obtain

$$p_f = \nabla \cdot \mathbf{S}' \qquad P_f = \oint\!\!\oint \mathbf{S}' \cdot d\mathbf{s}$$

for $\nabla \cdot \nabla \times \mathbf{A} = 0$ and $\oint\!\!\oint \nabla \times \mathbf{A} \cdot d\mathbf{s} = 0$ are vector identities. Thus, *the vector density of power flow is unique only because we have defined it so.* Poynting's theorem cannot be proved or disproved, for it rests on the previous hypothesis that electric and magnetic energies are spatially distributed, which in itself cannot be proved or disproved. The fact that more than one density of power flow can be defined should not be surprising, for we have a similar situation in circuit theory. If there are three or more wires connecting a source to a load, there is no unique power associated with each pair of wires. It is only the *total* power from source to load that is unique. In static fields, where **E** and **H** are not interrelated, the Poynting vector can exist even though there is no flow of energy. However, in such cases, $\nabla \cdot \mathbf{S} = 0$ everywhere, indicating no spatial change in energy distribution.

As an example, consider the coaxial line of Fig. 10-10. We have the current flowing in the $+z$ direction on the inner cylinder and in the $-z$ direction on the outer cylinder. For simplicity, we shall assume that the current is d-c. Between the two cylinders, the **H** field is the line-current field

$$H_\phi = \frac{i}{2\pi\rho}$$

where i is the total current on the inner cylinder. The **E** field is that of the line charge

$$E_\rho = \frac{q_l}{2\pi\epsilon}$$

where q_l is the charge per unit length on the inner cylinder. The voltage between the cylinders is related to the charge according to $q_l = vC/l$,

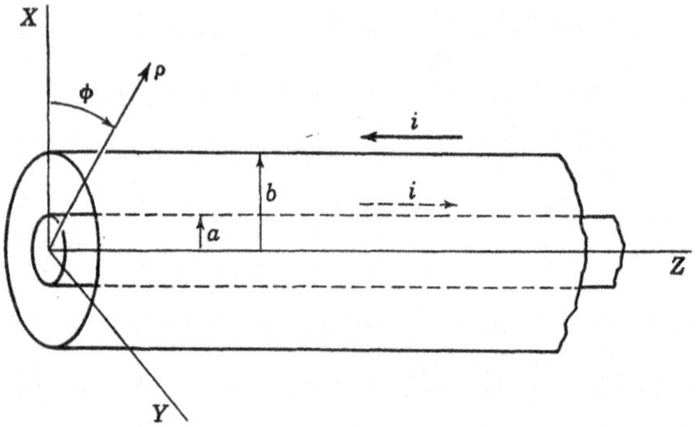

FIG. 10-10. A coaxial transmission line.

where C/l for concentric cylinders is given by Eq. (9-18). Thus,

$$E_\rho = \frac{v}{\rho \log (b/a)}$$

where a and b are the radii of the inner and outer cylinders. According to Poynting's theorem, the power flow along the coax is

$$P_f = \iint \mathbf{E} \times \mathbf{H} \cdot ds = \iint E_\rho H_\phi \, ds$$

where the integration is taken over the cross section of the line. Substituting for E_ρ and H_ϕ, we have the power flow along the coax given by

$$P_f = \int_a^b d\rho \int_0^{2\pi} \rho \, d\phi \left[\frac{v}{\rho \log (b/a)} \frac{i}{2\pi\rho} \right] = vi$$

which is in agreement with circuit concepts.

10-5. Electromagnetic Potentials. Consider Maxwell's equations in their original form

$$\begin{aligned} \nabla \times \mathbf{E} &= -\frac{\partial \mathbf{B}}{\partial t} \qquad & \nabla \cdot \mathbf{B} &= 0 \\ \nabla \times \mathbf{H} &= \frac{\partial \mathbf{D}}{\partial t} + \mathbf{J} \qquad & \nabla \cdot \mathbf{D} &= q_v \end{aligned} \tag{10-54}$$

We shall treat the extension to impressed magnetic currents later. In view of the divergenceless character of **B**, we can express it in terms of a magnetic vector potential according to

$$\mathbf{B} = \nabla \times \mathbf{A} \tag{10-55}$$

Substituting this into the first of Eqs. (10-54), we have

$$\nabla \times \left(\mathbf{E} + \frac{\partial \mathbf{A}}{\partial t}\right) = 0$$

The quantity in the parentheses is irrotational and can be expressed in terms of a scalar potential; so

$$\mathbf{E} = -\frac{\partial \mathbf{A}}{\partial t} - \nabla V \tag{10-56}$$

Note that in the static case, $\partial \mathbf{A}/\partial t \rightarrow 0$ and V reduces to a static electric potential. Thus, knowledge of a magnetic vector potential and a scalar electric potential is sufficient to determine **B** and **E** if there are only electric currents.

If we now proceed to linear media, for which

$$\mathbf{D} = \epsilon \mathbf{E} \qquad \mathbf{B} = \mu \mathbf{H}$$

we can obtain explicit equations for the potentials. Substituting $\mu \mathbf{H} = \nabla \times \mathbf{A}$ into the second of Eqs. (10-54), we have

$$\frac{1}{\mu} \nabla \times \nabla \times \mathbf{A} = \epsilon \frac{\partial \mathbf{E}}{\partial t} + \mathbf{J}$$

Substituting for **E** from Eq. (10-56), we have

$$\frac{1}{\mu} \nabla \times \nabla \times \mathbf{A} = -\epsilon \frac{\partial^2 \mathbf{A}}{\partial t^2} - \epsilon \nabla \frac{\partial V}{\partial t} + \mathbf{J}$$

Using the vector relationship $\nabla \times \nabla \times \mathbf{A} = \nabla(\nabla \cdot \mathbf{A}) - \nabla^2 \mathbf{A}$, we can write the above equation as

$$\nabla^2 \mathbf{A} - \nabla(\nabla \cdot \mathbf{A}) - \epsilon\mu \frac{\partial^2 \mathbf{A}}{\partial t^2} - \epsilon\mu \nabla \frac{\partial V}{\partial t} = -\mu \mathbf{J} \tag{10-57}$$

Returning now to the last of Eqs. (10-54), we substitute $\mathbf{D} = \epsilon \mathbf{E}$ with **E** given by Eq. (10-56), and obtain

$$-\epsilon \nabla \cdot \left(\frac{\partial \mathbf{A}}{\partial t} + \nabla V\right) = q_v$$

This can be rearranged to read

$$\nabla^2 V + \frac{\partial}{\partial t} \nabla \cdot \mathbf{A} = -\frac{q_v}{\epsilon} \tag{10-58}$$

Eqs. (10-57) and (10-58) are the general equations for the electromagnetic potentials of electric current and charge in homogeneous linear media.

We have interpreted only the curl of **A** according to Eq. (10-55), and we have as yet made no restrictions on its divergence. We may, *if we so desire*, impose the auxiliary condition

$$\nabla \cdot \mathbf{A} = -\epsilon\mu \frac{\partial V}{\partial t} \tag{10-59}$$

This reduces Eqs. (10-57) and (10-58) to

$$\begin{aligned} \nabla^2 \mathbf{A} - \epsilon\mu \frac{\partial^2 \mathbf{A}}{\partial t^2} &= -\mu \mathbf{J} \\ \nabla^2 V - \epsilon\mu \frac{\partial^2 V}{\partial t^2} &= -\frac{q_v}{\epsilon} \end{aligned} \tag{10-60}$$

In regions where **J** and q_v are zero these are the same wave equations satisfied by **E** and **H**. Potentials determined according to Eqs. (10-60) are called *wave potentials*. Thus, even though it is not necessary to impose Eq. (10-59), it is usually desirable to do so in order to simplify the equations for the potentials.

There are several advantages in using potentials in an electromagnetic field problem instead of solving for the field vectors directly. In addition to satisfying the vector wave equations, **E** and **H** must be divergenceless, whereas *any* solution to the vector wave equation is a possible **A**. Another advantage lies in the arbitrariness of the potentials, there being no unique set of potentials for a given field. Perhaps the greatest advantage in using potentials is the simple relationship of the potentials to **J** and q_v as sources. Expanding the first of Eqs. (10-60) in rectangular components, we have

$$\begin{aligned} \nabla^2 A_x - \epsilon\mu \frac{\partial^2 A_x}{\partial t^2} &= -\mu J_x \\ \nabla^2 A_y - \epsilon\mu \frac{\partial^2 A_y}{\partial t^2} &= -\mu J_y \\ \nabla^2 A_z - \epsilon\mu \frac{\partial^2 A_z}{\partial t^2} &= -\mu J_z \end{aligned} \tag{10-61}$$

Thus, the rectangular components of **A** have corresponding rectangular components of **J** as their sources. No such simple relationship exists between the field vectors and current.

Let us now consider the field equations amended to include magnetic current. We shall treat all "charge-flow" current as a source, dropping the superscript i. Thus, we start with

$$\begin{aligned} \nabla \times \mathbf{E} &= -\frac{\partial \mathbf{B}}{\partial t} - \mathbf{K} \qquad & \nabla \cdot \mathbf{B} &= m_v \\ \nabla \times \mathbf{H} &= \frac{\partial \mathbf{D}}{\partial t} + \mathbf{J} \qquad & \nabla \cdot \mathbf{D} &= q_v \end{aligned} \tag{10-62}$$

where m_v is the magnetic charge density. Using superposition, we may let the total field be the sum of the field produced by the electric sources (primed) and the field produced by the magnetic sources (double-primed); that is

$$\begin{aligned} \mathbf{E} &= \mathbf{E}' + \mathbf{E}'' \qquad & \mathbf{D} &= \mathbf{D}' + \mathbf{D}'' \\ \mathbf{H} &= \mathbf{H}' + \mathbf{H}'' \qquad & \mathbf{B} &= \mathbf{B}' + \mathbf{B}'' \end{aligned} \tag{10-63}$$

The primed quantities satisfy Eqs. (10-54), and the double-primed equations satisfy

$$\begin{aligned} \nabla \times \mathbf{E}'' &= -\frac{\partial \mathbf{B}''}{\partial t} - \mathbf{K}'' \qquad & \nabla \cdot \mathbf{B}'' &= m_v \\ \nabla \times \mathbf{H}'' &= \frac{\partial \mathbf{D}''}{\partial t} \qquad & \nabla \cdot \mathbf{D}'' &= 0 \end{aligned} \tag{10-64}$$

The determination of potentials for the primed field is just the analysis followed in the earlier part of this section; that is

$$\mathbf{B}' = \nabla \times \mathbf{A} \qquad \mathbf{E}' = -\frac{\partial \mathbf{A}}{\partial t} - \nabla V \tag{10-65}$$

where **A** and V satisfy Eqs. (10-60). The equations for the double-primed field, Eqs. (10-64), are dual to those for the primed field, Eqs. (10-54), so all ensuing equations must be dual. Taking into account the minus-sign difference, we have equations dual to Eqs. (10-65)

$$\mathbf{D}'' = -\nabla \times \mathbf{F} \qquad \mathbf{H}'' = -\frac{\partial \mathbf{F}}{\partial t} - \nabla U \tag{10-66}$$

where **F** is called an *electric vector potential*, and U a *magnetic scalar potential.* For these to be wave potentials, they must satisfy equations dual to Eqs. (10-60), or

$$\begin{aligned} \nabla^2 \mathbf{F} - \epsilon\mu \frac{\partial^2 \mathbf{F}}{\partial t^2} &= -\epsilon \mathbf{K} \\ \nabla^2 U - \epsilon\mu \frac{\partial^2 U}{\partial t^2} &= -\frac{m_v}{\mu} \end{aligned} \tag{10-67}$$

The complete field is the sum of the primed and double-primed fields, Eqs. (10-65) and Eqs. (10-66). Thus

$$\begin{aligned} \mathbf{E} &= -\frac{\partial \mathbf{A}}{\partial t} - \nabla V - \frac{1}{\epsilon} \nabla \times \mathbf{F} \\ \mathbf{H} &= -\frac{\partial \mathbf{F}}{\partial t} - \nabla U + \frac{1}{\mu} \nabla \times \mathbf{A} \end{aligned} \tag{10-68}$$

This is the general form for an electromagnetic field in linear homogeneous media in terms of potentials.

TABLE 10-1. DUAL QUANTITIES FOR ELECTROMAGNETIC FIELDS

Electric sources only	*Magnetic sources only*
E	**H**
H	$-$**E**
D	**B**
B	$-$**D**
J	**K**
q	m
ϵ	μ
μ	ϵ
A	**F**
V	U

To summarize the duality relationship used, consider Table 10-1. If we replace the quantities of the first column in an equation pertaining to electric currents and charges by the corresponding quantities of the second column, we obtain the dual equation for magnetic currents and charges. The reader may check this for the equations of this section. Similarly, any solution to a problem containing electric sources can be converted to the dual solution of the problem containing magnetic sources. Our work is thus materially reduced by the concept of duality, for two field problems correspond to one mathematical problem. It is for this reason that we prefer to use the concept of impressed magnetic currents rather than impressed electric intensities.

10-6. Radiation. We must now relate the wave potentials to their sources, charge and current. The scalar potential V and the rectangular components of **A** all satisfy the scalar wave equation,

$$\nabla^2 g - \frac{1}{\nu^2}\frac{\partial^2 g}{\partial t^2} = -w \tag{10-69}$$

We can construct solutions to this equation as we constructed solutions to the potential integral equations in electrostatics.

Suppose we have a point source denoted by w_0, situated at the coordinate origin. The potential g must satisfy the equation

$$\nabla^2 g - \frac{1}{\nu^2}\frac{\partial^2 g}{\partial t^2} = 0 \tag{10-70}$$

everywhere except at $r = 0$. The problem has complete spherical symmetry; so the solution can depend only upon the distance from the source and, of course, on time. We therefore assume

$$g = g(r,t)$$

and obtain

$$\nabla^2 g = \frac{1}{r^2}\frac{\partial}{\partial r}\left(r^2 \frac{\partial g}{\partial r}\right)$$

Note that the form $\nabla^2 g$ can be simplified by the substitution

$$g = \frac{f}{r} \tag{10-71}$$

for then we have

$$\nabla^2 g = \frac{1}{r}\frac{\partial^2 f}{\partial r^2}$$

Substituting this into Eq. (10-70) and multiplying through by r, we obtain

$$\frac{\partial^2 f}{\partial r^2} - \frac{1}{\nu^2}\frac{\partial^2 f}{\partial t^2} = 0$$

This is the same wave equation that we encountered in Sec. 10-1, solutions to which are

$$f\left(t - \frac{r}{\nu}\right) \qquad f\left(t + \frac{r}{\nu}\right)$$

The first form is a wave traveling outward from the origin, and the second is a wave traveling inward toward the origin. There is no mathematical reason for choosing one form of solution over the other. However, physically we view w_0 as the source; so waves should originate at the source and travel outward. We therefore choose the outward-traveling wave, which, substituted into Eq. (10-71), gives

$$g = \frac{1}{r} f\left(t - \frac{r}{\nu}\right) \tag{10-72}$$

This solution must be valid for all well-behaved f, even f = constant. If f is constant, then g is independent of time, and Eq. (10-69) reduces to Poisson's equation. We know that the point-source solution to Poisson's equation is

$$g = \frac{w_0}{4\pi r}$$

A comparison of this with Eq. (10-72) shows that $f = w_0/4\pi$. This must remain valid even if w_0 is time-varying. Thus

$$f(t) = \frac{w_0(t)}{4\pi} \tag{10-73}$$

which, upon substitution into Eq. (10-72), gives

$$g = \frac{w_0(t - r/\nu)}{4\pi r} \tag{10-74}$$

This is the wave potential of a point source. Note that it is an expanding wave, starting at the source and traveling radially outward with the velocity v. Its value depends on the magnitude of the source *at the time the wave starts.* The potential at a distance r depends therefore upon w_0 at the earlier time $(t - r/v)$ and is called a *retarded potential.*

The potential from a point source not at the origin is obtained by substituting $|\mathbf{r} - \mathbf{r}'|$ for r in Eq. (10-74). The potential at $\mathbf{r}$ due to a point source at $\mathbf{r}'$ is therefore given by

$$g(\mathbf{r}) = \frac{w_0(\mathbf{r}', t - |\mathbf{r} - \mathbf{r}'|/v)}{4\pi|\mathbf{r} - \mathbf{r}'|} \tag{10-75}$$

where we have shown the coordinates explicitly. To generalize to an arbitrary distribution of source, we replace w_0 in Eq. (10-75) by $w\,d\tau'$ and integrate over all elements of source. This gives

$$g(\mathbf{r}) = \frac{1}{4\pi}\iiint \frac{w(\mathbf{r}', t - |\mathbf{r} - \mathbf{r}'|/v)}{|\mathbf{r} - \mathbf{r}'|}\,d\tau' \tag{10-76}$$

Thus, the potential from each element of source is retarded a different amount, depending upon its distance from the field point.

Comparing Eq. (10-69) with Eqs. (10-60), we see that g corresponds to V and w to q_v/ϵ; so the *retarded scalar electric potential integral* is given by

$$V(\mathbf{r}) = \frac{1}{4\pi\epsilon}\iiint \frac{q_v(\mathbf{r}', t - |\mathbf{r} - \mathbf{r}'|/v)}{|\mathbf{r} - \mathbf{r}'|}\,d\tau' \tag{10-77}$$

where $v = 1/\sqrt{\epsilon\mu}$. The rectangular components of $\mathbf{A}$ and $\mu\mathbf{J}$ correspond to g and w, respectively, and the component equations can be recombined to give a *retarded magnetic vector potential integral*

$$\mathbf{A}(\mathbf{r}) = \frac{\mu}{4\pi}\iiint \frac{\mathbf{J}(\mathbf{r}', t - |\mathbf{r} - \mathbf{r}'|/v)}{|\mathbf{r} - \mathbf{r}'|}\,d\tau' \tag{10-78}$$

where $v = 1/\sqrt{\epsilon\mu}$. Eqs. (10-77) and (10-78) give the solution for electric charges and currents in unbounded, homogeneous, linear, isotropic media, such as free space. They can be specialized to surface, line, and point distributions in the usual manner. We should keep in mind that Eq. (10-78) relates vectors at different points in space; so only rectangular component equations should be used. The retarded potentials reduce to the usual static potentials when there is no time variation, for then q_v and $\mathbf{J}$ are the same at all instants of time.

The solution for the potentials from magnetic sources would be equations dual to Eqs. (10-77) and (10-78). Substituting from our table of duals, Table 10-1, we have a *retarded magnetic scalar potential integral*

$$U(\mathbf{r}) = \frac{1}{4\pi\mu}\iiint \frac{m_v(\mathbf{r}', t - |\mathbf{r} - \mathbf{r}'|/v)}{|\mathbf{r} - \mathbf{r}'|}\,d\tau' \tag{10-79}$$

and a *retarded electric vector potential integral*

$$\mathbf{F}(\mathbf{r}) = \frac{\epsilon}{4\pi} \iiint \frac{\mathbf{K}(\mathbf{r}', t - |\mathbf{r} - \mathbf{r}'|/v)}{|\mathbf{r} - \mathbf{r}'|} d\tau' \tag{10-80}$$

Again these give the solution for unbounded, homogeneous, linear, isotropic media, usually taken to be free space. Specialization to surface, line, and point distributions is straightforward. The electromagnetic field from a general distribution of electric and magnetic sources in free space is given by substitution of Eqs. (10-77) to (10-80) into Eqs. (10-68). Note that the potential integrals are wave potentials, subject to Eq. (10-59) and its dual.

The reader may have noted that q_v, V, and $\mathbf{E}$ are related to $\mathbf{J}$, $\mathbf{A}$, and $\mathbf{H}$ by a relationship involving a time derivative, the dual concept holding for magnetic sources. Explicitly, these equations are

$$\nabla \cdot \mathbf{J} = -\frac{\partial q_v}{\partial t} \qquad \nabla \cdot \mathbf{K} = -\frac{\partial m_v}{\partial t}$$

$$\nabla \cdot \mathbf{A} = -\epsilon\mu \frac{\partial V}{\partial t} \qquad \nabla \cdot \mathbf{F} = -\epsilon\mu \frac{\partial U}{\partial t}$$

$$\nabla \times \mathbf{H} = \epsilon \frac{\partial \mathbf{E}}{\partial t} \qquad \nabla \times \mathbf{E} = -\mu \frac{\partial \mathbf{H}}{\partial t}$$

Thus, if we know $\mathbf{J}$, we know q_v except for a term independent of time, that is, a static charge. If we know $\mathbf{A}$, we know V except for a term independent of time, that is, a static potential. If we know $\mathbf{H}$, we know $\mathbf{E}$ except for a term independent of time, that is, a static field. Similar comments hold for the magnetic-source case. *We therefore need evaluate only the vector potential integrals if all sources vary with time.* If both static and time-varying sources are present, we can use the principle of superposition and treat the electromagnetic (time-varying) field separately from the static field.

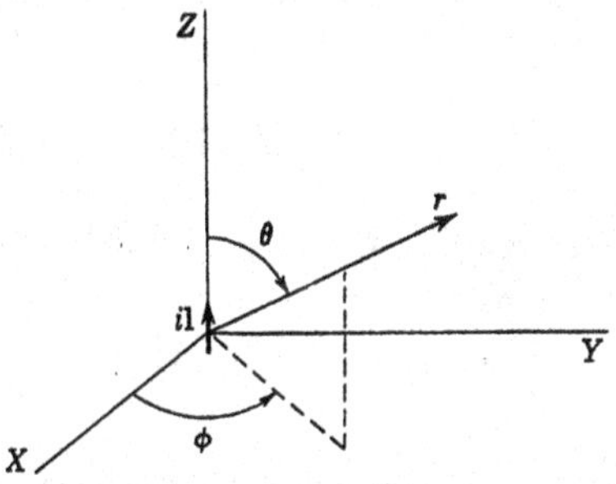

FIG. 10-11. An a-c current element.

As an example, consider the filament of a-c current

$$i = I \sin \omega t \tag{10-81}$$

extending over the incremental length l, as shown in Fig. 10-11. In the limits as $l \to 0$, $I \to \infty$, this source becomes an *electric dipole.* In Eq. (10-78), the integration reduces to a single volume element containing a moment $\mathbf{J}\,\Delta\tau = il$, pointing in the z direction, and the distance from the element becomes simply r. Therefore, there is only a z component of **A** given by

$$A_z(r,\theta,\phi) = \frac{\mu i(t - r/v)l}{4\pi r}$$

Using Eqs. (10-81) and (10-17), we have

$$i\left(t - \frac{r}{v}\right) = I \sin\left[\omega\left(t - \frac{r}{v}\right)\right] = I \sin\left(\omega t - \frac{\omega}{v} r\right)$$
$$= I \sin(\omega t - \beta r)$$

Therefore, the vector potential is specified by

$$A_z = \frac{\mu Il}{4\pi r} \sin(\omega t - \beta r) \tag{10-82}$$

This is a wave traveling radially outward with a phase velocity v, its amplitude being inversely proportional to r. The surfaces of constant phase (βr constant) are spheres concentric with the current element. Such a wave is called a *spherical wave.*

To obtain the magnetic field, we apply Eq. (10-55), which is

$$\mathbf{H} = \frac{1}{\mu}\mathbf{B} = \frac{1}{\mu}\nabla \times \mathbf{A}$$

Expressing **A** in spherical components and expanding the curl in spherical coordinates taking into account that A_z is a function only of r, we have

$$H_\phi = \frac{1}{r}\left[\frac{\partial}{\partial r}(-rA_z \sin\theta) - \frac{\partial}{\partial\theta}(A_z \cos\theta)\right]$$

Substituting from Eq. (10-82) and performing the indicated differentiation, we obtain

$$H_\phi = \frac{Il}{4\pi} \sin\theta \left[\frac{1}{r^2}\sin(\omega t - \beta r) + \frac{\beta}{r}\cos(\omega t - \beta r)\right] \tag{10-83}$$

This is the only component of **H**; so **H** lies in circles concentric with the axis of the current element. If $1/r \ll \beta = 2\pi/\lambda$ or, taking the reciprocal, if $r \gg \lambda/2\pi$, the first term of Eq. (10-83) becomes negligible with respect to the second, and we have

$$H_\phi \approx \frac{\beta Il}{4\pi r}\cos(\omega t - \beta r)\sin\theta \qquad r \gg \frac{\lambda}{2\pi} \tag{10-84}$$

This is called the *radiation field,* and we shall presently see that there is an outward flow of power associated with it. Note that amplitude of **H** varies as $1/r$ in contrast to the $1/r^2$ variation of **H** from a constant current element, Eq. (6-22). If we consider the region close to the current element, the first term of Eq. (10-83) becomes dominant, and we have

$$H_\phi \approx \frac{Il}{4\pi r^2} \sin(\omega t - \beta r) \sin\theta \qquad r \ll \frac{\lambda}{2\pi} \tag{10-85}$$

This is called the *induction field*, and is associated with a storage of magnetic energy. Comparing Eq. (10-85) with the field of a steady current element, Eq. (6-22), we see that they are similar in form. If we let r become sufficiently small, we can neglect the βr term, that is, neglect retardation. With $i = I\sin(\omega t)$ and retardation neglected, Eq. (10-85) becomes identical with Eq. (6-22), and we call it the *quasi-static field.*

There are a number of ways to determine the **E** field. We can determine the charge according to the equation of continuity, Eq. (2-45), the scalar potential integral according to Eq. (10-77), and the electric field according to Eq. (10-56). We can short-cut the first step and, if we wish, determine the scalar potential directly from Eq. (10-59). Finally, we can take the shortest route and determine the electric field directly from the Maxwell-Ampère equation

$$\nabla \times \mathbf{H} = \epsilon \frac{\partial \mathbf{E}}{\partial t}$$

We shall use the latter procedure. Evaluating the left-hand side in spherical coordinates, using **H** as given by Eq. (10-83), we have

$$\begin{aligned}\nabla \times \mathbf{H} = \frac{Il}{4\pi} \Bigg\{ &\mathbf{u}_r 2\cos\theta \left[\frac{1}{r^3}\sin(\omega t - \beta r) + \frac{\beta}{r^2}\cos(\omega t - \beta r)\right] \\ &+ \mathbf{u}_\theta \sin\theta \left[\frac{1}{r^3}\sin(\omega t - \beta r) + \frac{\beta}{r^2}\cos(\omega t - \beta r) - \frac{\beta^2}{r}\sin(\omega t - \beta r)\right]\Bigg\}\end{aligned}$$

Substituting this into the previous equation and taking the antiderivative (all integration constants are zero, for the field is a-c), we obtain

$$\begin{aligned}\mathbf{E} = \frac{Il}{4\pi\epsilon} \Bigg\{ &\mathbf{u}_r 2\cos\theta \left[\frac{-1}{\omega r^3}\cos(\omega t - \beta r) + \frac{\beta}{\omega r^2}\sin(\omega t - \beta r)\right] \\ &+ \mathbf{u}_\theta \sin\theta \bigg[\frac{-1}{\omega r^3}\cos(\omega t - \beta r) + \frac{\beta}{\omega r^2}\sin(\omega t - \beta r) \\ &+ \frac{\beta^2}{\omega r}\cos(\omega t - \beta r)\bigg]\Bigg\}\end{aligned} \tag{10-86}$$

If $r \gg \lambda/2\pi$, the $1/r$ term becomes the dominant term, and we have

$$E_\theta \approx \frac{\beta^2 Il}{4\pi\epsilon\omega r} \cos(\omega t - \beta r) \sin\theta \qquad r \gg \frac{\lambda}{2\pi} \tag{10-87}$$

This is the electric *radiation field* associated with the magnetic radiation field of Eq. (10-84). All other terms of Eq. (10-86) constitute the *induction field*, associated with the **H** of Eq. (10-85). If we allow r to become sufficiently small so that only the $1/r^3$ terms are appreciable, and $\beta r \to 0$, we have the *quasi-static field*

$$\mathbf{E}_r \xrightarrow[r \to 0]{} \frac{-Il \cos \omega t}{4\pi\epsilon\omega r^3} (\mathbf{u}_r 2 \cos \theta + \mathbf{u}_\theta \sin \theta) \tag{10-88}$$

Noting that the charge collecting at the end of the current element is

$$q = \int i \, dt = I \int \sin \omega t \, dt = \frac{-I}{\omega} \cos \omega t$$

and substituting this into Eq. (10-88), we can write

$$\mathbf{E} \xrightarrow[r \to 0]{} \frac{ql}{4\pi\epsilon r^3} (\mathbf{u}_r 2 \cos \theta + \mathbf{u}_\theta \sin \theta)$$

This is identical with the static charge-dipole field, Eq. (5-80).

To summarize, the field of an a-c current element is characterized as follows. (*a*) Very near the element the magnetic field has the same form as the field due to a steady current element, and the electric field has the same form as the field due to a static charge dipole. (*b*) At intermediate distances, the field is quite complicated and is called the induction field. (*c*) At large distances from the element, we have the relatively simple radiation field, given by Eqs. (10-84) and (10-87). We shall now consider further this radiation field.

If the substitutions

$$\frac{\beta}{\omega\epsilon} = \frac{\omega\sqrt{\epsilon\mu}}{\omega\epsilon} = \sqrt{\frac{\mu}{\epsilon}} \qquad \beta = \frac{2\pi}{\lambda}$$

are made, the radiation field of a current element can be written as

$$\left.\begin{aligned} H_\phi &= \frac{Il}{2\lambda r} \cos (\omega t - \beta r) \sin \theta \\ E_\theta &= \sqrt{\frac{\mu}{\epsilon}} \frac{Il}{2\lambda r} \cos (\omega t - \beta r) \sin \theta \end{aligned}\right\} r \gg \frac{\lambda}{2\pi} \tag{10-89}$$

The Poynting vector is given by $\mathbf{S} = \mathbf{E} \times \mathbf{H}$, which in the radiation zone has only an r component given by

$$S_r = E_\theta H_\phi = \sqrt{\frac{\mu}{\epsilon}} \left(\frac{Il}{2\lambda r}\right)^2 \cos^2 (\omega t - \beta r) \sin^2 \theta \tag{10-90}$$

Note that this is never negative, indicating a flow of power radially outward from the current element. This is called *radiated power*. Note that **E**, **H**, and **S** form a right-handed triad of mutually orthogonal vectors at every point in the radiation field, as suggested by Fig. 10-12. If we consider a small section of the radiation zone, the factor $1/r$ is very nearly constant, and Eqs. (10-89) are essentially the equations for a plane wave. Therefore we say that *the radiation field consists of outward traveling plane waves.* In dual manner, we note that the same holds true of a magnetic current element. Since any source can be represented in terms of electric and magnetic currents and since the total field is the superposition of the fields from all current elements, it follows that the radiation field from any source of finite size consists of outward-traveling plane waves. Note that the ratio of E to H in Eqs. (10-89) is the intrinsic impedance of the medium, $\eta = \sqrt{\mu/\epsilon}$. This is to be expected, for within a restricted region the wave is indistinguishable from a single traveling plane wave.

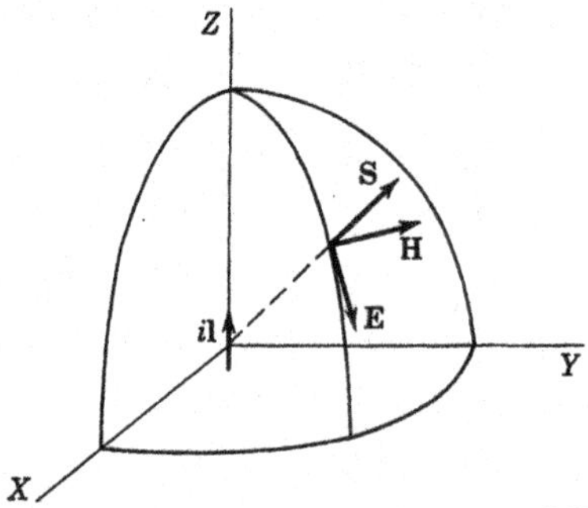

FIG. 10-12. Relationship of **E**, **H**, and **S** in the radiation field.

A plot of the amplitude (that quantity multiplying the $\cos(\omega t - \beta r)$ term) of E_θ for constant r in the radiation zone is called a *radiation field pattern.* This is also a plot of H_ϕ in the radiation zone, for the two are related by the intrinsic impedance $\eta = \sqrt{\mu/\epsilon}$ as shown by Eqs. (10-89). The radiation pattern of the current element varies as $\sin\theta$, as illustrated

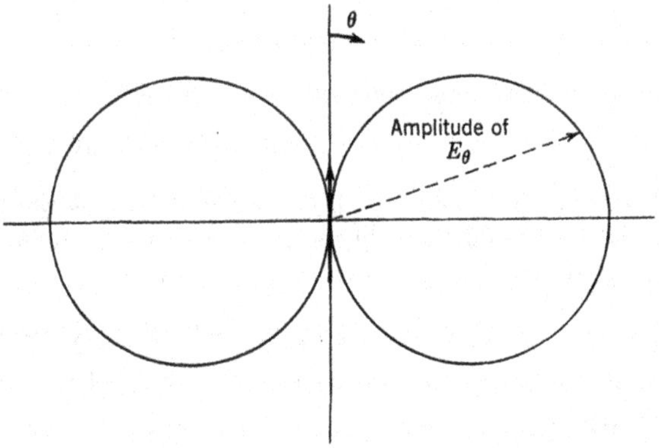

FIG. 10-13. The radiation field pattern of a current element.

by Fig. 10-13. A three-dimensional picture of the radiation pattern is obtained by rotating the pattern of Fig. 10-13 about the axis of the current element, giving a doughnut-shaped pattern. A plot of the time-average Poynting vector in the radiation zone is called a *radiation power pattern.* From Eq. (10-90) it is apparent that this would vary as $\sin^2\theta$, that is, as the field pattern squared.

10-7. Guided Waves. Under certain conditions it is possible for electromagnetic waves to be propagated through hollow conducting cylinders. In this section we shall illustrate such propagation by an example.

Let us first consider a uniform x-polarized plane wave propagating at an angle θ with respect to the z axis. Fig. 10-14 represents the geometry of the situation. Orienting a primed set of axes such that Z' points in

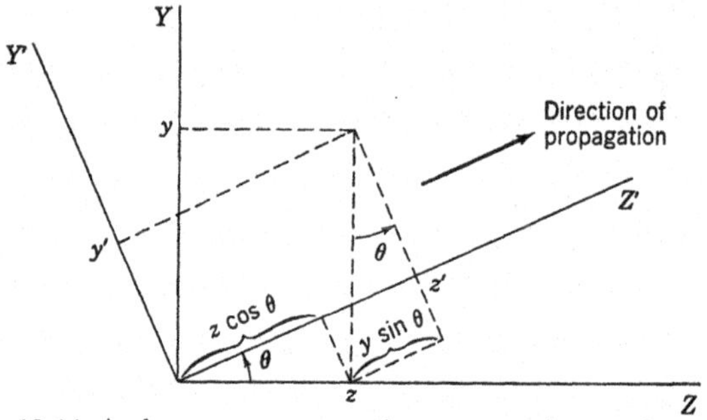

FIG. 10-14. A plane wave propagating at an angle θ to the z axis.

the direction of propagation and X' lies along X, we have the plane wave given by

$$E_x = C \sin (\omega t - \beta z')$$

Referring to Fig. 10-14, we see that

$$z' = z \cos \theta + y \sin \theta$$

so in terms of the unprimed coordinates, we have

$$E_x = C \sin [\omega t - \beta(z \cos \theta + y \sin \theta)]$$

Suppose we have two such waves, one propagating at the angle θ with respect to the z axis and the other propagating at the angle $-\theta$. If the two waves are of equal amplitude, the total field is

$$E_x = C \sin [\omega t - \beta(z \cos \theta + y \sin \theta)] + C \sin [\omega t - \beta(z \cos \theta - y \sin \theta)]$$

since $\cos(-\theta) = \cos \theta$ and $\sin(-\theta) = -\sin \theta$. Using the trigonometric identity

$$\sin a + \sin b = 2 \cos \tfrac{1}{2}(a - b) \sin \tfrac{1}{2}(a + b)$$

we can write the field of the two waves as

$$E_x = 2C \cos (\beta y \sin \theta) \sin (\omega t - \beta z \cos \theta) \tag{10-91}$$

Note that we have a standing-wave phenomenon in the y direction and a traveling-wave phenomenon in the z direction. The field is zero along the set of infinite planes $\beta y \sin \theta = \pm\pi/2, +3\pi/2, \pm 5\pi/2, \ldots$.

Now consider the hollow rectangular tube formed of conducting sheets, as represented by Fig. 10-15. We shall consider the conductor to be perfect, so the tangential components of **E** must be zero at the conducting sheets. Viewing the "possible field" of Eq. (10-91), we note that there is no component of **E** tangential to the top and bottom conductors of the tube. Furthermore, if the side walls lie at a position of zero field, we satisfy the condition of no tangential component of **E** along these walls.

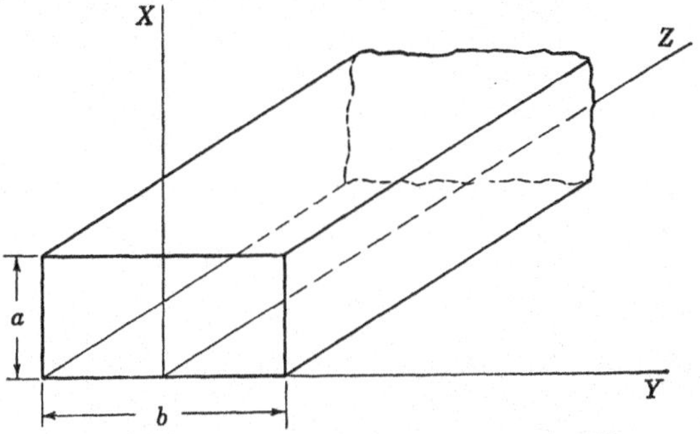

FIG. 10-15. A section of rectangular waveguide.

Therefore, a possible field within the tube is given by Eq. (10-91) subject to the restriction

$$\beta \frac{b}{2} \sin \theta = \frac{\pi}{2}, \frac{3\pi}{2}, \frac{5\pi}{2}, \ldots$$

The smallest value of b corresponds to the choice of $\pi/2$, which, upon substitution of $\beta = 2\pi/\lambda$, gives

$$b = \frac{\lambda}{2 \sin \theta} \tag{10-92}$$

For a given width b, this can be used to determine the angle θ that the partial plane waves make with respect to the z axis. We can picture the field inside the rectangular tube of Fig. 10-15 as being two x-polarized plane waves bouncing back and forth between the two side walls as they progress down the tube in the z direction. The wave is said to be "trapped" within the tube, and the tube is called a *waveguide*. Note that if $b < \lambda/2$, there is no "real" value of θ according to Eq. (10-92), and the bouncing wave picture does not apply. In this case, the waveguide is said to be *cut off*, and no wave can be propagated down the guide. The *cut-off wavelength* λ_c is defined as

$$\lambda_c = 2b \tag{10-93}$$

All frequencies having a shorter wavelength are propagated, and those having a longer wavelength are cut off. Note that the field, Eq. (10-91), has the form of a single wave traveling in the $+z$ direction. Another wave can be propagated in the $-z$ direction. Two waves of equal amplitude but traveling in opposite directions would form a standing wave in a manner similar to the plane-wave case. Waves of unequal amplitudes would form partial standing waves. The field that we have been considering is only one of many possible fields. Each possible field is called a *mode.* The one that we have obtained is called the *dominant mode,* for at a given frequency it can propagate in a smaller waveguide than any other mode.

So that we may better visualize the waveguide field as a whole, let us determine the associated **H** field. This may be done by determining the **H** associated with each of the two plane waves and then adding them together, or by determining **H** directly from the field equations. We shall use the latter approach. Taking **E** as given by Eq. (10-91) and using

$$\nabla \times \mathbf{E} = -\mu \frac{\partial \mathbf{H}}{\partial t}$$

we obtain
$$\begin{aligned} H_y &= \frac{\omega\mu}{\beta \cos \theta} E_x \\ H_z &= \frac{2C}{\omega\mu} \beta \sin \theta \sin (\beta y \sin \theta) \cos (\omega t - \beta z \cos \theta) \end{aligned} \tag{10-94}$$

A plot of the field lines at some instant of time would look like Fig. 10-16. Since this is a traveling wave, the entire picture would appear to move

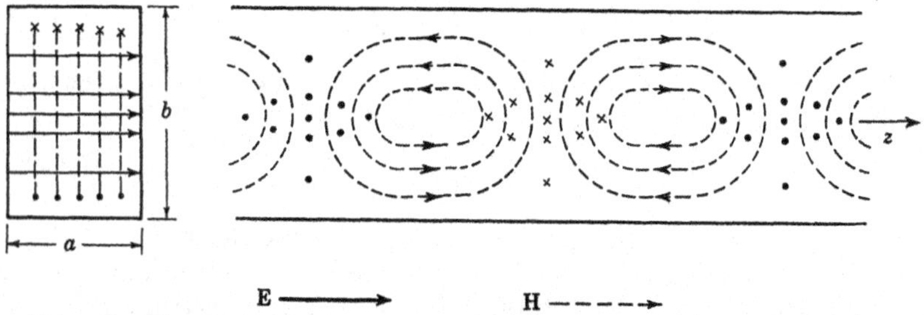

FIG. 10-16. Instantaneous picture of the field lines in a rectangular waveguide.

in the $+z$ direction as time elapses. Note that H_y is proportional to E_x, a fact which allows us to extend the concept of characteristic impedance to waveguide phenomena. A picture of standing waves in a waveguide would be the same as the picture for the resonant cavity considered in the next section.

10-8. Resonance. Under certain conditions an electromagnetic field can "resonate" in a dielectric region surrounded by a conductor. This phenomenon is closely related to the behavior of an interconnected capacitor and inductor in circuit theory. We shall use the "rectangular cavity" to illustrate electromagnetic resonance.

The right parallelopiped enclosed by the six conducting walls shown in Fig. 10-17 is called the rectangular cavity. A field existing within this cavity must be characterized by vanishing tangential components of **E** over all walls. The rectangular waveguide field already satisfies this condition over four of the walls. Recalling that standing waves have stationary planes of zero field, we consider the standing-wave rectangular waveguide field. This consists of two waves of equal amplitude traveling in opposite directions. Each wave would have the form of Eq. (10-91), the sign of z being reversed for the $-z$-traveling wave. The sum of the two waves is therefore

$$E_x = 2C \cos (\beta y \sin \theta)[\sin (\omega t - \beta z \cos \theta) + \sin (\omega t + \beta z \cos \theta)]$$

Using the trigonometric identity preceding Eq. (10-91), we can rewrite this as

$$E_x = 4C \cos (\beta y \sin \theta) \cos (\beta z \cos \theta) \sin \omega t \tag{10-95}$$

Note that this is the sum of four uniform plane waves, making angles θ, $-\theta$, $\pi + \theta$, and $\pi - \theta$ with respect to the z axis. The four side walls of the cavity must coincide with positions of zero **E**. Thus, we require that

$$\beta \frac{b}{2} \sin \theta = \frac{\pi}{2}, \frac{3\pi}{2}, \frac{5\pi}{2}, \ldots$$

$$\beta \frac{c}{2} \cos \theta = \frac{\pi}{2}, \frac{3\pi}{2}, \frac{5\pi}{2}, \ldots$$

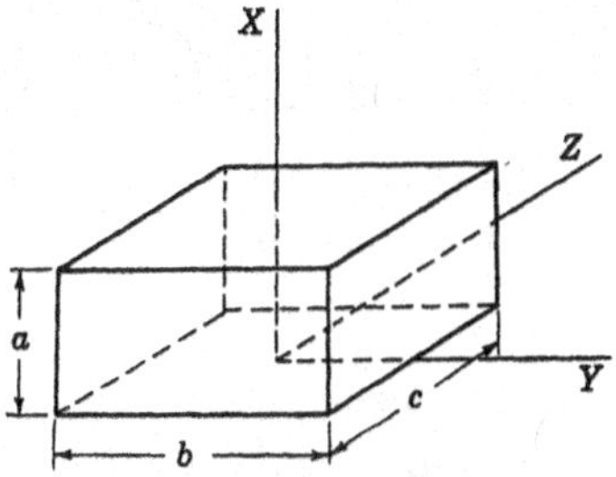

FIG. 10-17. A rectangular cavity.

The dimension a is arbitrary, for there is no component of **E** tangential to the top and bottom walls. The smallest possible dimensions b and c correspond to the first planes of zero field. Choosing $\pi/2$ in each of the above equations we have

$$\sin \theta = \frac{\pi}{\beta b} \qquad \cos \theta = \frac{\pi}{\beta c}$$

The trigonometric identity

$$\sin^2 \theta + \cos^2 \theta = \left(\frac{\pi}{\beta b}\right)^2 + \left(\frac{\pi}{\beta c}\right)^2 = 1$$

must be satisfied. Thus, for specific values of b and c, β must satisfy

$$\beta^2 = \left(\frac{\pi}{b}\right)^2 + \left(\frac{\pi}{c}\right)^2$$

or, substituting $\beta = \omega \sqrt{\mu\epsilon}$, the resonant frequency is specified by

$$2\pi f = \omega = \frac{\pi}{\sqrt{\epsilon\mu}} \sqrt{\frac{1}{b^2} + \frac{1}{c^2}} \tag{10-96}$$

This is the only frequency for which the field can exist, assuming no dissipation. However, other possible field configurations can exist, each possible field being called a *mode*. Associated with each mode is a particular resonant frequency. The one that we have been considering is the lowest possible frequency for a given size cavity (assuming a is the smallest dimension), and the field is called the *dominant mode*.

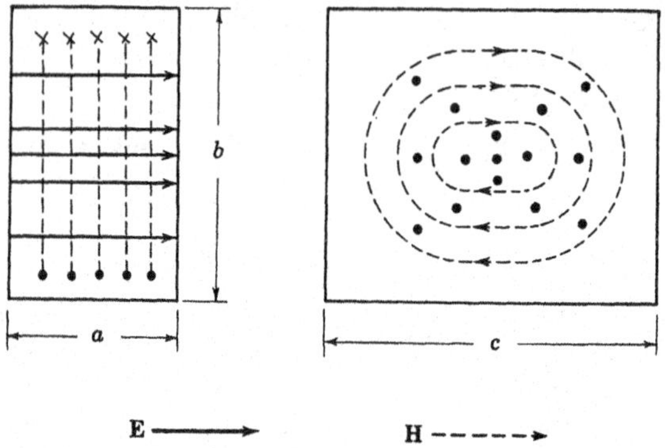

FIG. 10-18. Instantaneous picture of the field lines in a rectangular cavity.

The **H** associated with the **E** field of the cavity can be found directly from the field equation

$$\nabla \times \mathbf{E} = -\mu \frac{\partial \mathbf{H}}{\partial t}$$

The result is

$$\begin{aligned} H_y &= \frac{-1}{\omega\mu} 4C\beta \cos\theta \cos(\beta y \sin\theta) \sin(\beta z \cos\theta) \cos\omega t \\ H_z &= \frac{1}{\omega\mu} 4C\beta \sin\theta \sin(\beta y \sin\theta) \cos(\beta z \cos\theta) \cos\omega t \end{aligned} \qquad (10\text{-}97)$$

Note that **H** is 90° out of phase with **E**. This means that **E** reaches its peak value when **H** is zero, and vice versa. An instantaneous picture of the field lines in the resonant cavity is shown in Fig. 10-18. The energy oscillates between the electric and magnetic forms. It can be shown that the total energy within the cavity is the same at all instants of time. This is a direct consequence of Poynting's theorem and the conservation of energy.

QUESTIONS FOR DISCUSSION

10-1. Why is Eq. (10-2) called a "wave equation"?

10-2. Why is it possible to define μ_0 arbitrarily as $4\pi \times 10^{-7}$? Does this choice fix the unit of charge, or can this also be chosen arbitrarily?

10-3. In establishing Eq. (10-9), it is assumed that no static field exists. Show that this assumption is permissible. What change in viewpoint could be made to obviate this assumption?

10-4. Under what restrictions does Eq. (10-10) apply?

10-5. Justify that Eq. (10-12) is the most general form for an a-c quantity.

10-6. Discuss the following a-c wave terminology: frequency, amplitude, phase, phase constant, wavelength, phase velocity, plane wave, uniform wave, linear polarization, traveling wave, standing wave.

10-7. Interpret each term of Eq. (10-31) and Eq. (10-38).

10-8. Discuss the field-theory representations of the voltage and current sources of circuit theory, Figs. 10-7 and 10-8.

10-9. It cannot be proved that $\mathbf{S} = \mathbf{E} \times \mathbf{H}$ is a vector density of power flow. What justification do we have for so defining it?

10-10. Interpret each term of Eqs. (10-48) and (10-53).

10-11. The power leaving a region is the integral of the normal component of $\mathbf{S}$ over the entire surface enclosing the region. What justification do we have for integrating only over the cross section of the coaxial line of Fig. 10-10?

10-12. Using the equations of Sec. 10-5, show that the electromagnetic potentials $\mathbf{A}$, $\mathbf{F}$, V, and U reduce to the static potentials when the time derivatives vanish.

10-13. Discuss the use of Table 10-1.

10-14. What is the meaning of the retarded-potential notation?

10-15. Is the following statement true? In a source-free region, the electromagnetic field can be specified either in terms of a magnetic vector potential only or in terms of an electric vector potential only.

10-16. Discuss the difference between the radiation field, the induction field, and the quasi-static field of an electric dipole.

10-17. Is the following statement reasonable? In the radiation field of an electric dipole, electromagnetic energy appears to be trapped within spherical shells of thickness $\lambda/2$, which expand radially outward with the velocity of light.

10-18. Give a verbal description of the dominant-mode field in a rectangular waveguide.

10-19. Give a verbal description of the dominant-mode field in a rectangular cavity.

PROBLEMS

10-1. Determine the velocity of propagation, the intrinsic impedance, and the magnetic intensity associated with an electric intensity of 1 volt per meter for a plane wave traveling in: (*a*) vacuum; (*b*) polystyrene; (*c*) fresh water. Assume $\sigma = 0$.

10-2. Suppose that the wave of Prob. 10-1 is a uniform-plane a-c wave. If the frequency is 1 megacycle, determine the phase constant and the wavelength in each material.

10-3. Determine the electromagnetic energy per unit cross section contained within each energy "pulse" of a uniform-plane a-c traveling wave. (A pulse is one-half wavelength thick.)

10-4. Establish Eq. (10-23) by direct substitution of Eq. (10-21) with $\Theta = \Phi = 0$ into Maxwell's equations.

10-5. Given the partial standing wave

$$E_x = B \cos (\omega t - \beta z) + C \cos (\omega t + \beta z)$$

where B and C are constants, determine the magnetic intensity: (*a*) by treating the $+z$ and $-z$ components as separate traveling waves; and (*b*) by direct substitution into Maxwell's equations.

10-6. The E_x of Prob. 10-5 must be expressible in the form of Eq. (10-12). Determine $A(z)$ and $\Phi(z)$ in this case. The *standing wave ratio* (S.W.R.) of a partial standing wave is defined as the ratio of the maximum amplitude of E to the minimum amplitude, that is

$$\text{S.W.R.} = \frac{A_{\max}}{A_{\min}}$$

Determine the S.W.R. of the above field.

10-7. A long straight wire with circular cross section of radius a carries a d-c current i uniformly distributed over its cross section. The conductivity of the wire is σ. Determine the **E** and **H** fields assuming that the wire is infinitely long. Show that the power dissipation per unit length of wire, evaluated according to Eqs. (10-51), equals the inward flow of power over its surface, evaluated according to Eq. (10-52).

10-8. In vacuum, there exists an electric intensity

$$\mathbf{E} = \mathbf{u}_x \sin \pi y \sin \pi z \sin \omega t$$

Assuming that all quantities are a-c, determine the magnetic intensity, the electric displacement current, and the magnetic displacement current.

10-9. For the field of Prob. 10-8, determine the total electric current through the square loop having vertices $(\frac{1}{2},0,0)$, $(\frac{1}{2},1,0)$, $(\frac{1}{2},1,\frac{1}{2})$, $(\frac{1}{2},0,\frac{1}{2})$: (*a*) by integrating the current density over the area enclosed; and (*b*) by integrating the magnetic intensity about the contour. Also (*c*) in terms of ϵ_0 and μ_0 determine the value of ω necessary to make results (*a*) and (*b*) equal.

10-10. For the field of Prob. 10-8, determine the electric-energy density, the magnetic-energy density, and the Poynting vector.

10-11. Suppose that an electric intensity $\mathbf{E} = \mathbf{u}_y E_0$, where E_0 is a constant, is associated with the problem of Fig. 6-6. Determine the power per unit area supplied by the current sheet. Show by Poynting's theorem that this energy divides equally, one-half flowing in the $+x$ direction and one-half flowing in the $-x$ direction.

10-12. Show that the magnetic vector potential

$$\mathbf{A} = \mathbf{u}_x \sin \beta y \sin \omega t$$

in a source-free medium is a wave potential. Determine the electric and magnetic intensities associated with this wave potential. Determine a possible electric vector potential which gives the same electric and magnetic fields.

10-13. Given the wave potentials in vacuum

$$\mathbf{A} = \mathbf{u}_z \cos \beta y \sin \omega t$$
$$\mathbf{F} = \mathbf{u}_z \cos \beta y \cos \omega t$$

find U, V, **E**, and **H** associated with them.

10-14. State the problem dual to that of Fig. 10-11. Write down the **E** and **H** fields of this dual problem.

10-15. Verify the derivation of Eq. (10-74) starting from Eqs. (10-72) and (10-73).

10-16. Evaluate the retarded potentials (**A** and V) for the d-c current element of Fig. 6-3. Determine **E**, **H**, and the Poynting vector **S**. Find the power crossing a sphere of radius R concentric with the current element.

10-17. For the electric dipole of Sec. 10-6 situated in free space, determine the instantaneous power crossing a sphere of radius R. Find the time-average power, and show that it is independent of R. Denoting the time-average power by $\bar{P}$, show that the radiation field can be expressed as

$$H_\phi = \frac{\sqrt{P}}{4\pi r \sqrt{10}} \sin\theta \cos(\omega t - \beta r) \qquad E_\theta = \sqrt{\frac{\mu}{\epsilon}}\, H_\phi$$

10-18. Suppose we have two z-directed electric dipoles, of equal magnitude and phase, located at $(0, -\lambda/4, 0)$ and $(0, \lambda/4, 0)$. Determine the vector potential and the radiation field ($r \to \infty$). Plot the radiation field patterns in the $x = 0$ plane, the $y = 0$ plane, and the $z = 0$ plane, as a function of θ or ϕ.

10-19. Show that when

$$\mathbf{E} = \hat{\mathbf{E}}(x,y) \sin(\omega t - \beta z)$$

where $\hat{\mathbf{E}} = \mathbf{u}_x \hat{E}_x + \mathbf{u}_y \hat{E}_y$, Eq. (10-3) reduces to

$$\frac{\partial^2 \hat{E}_x}{\partial x^2} + \frac{\partial^2 \hat{E}_x}{\partial y^2} = 0 \qquad \frac{\partial^2 \hat{E}_y}{\partial x^2} + \frac{\partial^2 \hat{E}_y}{\partial y^2} = 0$$

Hence, $\hat{\mathbf{E}}$ is a solution to the two-dimensional Laplace equation. Thus, the solution to a two-dimensional electrostatic problem can be converted into a solution to a two-dimensional a-c problem according to the first of the above equations. This is the basis of the analysis of two-wire transmission lines.

10-20. Determine the Poynting vector over a cross section of a rectangular waveguide for the dominant mode. Integrate the Poynting vector over the cross section, and determine the power flow along the guide. Determine the time-average power flow.

10-21. Assume that the field outside the rectangular waveguide is zero. Using Eq. (4-22), determine the surface current on the waveguide walls. Sketch the current distribution in the top and side walls of the guide for some instant of time. What happens to this current as time increases?

10-22. Determine the total z-directed current in the top wall of the rectangular waveguide ($x = a$ of Fig. 10-15). Determine the voltage across the center of the guide ($y = 0$ of Fig. 10-15), reference positive at $x = a$, $y = 0$. Note that this current and voltage are in phase at each value of z. Hence, the ratio of their magnitudes is a resistance. Determine this "characteristic impedance." Note that the product vi of the above determined current and voltage is not equal to the answer to Prob. 10-20. How is this possible?

10-23. It is desired to construct a cubic cavity resonating in the dominant mode at 1000 megacycles. Determine its dimensions.

10-24. Determine the electric and magnetic energy densities for the dominant mode in a rectangular cavity. Integrate, and obtain the total electric and magnetic energies within the cavity as a function of time. Note that they reach peak values at different instants of time. Show that the time-average electric energy is equal to the time-average magnetic energy.

APPENDIX A

UNITS AND DIMENSIONS

A table of the principal quantities used in this book, their symbolic representation, their units, their dimensions, and the sections where they are introduced, listed in order of introduction.

Quantity	Symbol	Unit	Dimensions	Section introduced
Charge (electric)	q	coulomb	Q	1-3
Current (electric)	i	ampere	Q/T	1-3
Voltage	v	volt	ML^2/T^2Q	1-3
Magnetic flux	ψ	weber	ML^2/TQ	1-3
Resistance	R	ohm	ML^2/TQ^2	1-4
Inductance	L	henry	ML^2/Q^2	1-4
Capacitance	C	farad	T^2Q^2/ML^2	1-4
Power	P	watt	ML^2/T^3	1-6
Energy	W	joule	ML^2/T^2	1-6
Mutual inductance	M	henry	ML^2/Q^2	1-8
Charge density (electric)				
Volume	q_v	coulomb/cubic meter	Q/L^3	2-1
Surface	q_s	coulomb/square meter	Q/L^2	2-1
Line	q_l	coulomb/meter	Q/L	2-1
Current density (electric)				
Volume	$\mathbf{J}$	ampere/square meter	Q/TL^2	2-4
Surface	$\mathbf{J}_s$	ampere/meter	Q/TL	2-4
Electric intensity	$\mathbf{E}$	volt/meter	ML/T^2Q	3-2
Conductivity	σ	mho/meter	TQ^2/ML^3	3-3
Power density	p	watt/cubic meter	M/T^3L	3-3
Magnetic flux density	$\mathbf{B}$	weber/square meter	M/TQ	3-4
Electric flux density	$\mathbf{D}$	coulomb/square meter	Q/L^2	4-2
Electric flux	ψ^e	coulomb	Q	4-2
Capacitivity (permittivity)	ϵ	farad/meter	T^2Q^2/ML^3	4-4
Relative capacitivity (dielectric constant)	ϵ_r	(numeric)	none	4-4
Magnetic intensity	$\mathbf{H}$	ampere/meter	Q/TL	4-5
Magnetomotive force	u	ampere	Q/T	4-5

Quantity	Symbol	Unit	Dimensions	Section introduced
Inductivity (permeability)	μ	henry/meter	ML/Q^2	4-7
Relative inductivity (relative permeability)	μ_r	(numeric)	none	4-7
Energy density	w	joule/cubic meter	M/LT^2	4-8
Electric scalar potential	V	volt	ML^2/T^2Q	5-4
Magnetic vector potential	$\mathbf{A}$	weber/meter	ML/TQ	6-2
Polarization (electric)	$\mathbf{P}$	coulomb/square meter	Q/L^2	7-3
Electric susceptibility	χ_e	(numeric)	none	7-3
Magnetization	$\mathbf{M}$	ampere/meter	Q/TL	7-4
Magnetic scalar potential	U	ampere	Q/T	7-4
Magnetic susceptibility	χ_m	(numeric)	none	7-4
Magnetic charge	m	weber	ML^2/TQ	7-5
Magnetic charge density				
Volume	m_v	weber/cubic meter	M/LTQ	7-5
Surface	m_s	weber/square meter	M/TQ	7-5
Line	m_l	weber/meter	ML/TQ	7-5
Magnetic current	k	volt	ML^2/T^2Q	7-5
Magnetic current density				
Volume	$\mathbf{K}$	volt/square meter	M/T^2Q	7-5
Surface	$\mathbf{K}_s$	volt/meter	ML/T^2Q	7-5
Electric vector potential	$\mathbf{F}$	coulomb/meter	Q/L	7-5
Coefficients of potential	$p_{i,j}$	volt/coulomb	ML^2/T^2Q^2	9-5
Coefficients of capacitance and electrostatic induction	$c_{i,j}$	farad	T^2Q^2/ML^2	9-5
Direct capacitances	$C_{i,j}$	farad	T^2Q^2/ML^2	9-5
Coefficients of resistance	$r_{i,j}$	ohm	ML^2/TQ^2	9-6
Coefficients of conductance	$g_{i,j}$	mho	TQ^2/ML^2	9-6
Direct conductances	$G_{i,j}$	mho	TQ^2/ML^2	9-6
Grounding resistance	R_g	ohm	ML^2/TQ^2	9-6
Reluctance	$\mathcal{R}$	ampere/weber	Q^2/ML^2	9-7
Velocity of propagation	v	meter/second	L/T	10-1
Velocity of light	c	meter/second	L/T	10-1
Intrinsic impedance	η	ohm	ML^2/TQ^2	10-1
Angular frequency	ω	1/second	$1/T$	10-2
Phase constant	β	1/meter	$1/L$	10-2
Wavelength	λ	meter	L	10-2
Poynting vector	$\mathbf{S}$	watt/square meter	M/T^3	10-4

APPENDIX B

SUMMARY OF VECTOR ANALYSIS

1. Coordinates

a. Whenever possible, we orient rectangular (x,y,z), cylindrical (ρ,ϕ,z), and spherical (r,θ,ϕ) coordinates as shown in Fig. B-1. Rules for

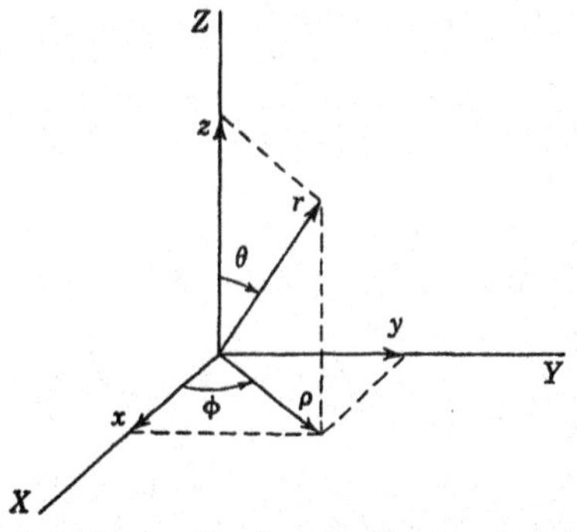

FIG. B-1. Convention for coordinate orientation.

transformation among coordinates are then

$$\begin{cases} x = \rho\cos\phi = r\sin\theta\cos\phi \\ y = \rho\sin\phi = r\sin\theta\sin\phi \\ z = r\cos\theta \end{cases}$$

$$\begin{cases} \rho = \sqrt{x^2+y^2} = r\sin\theta \\ \phi = \tan^{-1}(y/x) \\ z = r\cos\theta \end{cases}$$

$$\begin{cases} r = \sqrt{x^2+y^2+z^2} = \sqrt{\rho^2+z^2} \\ \theta = \tan^{-1}(\sqrt{x^2+y^2}/z) = \tan^{-1}(\rho/z) \\ \phi = \tan^{-1}(y/x) \end{cases}$$

b. The coordinate unit vectors are defined as vectors of unit length pointing along coordinate lines in the direction of increasing coordi-

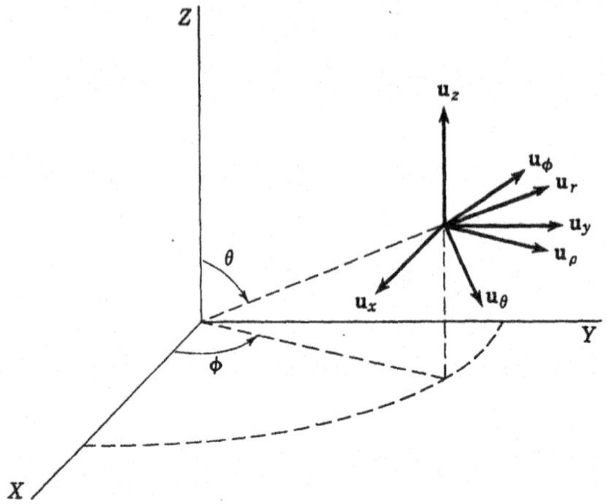

FIG. B-2. The coordinate unit vectors.

nate variables. Fig. B-2 illustrates coordinate unit vectors at a point. Rules for transformation among unit coordinate vectors are then

$$\begin{cases} \mathbf{u}_x = \mathbf{u}_\rho \cos\phi - \mathbf{u}_\phi \sin\phi \\ \quad\;\, = \mathbf{u}_r \sin\theta\cos\phi + \mathbf{u}_\theta\cos\theta\cos\phi - \mathbf{u}_\phi\sin\phi \\ \mathbf{u}_y = \mathbf{u}_\rho\sin\phi + \mathbf{u}_\phi\cos\phi \\ \quad\;\, = \mathbf{u}_r\sin\theta\sin\phi + \mathbf{u}_\theta\cos\theta\sin\phi + \mathbf{u}_\phi\cos\phi \\ \mathbf{u}_z = \mathbf{u}_r\cos\theta - \mathbf{u}_\theta\sin\theta \end{cases}$$

$$\begin{cases} \mathbf{u}_\rho = \mathbf{u}_x\cos\phi + \mathbf{u}_y\sin\phi = \mathbf{u}_r\sin\theta + \mathbf{u}_\theta\cos\theta \\ \mathbf{u}_\phi = -\mathbf{u}_x\sin\phi + \mathbf{u}_y\cos\phi \\ \mathbf{u}_z = \mathbf{u}_r\cos\theta - \mathbf{u}_\theta\sin\theta \end{cases}$$

$$\begin{cases} \mathbf{u}_r = \mathbf{u}_x\sin\theta\cos\phi + \mathbf{u}_y\sin\theta\sin\phi + \mathbf{u}_z\cos\theta \\ \quad\;\, = \mathbf{u}_\rho\sin\theta + \mathbf{u}_z\cos\theta \\ \mathbf{u}_\theta = \mathbf{u}_x\cos\theta\cos\phi + \mathbf{u}_y\cos\theta\sin\phi - \mathbf{u}_z\sin\theta \\ \quad\;\, = \mathbf{u}_\rho\cos\theta - \mathbf{u}_z\sin\theta \\ \mathbf{u}_\phi = -\mathbf{u}_x\sin\phi + \mathbf{u}_y\cos\phi \end{cases}$$

2. Addition, scalar multiplication, vector multiplication

a. General

$\mathbf{F} + \mathbf{G}$ formed according to parallelogram rule

$\mathbf{F} \cdot \mathbf{G} = FG\cos\eta$ (η is the angle between **F** and **G**)

$\mathbf{F} \times \mathbf{G} = \mathbf{u}_n FG\sin\eta$ ($\mathbf{u}_n$ is the unit vector perpendicular to **F** and **G** according to the right-hand rule)

b. Rectangular components (always valid)

$$\mathbf{F} + \mathbf{G} = \mathbf{u}_x(F_x + G_x) + \mathbf{u}_y(F_y + G_y) + \mathbf{u}_z(F_z + G_z)$$
$$\mathbf{F} \cdot \mathbf{G} = F_xG_x + F_yG_y + F_zG_z$$
$$\mathbf{F} \times \mathbf{G} = \begin{vmatrix} \mathbf{u}_x & \mathbf{u}_y & \mathbf{u}_z \\ F_x & F_y & F_z \\ G_x & G_y & G_z \end{vmatrix}$$

c. Cylindrical components (valid if **F** and **G** are at the same point)

$$\mathbf{F} + \mathbf{G} = \mathbf{u}_\rho(F_\rho + G_\rho) + \mathbf{u}_\phi(F_\phi + G_\phi) + \mathbf{u}_z(F_z + G_z)$$
$$\mathbf{F} \cdot \mathbf{G} = F_\rho G_\rho + F_\phi G_\phi + F_zG_z$$
$$\mathbf{F} \times \mathbf{G} = \begin{vmatrix} \mathbf{u}_\rho & \mathbf{u}_\phi & \mathbf{u}_z \\ F_\rho & F_\phi & F_z \\ G_\rho & G_\phi & G_z \end{vmatrix}$$

d. Spherical components (valid if **F** and **G** are at the same point)

$$\mathbf{F} + \mathbf{G} = \mathbf{u}_r(F_r + G_r) + \mathbf{u}_\theta(F_\theta + G_\theta) + \mathbf{u}_\phi(F_\phi + G_\phi)$$
$$\mathbf{F} \cdot \mathbf{G} = F_rG_r + F_\theta G_\theta + F_\phi G_\phi$$
$$\mathbf{F} \times \mathbf{G} = \begin{vmatrix} \mathbf{u}_r & \mathbf{u}_\theta & \mathbf{u}_\phi \\ F_r & F_\theta & F_\phi \\ G_r & G_\theta & G_\phi \end{vmatrix}$$

3. Line integral and surface integral

Coordinates	$d\mathbf{l}$ or $d\mathbf{s}$ points in	action of **F**	flow of **F**
a. General	Arbitrary direction	$\int \mathbf{F} \cdot d\mathbf{l}$	$\int\int \mathbf{F} \cdot d\mathbf{s}$
b. Rectangular coordinates	x direction y direction z direction	$\int F_x\, dx$ $\int F_y\, dy$ $\int F_z\, dz$	$\int\int F_x\, dy\, dz$ $\int\int F_y\, dx\, dz$ $\int\int F_z\, dx\, dy$
c. Cylindrical coordinates	ρ direction ϕ direction z direction	$\int F_\rho\, d\rho$ $\int F_\phi\, \rho\, d\phi$ $\int F_z\, dz$	$\int\int F_\rho\, \rho\, d\phi\, dz$ $\int\int F_\phi\, d\rho\, dz$ $\int\int F_z\, \rho\, d\phi\, d\rho$
d. Spherical coordinates	r direction θ direction ϕ direction	$\int F_r\, dr$ $\int F_\theta\, r\, d\theta$ $\int F_\phi\, r \sin\theta\, d\phi$	$\int\int F_r\, r^2 \sin\theta\, d\theta\, d\phi$ $\int\int F_\theta\, r \sin\theta\, dr\, d\phi$ $\int\int F_\phi\, r\, dr\, d\theta$

4. Gradient, divergence, curl, and Laplacian

a. General

$$\text{grad } w = \nabla w = \left(\frac{dw}{dl}\mathbf{u}_l\right)_{\max} \quad \text{(maximum directional derivative)}$$

$$\text{div } \mathbf{F} = \nabla \cdot \mathbf{F} = \lim_{\Delta\tau \to 0} \frac{1}{\Delta\tau} \oiint \mathbf{F} \cdot d\mathbf{s} \quad \text{(outward flow per unit volume)}$$

$$\text{curl } \mathbf{F} = \nabla \times \mathbf{F} = \lim_{\Delta s \to 0} \left[\frac{\mathbf{u}_n}{\Delta s} \oint \mathbf{F} \cdot d\mathbf{l}\right]_{\max} \quad \text{(maximum circulation per unit area)}$$

$$\nabla^2 w = \text{div (grad } w) = \nabla \cdot \nabla w$$

b. Rectangular coordinates $\left(\nabla = \mathbf{u}_x \frac{\partial}{\partial x} + \mathbf{u}_y \frac{\partial}{\partial y} + \mathbf{u}_z \frac{\partial}{\partial z}\right)$

$$\nabla w = \mathbf{u}_x \frac{\partial w}{\partial x} + \mathbf{u}_y \frac{\partial w}{\partial y} + \mathbf{u}_z \frac{\partial w}{\partial z}$$

$$\nabla \cdot \mathbf{F} = \frac{\partial F_x}{\partial x} + \frac{\partial F_y}{\partial y} + \frac{\partial F_z}{\partial z}$$

$$\nabla \times \mathbf{F} = \begin{vmatrix} \mathbf{u}_x & \mathbf{u}_y & \mathbf{u}_z \\ \frac{\partial}{\partial x} & \frac{\partial}{\partial y} & \frac{\partial}{\partial z} \\ F_x & F_y & F_z \end{vmatrix}$$

$$\nabla^2 w = \frac{\partial^2 w}{\partial x^2} + \frac{\partial^2 w}{\partial y^2} + \frac{\partial^2 w}{\partial z^2}$$

c. Cylindrical coordinates (no single differential operator)

$$\nabla w = \mathbf{u}_\rho \frac{\partial w}{\partial \rho} + \mathbf{u}_\phi \frac{1}{\rho}\frac{\partial w}{\partial \phi} + \mathbf{u}_z \frac{\partial w}{\partial z}$$

$$\nabla \cdot \mathbf{F} = \frac{1}{\rho}\frac{\partial}{\partial \rho}(\rho F_\rho) + \frac{1}{\rho}\frac{\partial F_\phi}{\partial \phi} + \frac{\partial F_z}{\partial z}$$

$$\nabla \times \mathbf{F} = \mathbf{u}_\rho \left(\frac{1}{\rho}\frac{\partial F_z}{\partial \phi} - \frac{\partial F_\phi}{\partial z}\right) + \mathbf{u}_\phi \left(\frac{\partial F_\rho}{\partial z} - \frac{\partial F_z}{\partial \rho}\right) + \mathbf{u}_z \left[\frac{1}{\rho}\frac{\partial}{\partial \rho}(\rho F_\phi) - \frac{1}{\rho}\frac{\partial F_\rho}{\partial \phi}\right]$$

$$\nabla^2 w = \frac{1}{\rho}\frac{\partial}{\partial \rho}\left(\rho \frac{\partial w}{\partial \rho}\right) + \frac{1}{\rho^2}\frac{\partial^2 w}{\partial \phi^2} + \frac{\partial^2 w}{\partial z^2}$$

d. Spherical coordinates (no single differential operator)

$$\nabla w = \mathbf{u}_r \frac{\partial w}{\partial r} + \mathbf{u}_\theta \frac{1}{r}\frac{\partial w}{\partial \theta} + \mathbf{u}_\phi \frac{1}{r \sin\theta}\frac{\partial w}{\partial \phi}$$

$$\nabla \cdot \mathbf{F} = \frac{1}{r^2}\frac{\partial}{\partial r}(r^2 F_r) + \frac{1}{r\sin\theta}\frac{\partial}{\partial\theta}(F_\theta \sin\theta) + \frac{1}{r\sin\theta}\frac{\partial F_\phi}{\partial\phi}$$

$$\nabla \times \mathbf{F} = \mathbf{u}_r \frac{1}{r\sin\theta}\left[\frac{\partial}{\partial\theta}(F_\phi \sin\theta) - \frac{\partial F_\theta}{\partial\phi}\right]$$
$$+ \mathbf{u}_\theta \frac{1}{r}\left[\frac{1}{\sin\theta}\frac{\partial F_r}{\partial\phi} - \frac{\partial}{\partial r}(rF_\phi)\right] + \mathbf{u}_\phi \frac{1}{r}\left[\frac{\partial}{\partial r}(rF_\theta) - \frac{\partial F_r}{\partial\theta}\right]$$

$$\nabla^2 w = \frac{1}{r^2}\frac{\partial}{\partial r}\left(r^2 \frac{\partial w}{\partial r}\right) + \frac{1}{r^2\sin\theta}\frac{\partial}{\partial\theta}\left(\sin\theta\frac{\partial w}{\partial\theta}\right) + \frac{1}{r^2\sin^2\theta}\frac{\partial^2 w}{\partial\phi^2}$$

5. Vector identities

a. Addition and multiplication

$$F = \sqrt{\mathbf{F}\cdot\mathbf{F}}$$
$$\mathbf{F} + \mathbf{G} = \mathbf{G} + \mathbf{F}$$
$$\mathbf{F}\cdot\mathbf{G} = \mathbf{G}\cdot\mathbf{F}$$
$$\mathbf{F}\times\mathbf{G} = -\mathbf{G}\times\mathbf{F}$$
$$(\mathbf{F}+\mathbf{G})\cdot\mathbf{H} = \mathbf{F}\cdot\mathbf{H} + \mathbf{G}\cdot\mathbf{H}$$
$$(\mathbf{F}+\mathbf{G})\times\mathbf{H} = \mathbf{F}\times\mathbf{H} + \mathbf{G}\times\mathbf{H}$$
$$\mathbf{F}\cdot\mathbf{G}\times\mathbf{H} = \mathbf{G}\cdot\mathbf{H}\times\mathbf{F} = \mathbf{H}\cdot\mathbf{F}\times\mathbf{G}$$
$$\mathbf{F}\times(\mathbf{G}\times\mathbf{H}) = (\mathbf{F}\cdot\mathbf{H})\mathbf{G} - (\mathbf{F}\cdot\mathbf{G})\mathbf{H}$$

b. Differentiation

$$\nabla(v + w) = \nabla v + \nabla w$$
$$\nabla\cdot(\mathbf{F}+\mathbf{G}) = \nabla\cdot\mathbf{F} + \nabla\cdot\mathbf{G}$$
$$\nabla\times(\mathbf{F}+\mathbf{G}) = \nabla\times\mathbf{F} + \nabla\times\mathbf{G}$$
$$\nabla(vw) = v\nabla w + w\nabla v$$
$$\nabla\cdot(w\mathbf{F}) = w\nabla\cdot\mathbf{F} + \mathbf{F}\cdot\nabla w$$
$$\nabla\times(w\mathbf{F}) = w\nabla\times\mathbf{F} + (\nabla w)\times\mathbf{F}$$
$$\nabla\cdot(\mathbf{F}\times\mathbf{G}) = \mathbf{G}\cdot\nabla\times\mathbf{F} - \mathbf{F}\cdot\nabla\times\mathbf{G}$$
$$\nabla^2\mathbf{F} = \nabla(\nabla\cdot\mathbf{F}) - \nabla\times\nabla\times\mathbf{F}$$
$$\nabla\times\nabla w = 0$$
$$\nabla\cdot\nabla\times\mathbf{F} = 0$$

c. Integration

$$\iiint \nabla\cdot\mathbf{F}\,d\tau = \oint\!\!\oint \mathbf{F}\cdot d\mathbf{s} \qquad \text{(Gauss's, or divergence, theorem)}$$
$$\iint \nabla\times\mathbf{F}\cdot d\mathbf{s} = \oint \mathbf{F}\cdot d\mathbf{l} \qquad \text{(Stokes', or circulation, theorem)}$$
$$\iiint \nabla\times\mathbf{F}\,d\tau = -\oint\!\!\oint \mathbf{F}\times d\mathbf{s}$$

6. Coordinate transformations for vector components (coordinate orientation as shown in Fig. B-1)

   a. Rectangular to cylindrical components

$$F_\rho = F_x \cos\phi + F_y \sin\phi$$
$$F_\phi = -F_x \sin\phi + F_y \cos\phi$$
$$F_z = F_z$$

   b. Rectangular to spherical components

$$F_r = F_x \sin\theta \cos\phi + F_y \sin\theta \sin\phi + F_z \cos\theta$$
$$F_\theta = F_x \cos\theta \cos\phi + F_y \cos\theta \sin\phi - F_z \sin\theta$$
$$F_\phi = -F_x \sin\phi + F_y \cos\phi$$

   c. Cylindrical to rectangular components

$$F_x = F_\rho \cos\phi - F_\phi \sin\phi$$
$$F_y = F_\rho \sin\phi + F_\phi \cos\phi$$
$$F_z = F_z$$

   d. Cylindrical to spherical components

$$F_r = F_\rho \sin\theta + F_z \cos\theta$$
$$F_\theta = F_\rho \cos\theta - F_z \sin\theta$$
$$F_\phi = F_\phi$$

   e. Spherical to rectangular components

$$F_x = F_r \sin\theta \cos\phi + F_\theta \cos\theta \cos\phi - F_\phi \sin\phi$$
$$F_y = F_r \sin\theta \sin\phi + F_\theta \cos\theta \sin\phi + F_\phi \cos\phi$$
$$F_z = F_r \cos\theta - F_\theta \sin\theta$$

   f. Spherical to cylindrical

$$F_\rho = F_r \sin\theta + F_\theta \cos\theta$$
$$F_\phi = F_\phi$$
$$F_z = F_r \cos\theta - F_\theta \sin\theta$$

7. Radius vectors

   a. General

   The radius vector $\mathbf{r}$ from the origin to the point $(\mathbf{r})$ is a vector pointing from the origin to the point $(\mathbf{r})$, of magnitude equal to the distance from the origin to the point $(\mathbf{r})$.
   The radius vector $\mathbf{r} - \mathbf{r}'$ from the point $(\mathbf{r}')$ to the point $(\mathbf{r})$ is a vector pointing from $(\mathbf{r}')$ to $(\mathbf{r})$, of magnitude equal to the distance from $(\mathbf{r}')$ to $(\mathbf{r})$.

b. Rectangular coordinates

$$
\begin{aligned}
\mathbf{r} &= \mathbf{u}_x x + \mathbf{u}_y y + \mathbf{u}_z z \\
|\mathbf{r}| &= \sqrt{x^2 + y^2 + z^2} \\
\mathbf{r} - \mathbf{r}' &= \mathbf{u}_x(x - x') + \mathbf{u}_y(y - y') + \mathbf{u}_z(z - z') \\
|\mathbf{r} - \mathbf{r}'| &= \sqrt{(x - x')^2 + (y - y')^2 + (z - z')^2}
\end{aligned}
$$

c. Cylindrical coordinates

$$
\begin{aligned}
\mathbf{r} &= \mathbf{u}_\rho \rho + \mathbf{u}_z z \\
|\mathbf{r}| &= \sqrt{\rho^2 + z^2} \\
\mathbf{r} - \mathbf{r}' &= \mathbf{u}_x(\rho \cos \phi - \rho' \cos \phi') + \mathbf{u}_y(\rho \sin \phi - \rho' \sin \phi') \\
&\qquad + \mathbf{u}_z(z - z') \\
|\mathbf{r} - \mathbf{r}'| &= \sqrt{\rho^2 + \rho'^2 - 2\rho\rho' \cos (\phi - \phi') + (z - z')^2}
\end{aligned}
$$

d. Spherical coordinates

$$
\begin{aligned}
\mathbf{r} &= \mathbf{u}_r r \\
|\mathbf{r}| &= r \\
\mathbf{r} - \mathbf{r}' &= \mathbf{u}_x(r \sin \theta \cos \phi - r' \sin \theta' \cos \phi') + \mathbf{u}_y(r \sin \theta \sin \phi \\
&\qquad - r' \sin \theta' \sin \phi') + \mathbf{u}_z(r \cos \theta - r' \cos \theta') \\
|\mathbf{r} - \mathbf{r}| &= \sqrt{r^2 + r'^2 - 2rr' \cos \psi} \\
\text{where } \cos \psi &= \cos \theta \cos \theta' + \sin \theta \sin \theta' \cos (\phi - \phi')
\end{aligned}
$$

e. Mixed coordinates (**r** in one system, **r**′ in another)—examples

$$
\begin{aligned}
\mathbf{r} - \mathbf{r}' &= \mathbf{u}_x(x - \rho' \cos \phi') + \mathbf{u}_y(y - \rho' \sin \phi') + \mathbf{u}_z(z - z') \\
&= \mathbf{u}_x(x - r' \sin \theta' \cos \phi') + \mathbf{u}_y(y - r' \sin \theta' \sin \phi') \\
&\qquad + \mathbf{u}_z(z - r' \cos \theta') \\
&= \mathbf{u}_x(\rho \cos \phi - r' \sin \theta' \cos \phi') \\
&\qquad + \mathbf{u}_y(\rho \sin \phi - r' \sin \theta' \sin \phi') + \mathbf{u}_z(z - r' \cos \theta') \\
|\mathbf{r} - \mathbf{r}'| &= \sqrt{x^2 + y^2 + \rho'^2 - 2\rho'(x \cos \phi' + y \sin \phi') + (z - z')^2} \\
&= \sqrt{x^2 + y^2 + z^2 + r'^2 - 2r'[\sin \theta'(x \cos \phi' + y \sin \phi') + z \cos \theta']} \\
&= \sqrt{\rho^2 + z^2 + r'^2 - 2r'[\rho \sin \theta' \cos (\phi - \phi') + z \cos \theta']}
\end{aligned}
$$

8. Sources and potential theory

a. Volume distributions of sources

If $\nabla \cdot \mathbf{C} = w$, then w is a flow source of $\mathbf{C}$.
If $\nabla \times \mathbf{C} = \mathbf{W}$, then $\mathbf{W}$ is a vortex source of $\mathbf{C}$.

b. Surface distributions of sources ($\mathbf{u}_n$ points into region (1))

If $\mathbf{u}_n \cdot (\mathbf{C}^{(1)} - \mathbf{C}^{(2)}) = C_n^{(1)} - C_n^{(2)} = w_s$, then w_s is a surface flow source of $\mathbf{C}$.

If $\mathbf{u}_n \times (\mathbf{C}^{(1)} - \mathbf{C}^{(2)}) = \mathbf{W}_s$, then $\mathbf{W}_s$ is a surface vortex source of $\mathbf{C}$.

c. Given the flow sources w and the vortex sources $\mathbf{W}$ of $\mathbf{C}$ everywhere in space, then

$$\mathbf{C} = -\nabla g + \nabla \times \mathbf{G}$$

where the scalar potential g is

$$g(\mathbf{r}) = \frac{1}{4\pi} \iiint \frac{w(\mathbf{r}')}{|\mathbf{r} - \mathbf{r}'|} \, d\tau'$$

and the vector potential $\mathbf{G}$ is

$$\mathbf{G}(\mathbf{r}) = \frac{1}{4\pi} \iiint \frac{\mathbf{W}(\mathbf{r}')}{|\mathbf{r} - \mathbf{r}'|} \, d\tau'$$

APPENDIX C

CONSTITUTIVE PARAMETERS OF MATTER

TABLE C-1. CONDUCTIVITY (APPROXIMATE)

Material	*Conductivity* σ *(mhos/meter)*
Silver	6.1×10^7
Copper	5.7×10^7
Gold	4.1×10^7
Aluminum	3.5×10^7
Nickel	1.3×10^7
Iron	1.0×10^7
Lead	4.8×10^6
German silver	3.0×10^6
Mercury	1.0×10^6
Nichrome	1.0×10^6
Graphite	1.1×10^5
Sea water	4
Ethyl alcohol	3×10^{-4}
Distilled water	2×10^{-4}
Celluloid	1×10^{-8}
Bakelite	1×10^{-9}
Glass	1×10^{-12}
Hard rubber	1×10^{-15}
Paraffin	1×10^{-15}
Mica	1×10^{-15}
Sulfur	1×10^{-15}
Quartz, fused	1×10^{-17}

TABLE C-2. RELATIVE CAPACITIVITY OR DIELECTRIC CONSTANT (APPROXIMATE)

$\epsilon_r = \epsilon/\epsilon_0 \qquad \epsilon_0 = 8.854 \times 10^{-12}$ farad/meter

Material	*Relative capacitivity* ϵ_r
Metals	1
Air	1.006
Paraffin	2.1
Mineral oil	2.2
Polyethylene	2.2
Polystyrene	2.5
Rubber	3
Linseed oil	3.4
Sulfur	4
Castor oil	4.7
Quartz	5
Bakelite	5
Porcelain	5.7
Plate glass	6
Mica	6
Cellulose acetate	7
Ethyl alcohol	26
Distilled water	81

TABLE C-3. RELATIVE INDUCTIVITY OR RELATIVE PERMEABILITY (APPROXIMATE)

$\mu_r = \mu/\mu_0 \qquad \mu_0 = 4\pi \times 10^{-7}$ henry/meter

Material	*Relative permeability* μ_r
Silver	0.99998
Copper	0.999991
Air	1.0000004
Aluminum	1.00002
Cobalt	250
Nickel	600
Steel	2,000
Iron	5,000
Permalloy	1×10^5
Supermalloy	1×10^6

BIBLIOGRAPHY

Abraham, A., and R. Becker: "The Classical Theory of Electricity," Blackie & Son, Ltd., Glasgow, 1932.

Attwood, S.: "Electric and Magnetic Fields," 3d ed., John Wiley & Sons, Inc., New York, 1949.

Harnwell, G. P.: "Principles of Electricity and Magnetism," McGraw-Hill Book Company, Inc., New York, 1938.

Jeans, J.: "Electricity and Magnetism," Cambridge University Press, London, 1933.

Jordan, E.: "Electromagnetic Waves and Radiating Systems," Prentice-Hall, Inc., Englewood Cliffs, N.J., 1950.

King, R. W. P.: "Electromagnetic Engineering," McGraw-Hill Book Company, Inc., New York, 1945.

Kraus, J. D.: "Electromagnetics," McGraw-Hill Book Company, Inc., New York, 1953.

Mason, M., and W. Weaver: "The Electromagnetic Field," University of Chicago Press, Chicago, 1929.

Maxwell, J. C.: "A Treatise on Electricity and Magnetism," Clarendon Press, Oxford, 1892, reprinted by Dover Publications, New York, 1954.

Page, L., and N. Adams: "Principles of Electricity," D. Van Nostrand Company, Inc., Princeton, N.J., 1931.

Peck, E. R.: "Electricity and Magnetism," McGraw-Hill Book Company, Inc., New York, 1953.

Ramo, S., and J. R. Whinnery, "Fields and Waves in Modern Radio," 2d ed., John Wiley & Sons, Inc., New York, 1953.

Rogers, W. E.: "Introduction to Electric Fields," McGraw-Hill Book Company, Inc., New York, 1954.

Schelkunoff, S. A.: "Electromagnetic Waves," D. Van Nostrand Company, Inc., Princeton, N.J., 1943.

Sears, F. W.: "Electricity and Magnetism," Addison-Wesley Publishing Company, Reading, Mass., 1946.

Shedd, P. C.: "Fundamentals of Electromagnetic Waves," Prentice-Hall, Inc., Englewood Cliffs, N.J., 1955.

Skilling, H. H.: "Fundamentals of Electric Waves," 2d ed., John Wiley & Sons, Inc., New York, 1948.

Slater, J. C., and N. H. Frank: "Electromagnetism," McGraw-Hill Book Company, Inc., New York, 1947.

Smythe, W. R.: "Static and Dynamic Electricity," 2d ed., McGraw-Hill Book Company, Inc., New York, 1950.

Stratton, J. A.: "Electromagnetic Theory," McGraw-Hill Book Company, Inc., New York, 1941.

Ware, L. A.: "Elements of Electromagnetic Waves," Pitman Publishing Corporation, New York, 1949.

Weber, E.: "Electromagnetic Fields," John Wiley & Sons, Inc., New York, 1950.

INDEX

Numbers in **boldface** type refer to problems

Astronomy

CHARIOTS FOR APOLLO: The NASA History of Manned Lunar Spacecraft to 1969, Courtney G. Brooks, James M. Grimwood, and Loyd S. Swenson, Jr. This illustrated history by a trio of experts is the definitive reference on the Apollo spacecraft and lunar modules. It traces the vehicles' design, development, and operation in space. More than 100 photographs and illustrations. 576pp. 6 3/4 x 9 1/4. 0-486-46756-2

EXPLORING THE MOON THROUGH BINOCULARS AND SMALL TELESCOPES, Ernest H. Cherrington, Jr. Informative, profusely illustrated guide to locating and identifying craters, rills, seas, mountains, other lunar features. Newly revised and updated with special section of new photos. Over 100 photos and diagrams. 240pp. 8 1/4 x 11. 0-486-24491-1

WHERE NO MAN HAS GONE BEFORE: A History of NASA's Apollo Lunar Expeditions, William David Compton. Introduction by Paul Dickson. This official NASA history traces behind-the-scenes conflicts and cooperation between scientists and engineers. The first half concerns preparations for the Moon landings, and the second half documents the flights that followed Apollo 11. 1989 edition. 432pp. 7 x 10. 0-486-47888-2

APOLLO EXPEDITIONS TO THE MOON: The NASA History, Edited by Edgar M. Cortright. Official NASA publication marks the 40th anniversary of the first lunar landing and features essays by project participants recalling engineering and administrative challenges. Accessible, jargon-free accounts, highlighted by numerous illustrations. 336pp. 8 3/8 x 10 7/8. 0-486-47175-6

ON MARS: Exploration of the Red Planet, 1958-1978--The NASA History, Edward Clinton Ezell and Linda Neuman Ezell. NASA's official history chronicles the start of our explorations of our planetary neighbor. It recounts cooperation among government, industry, and academia, and it features dozens of photos from Viking cameras. 560pp. 6 3/4 x 9 1/4. 0-486-46757-0

ARISTARCHUS OF SAMOS: The Ancient Copernicus, Sir Thomas Heath. Heath's history of astronomy ranges from Homer and Hesiod to Aristarchus and includes quotes from numerous thinkers, compilers, and scholasticists from Thales and Anaximander through Pythagoras, Plato, Aristotle, and Heraclides. 34 figures. 448pp. 5 3/8 x 8 1/2. 0-486-43886-4

AN INTRODUCTION TO CELESTIAL MECHANICS, Forest Ray Moulton. Classic text still unsurpassed in presentation of fundamental principles. Covers rectilinear motion, central forces, problems of two and three bodies, much more. Includes over 200 problems, some with answers. 437pp. 5 3/8 x 8 1/2. 0-486-64687-4

BEYOND THE ATMOSPHERE: Early Years of Space Science, Homer E. Newell. This exciting survey is the work of a top NASA administrator who chronicles technological advances, the relationship of space science to general science, and the space program's social, political, and economic contexts. 528pp. 6 3/4 x 9 1/4. 0-486-47464-X

STAR LORE: Myths, Legends, and Facts, William Tyler Olcott. Captivating retellings of the origins and histories of ancient star groups include Pegasus, Ursa Major, Pleiades, signs of the zodiac, and other constellations. "Classic." – *Sky & Telescope*. 58 illustrations. 544pp. 5 3/8 x 8 1/2. 0-486-43581-4

A COMPLETE MANUAL OF AMATEUR ASTRONOMY: Tools and Techniques for Astronomical Observations, P. Clay Sherrod with Thomas L. Koed. Concise, highly readable book discusses the selection, set-up, and maintenance of a telescope; amateur studies of the sun; lunar topography and occultations; and more. 124 figures. 26 halftones. 37 tables. 335pp. 6 1/2 x 9 1/4. 0-486-42820-6

Chemistry

MOLECULAR COLLISION THEORY, M. S. Child. This high-level monograph offers an analytical treatment of classical scattering by a central force, quantum scattering by a central force, elastic scattering phase shifts, and semi-classical elastic scattering. 1974 edition. 310pp. 5 3/8 x 8 1/2. 0-486-69437-2

HANDBOOK OF COMPUTATIONAL QUANTUM CHEMISTRY, David B. Cook. This comprehensive text provides upper-level undergraduates and graduate students with an accessible introduction to the implementation of quantum ideas in molecular modeling, exploring practical applications alongside theoretical explanations. 1998 edition. 832pp. 5 3/8 x 8 1/2. 0-486-44307-8

RADIOACTIVE SUBSTANCES, Marie Curie. The celebrated scientist's thesis, which directly preceded her 1903 Nobel Prize, discusses establishing atomic character of radioactivity; extraction from pitchblende of polonium and radium; isolation of pure radium chloride; more. 96pp. 5 3/8 x 8 1/2. 0-486-42550-9

CHEMICAL MAGIC, Leonard A. Ford. Classic guide provides intriguing entertainment while elucidating sound scientific principles, with more than 100 unusual stunts: cold fire, dust explosions, a nylon rope trick, a disappearing beaker, much more. 128pp. 5 3/8 x 8 1/2. 0-486-67628-5

ALCHEMY, E. J. Holmyard. Classic study by noted authority covers 2,000 years of alchemical history: religious, mystical overtones; apparatus; signs, symbols, and secret terms; advent of scientific method, much more. Illustrated. 320pp. 5 3/8 x 8 1/2. 0-486-26298-7

CHEMICAL KINETICS AND REACTION DYNAMICS, Paul L. Houston. This text teaches the principles underlying modern chemical kinetics in a clear, direct fashion, using several examples to enhance basic understanding. Solutions to selected problems. 2001 edition. 352pp. 8 3/8 x 11. 0-486-45334-0

PROBLEMS AND SOLUTIONS IN QUANTUM CHEMISTRY AND PHYSICS, Charles S. Johnson and Lee G. Pedersen. Unusually varied problems, with detailed solutions, cover of quantum mechanics, wave mechanics, angular momentum, molecular spectroscopy, scattering theory, more. 280 problems, plus 139 supplementary exercises. 430pp. 6 1/2 x 9 1/4. 0-486-65236-X

ELEMENTS OF CHEMISTRY, Antoine Lavoisier. Monumental classic by the founder of modern chemistry features first explicit statement of law of conservation of matter in chemical change, and more. Facsimile reprint of original (1790) Kerr translation. 539pp. 5 3/8 x 8 1/2. 0-486-64624-6

MAGNETISM AND TRANSITION METAL COMPLEXES, F. E. Mabbs and D. J. Machin. A detailed view of the calculation methods involved in the magnetic properties of transition metal complexes, this volume offers sufficient background for original work in the field. 1973 edition. 240pp. 5 3/8 x 8 1/2. 0-486-46284-6

GENERAL CHEMISTRY, Linus Pauling. Revised third edition of classic first-year text by Nobel laureate. Atomic and molecular structure, quantum mechanics, statistical mechanics, thermodynamics correlated with descriptive chemistry. Problems. 992pp. 5 3/8 x 8 1/2. 0-486-65622-5

ELECTROLYTE SOLUTIONS: Second Revised Edition, R. A. Robinson and R. H. Stokes. Classic text deals primarily with measurement, interpretation of conductance, chemical potential, and diffusion in electrolyte solutions. Detailed theoretical interpretations, plus extensive tables of thermodynamic and transport properties. 1970 edition. 590pp. 5 3/8 x 8 1/2. 0-486-42225-9

Engineering

FUNDAMENTALS OF ASTRODYNAMICS, Roger R. Bate, Donald D. Mueller, and Jerry E. White. Teaching text developed by U.S. Air Force Academy develops the basic two-body and n-body equations of motion; orbit determination; classical orbital elements, coordinate transformations; differential correction; more. 1971 edition. 455pp. 5 3/8 x 8 1/2. 0-486-60061-0

INTRODUCTION TO CONTINUUM MECHANICS FOR ENGINEERS: Revised Edition, Ray M. Bowen. This self-contained text introduces classical continuum models within a modern framework. Its numerous exercises illustrate the governing principles, linearizations, and other approximations that constitute classical continuum models. 2007 edition. 320pp. 6 1/8 x 9 1/4. 0-486-47460-7

ENGINEERING MECHANICS FOR STRUCTURES, Louis L. Bucciarelli. This text explores the mechanics of solids and statics as well as the strength of materials and elasticity theory. Its many design exercises encourage creative initiative and systems thinking. 2009 edition. 320pp. 6 1/8 x 9 1/4. 0-486-46855-0

FEEDBACK CONTROL THEORY, John C. Doyle, Bruce A. Francis and Allen R. Tannenbaum. This excellent introduction to feedback control system design offers a theoretical approach that captures the essential issues and can be applied to a wide range of practical problems. 1992 edition. 224pp. 6 1/2 x 9 1/4. 0-486-46933-6

THE FORCES OF MATTER, Michael Faraday. These lectures by a famous inventor offer an easy-to-understand introduction to the interactions of the universe's physical forces. Six essays explore gravitation, cohesion, chemical affinity, heat, magnetism, and electricity. 1993 edition. 96pp. 5 3/8 x 8 1/2. 0-486-47482-8

DYNAMICS, Lawrence E. Goodman and William H. Warner. Beginning engineering text introduces calculus of vectors, particle motion, dynamics of particle systems and plane rigid bodies, technical applications in plane motions, and more. Exercises and answers in every chapter. 619pp. 5 3/8 x 8 1/2. 0-486-42006-X

ADAPTIVE FILTERING PREDICTION AND CONTROL, Graham C. Goodwin and Kwai Sang Sin. This unified survey focuses on linear discrete-time systems and explores natural extensions to nonlinear systems. It emphasizes discrete-time systems, summarizing theoretical and practical aspects of a large class of adaptive algorithms. 1984 edition. 560pp. 6 1/2 x 9 1/4. 0-486-46932-8

INDUCTANCE CALCULATIONS, Frederick W. Grover. This authoritative reference enables the design of virtually every type of inductor. It features a single simple formula for each type of inductor, together with tables containing essential numerical factors. 1946 edition. 304pp. 5 3/8 x 8 1/2. 0-486-47440-2

THERMODYNAMICS: Foundations and Applications, Elias P. Gyftopoulos and Gian Paolo Beretta. Designed by two MIT professors, this authoritative text discusses basic concepts and applications in detail, emphasizing generality, definitions, and logical consistency. More than 300 solved problems cover realistic energy systems and processes. 800pp. 6 1/8 x 9 1/4. 0-486-43932-1

THE FINITE ELEMENT METHOD: Linear Static and Dynamic Finite Element Analysis, Thomas J. R. Hughes. Text for students without in-depth mathematical training, this text includes a comprehensive presentation and analysis of algorithms of time-dependent phenomena plus beam, plate, and shell theories. Solution guide available upon request. 672pp. 6 1/2 x 9 1/4. 0-486-41181-8

Mathematics–Bestsellers

HANDBOOK OF MATHEMATICAL FUNCTIONS: with Formulas, Graphs, and Mathematical Tables, Edited by Milton Abramowitz and Irene A. Stegun. A classic resource for working with special functions, standard trig, and exponential logarithmic definitions and extensions, it features 29 sets of tables, some to as high as 20 places. 1046pp. 8 x 10 1/2. 0-486-61272-4

ABSTRACT AND CONCRETE CATEGORIES: The Joy of Cats, Jiri Adamek, Horst Herrlich, and George E. Strecker. This up-to-date introductory treatment employs category theory to explore the theory of structures. Its unique approach stresses concrete categories and presents a systematic view of factorization structures. Numerous examples. 1990 edition, updated 2004. 528pp. 6 1/8 x 9 1/4. 0-486-46934-4

MATHEMATICS: Its Content, Methods and Meaning, A. D. Aleksandrov, A. N. Kolmogorov, and M. A. Lavrent'ev. Major survey offers comprehensive, coherent discussions of analytic geometry, algebra, differential equations, calculus of variations, functions of a complex variable, prime numbers, linear and non-Euclidean geometry, topology, functional analysis, more. 1963 edition. 1120pp. 5 3/8 x 8 1/2. 0-486-40916-3

INTRODUCTION TO VECTORS AND TENSORS: Second Edition--Two Volumes Bound as One, Ray M. Bowen and C.-C. Wang. Convenient single-volume compilation of two texts offers both introduction and in-depth survey. Geared toward engineering and science students rather than mathematicians, it focuses on physics and engineering applications. 1976 edition. 560pp. 6 1/2 x 9 1/4. 0-486-46914-X

AN INTRODUCTION TO ORTHOGONAL POLYNOMIALS, Theodore S. Chihara. Concise introduction covers general elementary theory, including the representation theorem and distribution functions, continued fractions and chain sequences, the recurrence formula, special functions, and some specific systems. 1978 edition. 272pp. 5 3/8 x 8 1/2. 0-486-47929-3

ADVANCED MATHEMATICS FOR ENGINEERS AND SCIENTISTS, Paul DuChateau. This primary text and supplemental reference focuses on linear algebra, calculus, and ordinary differential equations. Additional topics include partial differential equations and approximation methods. Includes solved problems. 1992 edition. 400pp. 7 1/2 x 9 1/4. 0-486-47930-7

PARTIAL DIFFERENTIAL EQUATIONS FOR SCIENTISTS AND ENGINEERS, Stanley J. Farlow. Practical text shows how to formulate and solve partial differential equations. Coverage of diffusion-type problems, hyperbolic-type problems, elliptic-type problems, numerical and approximate methods. Solution guide available upon request. 1982 edition. 414pp. 6 1/8 x 9 1/4. 0-486-67620-X

VARIATIONAL PRINCIPLES AND FREE-BOUNDARY PROBLEMS, Avner Friedman. Advanced graduate-level text examines variational methods in partial differential equations and illustrates their applications to free-boundary problems. Features detailed statements of standard theory of elliptic and parabolic operators. 1982 edition. 720pp. 6 1/8 x 9 1/4. 0-486-47853-X

LINEAR ANALYSIS AND REPRESENTATION THEORY, Steven A. Gaal. Unified treatment covers topics from the theory of operators and operator algebras on Hilbert spaces; integration and representation theory for topological groups; and the theory of Lie algebras, Lie groups, and transform groups. 1973 edition. 704pp. 6 1/8 x 9 1/4. 0-486-47851-3

Mathematics–Logic and Problem Solving

PERPLEXING PUZZLES AND TANTALIZING TEASERS, Martin Gardner. Ninety-three riddles, mazes, illusions, tricky questions, word and picture puzzles, and other challenges offer hours of entertainment for youngsters. Filled with rib-tickling drawings. Solutions. 224pp. 5 3/8 x 8 1/2. 0-486-25637-5

MY BEST MATHEMATICAL AND LOGIC PUZZLES, Martin Gardner. The noted expert selects 70 of his favorite "short" puzzles. Includes The Returning Explorer, The Mutilated Chessboard, Scrambled Box Tops, and dozens more. Complete solutions included. 96pp. 5 3/8 x 8 1/2. 0-486-28152-3

THE LADY OR THE TIGER?: and Other Logic Puzzles, Raymond M. Smullyan. Created by a renowned puzzle master, these whimsically themed challenges involve paradoxes about probability, time, and change; metapuzzles; and self-referentiality. Nineteen chapters advance in difficulty from relatively simple to highly complex. 1982 edition. 240pp. 5 3/8 x 8 1/2. 0-486-47027-X

SATAN, CANTOR AND INFINITY: Mind-Boggling Puzzles, Raymond M. Smullyan. A renowned mathematician tells stories of knights and knaves in an entertaining look at the logical precepts behind infinity, probability, time, and change. Requires a strong background in mathematics. Complete solutions. 288pp. 5 3/8 x 8 1/2. 0-486-47036-9

THE RED BOOK OF MATHEMATICAL PROBLEMS, Kenneth S. Williams and Kenneth Hardy. Handy compilation of 100 practice problems, hints and solutions indispensable for students preparing for the William Lowell Putnam and other mathematical competitions. Preface to the First Edition. Sources. 1988 edition. 192pp. 5 3/8 x 8 1/2. 0-486-69415-1

KING ARTHUR IN SEARCH OF HIS DOG AND OTHER CURIOUS PUZZLES, Raymond M. Smullyan. This fanciful, original collection for readers of all ages features arithmetic puzzles, logic problems related to crime detection, and logic and arithmetic puzzles involving King Arthur and his Dogs of the Round Table. 160pp. 5 3/8 x 8 1/2. 0-486-47435-6

UNDECIDABLE THEORIES: Studies in Logic and the Foundation of Mathematics, Alfred Tarski in collaboration with Andrzej Mostowski and Raphael M. Robinson. This well-known book by the famed logician consists of three treatises: "A General Method in Proofs of Undecidability," "Undecidability and Essential Undecidability in Mathematics," and "Undecidability of the Elementary Theory of Groups." 1953 edition. 112pp. 5 3/8 x 8 1/2. 0-486-47703-7

LOGIC FOR MATHEMATICIANS, J. Barkley Rosser. Examination of essential topics and theorems assumes no background in logic. "Undoubtedly a major addition to the literature of mathematical logic." – *Bulletin of the American Mathematical Society.* 1978 edition. 592pp. 6 1/8 x 9 1/4. 0-486-46898-4

INTRODUCTION TO PROOF IN ABSTRACT MATHEMATICS, Andrew Wohlgemuth. This undergraduate text teaches students what constitutes an acceptable proof, and it develops their ability to do proofs of routine problems as well as those requiring creative insights. 1990 edition. 384pp. 6 1/2 x 9 1/4. 0-486-47854-8

FIRST COURSE IN MATHEMATICAL LOGIC, Patrick Suppes and Shirley Hill. Rigorous introduction is simple enough in presentation and context for wide range of students. Symbolizing sentences; logical inference; truth and validity; truth tables; terms, predicates, universal quantifiers; universal specification and laws of identity; more. 288pp. 5 3/8 x 8 1/2. 0-486-42259-3

Mathematics–Algebra and Calculus

VECTOR CALCULUS, Peter Baxandall and Hans Liebeck. This introductory text offers a rigorous, comprehensive treatment. Classical theorems of vector calculus are amply illustrated with figures, worked examples, physical applications, and exercises with hints and answers. 1986 edition. 560pp. 5 3/8 x 8 1/2. 0-486-46620-5

ADVANCED CALCULUS: An Introduction to Classical Analysis, Louis Brand. A course in analysis that focuses on the functions of a real variable, this text introduces the basic concepts in their simplest setting and illustrates its teachings with numerous examples, theorems, and proofs. 1955 edition. 592pp. 5 3/8 x 8 1/2. 0-486-44548-8

ADVANCED CALCULUS, Avner Friedman. Intended for students who have already completed a one-year course in elementary calculus, this two-part treatment advances from functions of one variable to those of several variables. Solutions. 1971 edition. 432pp. 5 3/8 x 8 1/2. 0-486-45795-8

METHODS OF MATHEMATICS APPLIED TO CALCULUS, PROBABILITY, AND STATISTICS, Richard W. Hamming. This 4-part treatment begins with algebra and analytic geometry and proceeds to an exploration of the calculus of algebraic functions and transcendental functions and applications. 1985 edition. Includes 310 figures and 18 tables. 880pp. 6 1/2 x 9 1/4. 0-486-43945-3

BASIC ALGEBRA I: Second Edition, Nathan Jacobson. A classic text and standard reference for a generation, this volume covers all undergraduate algebra topics, including groups, rings, modules, Galois theory, polynomials, linear algebra, and associative algebra. 1985 edition. 528pp. 6 1/8 x 9 1/4. 0-486-47189-6

BASIC ALGEBRA II: Second Edition, Nathan Jacobson. This classic text and standard reference comprises all subjects of a first-year graduate-level course, including in-depth coverage of groups and polynomials and extensive use of categories and functors. 1989 edition. 704pp. 6 1/8 x 9 1/4. 0-486-47187-X

CALCULUS: An Intuitive and Physical Approach (Second Edition), Morris Kline. Application-oriented introduction relates the subject as closely as possible to science with explorations of the derivative; differentiation and integration of the powers of x; theorems on differentiation, antidifferentiation; the chain rule; trigonometric functions; more. Examples. 1967 edition. 960pp. 6 1/2 x 9 1/4. 0-486-40453-6

ABSTRACT ALGEBRA AND SOLUTION BY RADICALS, John E. Maxfield and Margaret W. Maxfield. Accessible advanced undergraduate-level text starts with groups, rings, fields, and polynomials and advances to Galois theory, radicals and roots of unity, and solution by radicals. Numerous examples, illustrations, exercises, appendixes. 1971 edition. 224pp. 6 1/8 x 9 1/4. 0-486-47723-1

AN INTRODUCTION TO THE THEORY OF LINEAR SPACES, Georgi E. Shilov. Translated by Richard A. Silverman. Introductory treatment offers a clear exposition of algebra, geometry, and analysis as parts of an integrated whole rather than separate subjects. Numerous examples illustrate many different fields, and problems include hints or answers. 1961 edition. 320pp. 5 3/8 x 8 1/2. 0-486-63070-6

LINEAR ALGEBRA, Georgi E. Shilov. Covers determinants, linear spaces, systems of linear equations, linear functions of a vector argument, coordinate transformations, the canonical form of the matrix of a linear operator, bilinear and quadratic forms, and more. 387pp. 5 3/8 x 8 1/2. 0-486-63518-X

Mathematics–Probability and Statistics

Mathematics–Geometry and Topology

PROBLEMS AND SOLUTIONS IN EUCLIDEAN GEOMETRY, M. N. Aref and William Wernick. Based on classical principles, this book is intended for a second course in Euclidean geometry and can be used as a refresher. More than 200 problems include hints and solutions. 1968 edition. 272pp. 5 3/8 x 8 1/2. 0-486-47720-7

TOPOLOGY OF 3-MANIFOLDS AND RELATED TOPICS, Edited by M. K. Fort, Jr. With a New Introduction by Daniel Silver. Summaries and full reports from a 1961 conference discuss decompositions and subsets of 3-space; n-manifolds; knot theory; the Poincaré conjecture; and periodic maps and isotopies. Familiarity with algebraic topology required. 1962 edition. 272pp. 6 1/8 x 9 1/4. 0-486-47753-3

POINT SET TOPOLOGY, Steven A. Gaal. Suitable for a complete course in topology, this text also functions as a self-contained treatment for independent study. Additional enrichment materials make it equally valuable as a reference. 1964 edition. 336pp. 5 3/8 x 8 1/2. 0-486-47222-1

INVITATION TO GEOMETRY, Z. A. Melzak. Intended for students of many different backgrounds with only a modest knowledge of mathematics, this text features self-contained chapters that can be adapted to several types of geometry courses. 1983 edition. 240pp. 5 3/8 x 8 1/2. 0-486-46626-4

TOPOLOGY AND GEOMETRY FOR PHYSICISTS, Charles Nash and Siddhartha Sen. Written by physicists for physics students, this text assumes no detailed background in topology or geometry. Topics include differential forms, homotopy, homology, cohomology, fiber bundles, connection and covariant derivatives, and Morse theory. 1983 edition. 320pp. 5 3/8 x 8 1/2. 0-486-47852-1

BEYOND GEOMETRY: Classic Papers from Riemann to Einstein, Edited with an Introduction and Notes by Peter Pesic. This is the only English-language collection of these 8 accessible essays. They trace seminal ideas about the foundations of geometry that led to Einstein's general theory of relativity. 224pp. 6 1/8 x 9 1/4. 0-486-45350-2

GEOMETRY FROM EUCLID TO KNOTS, Saul Stahl. This text provides a historical perspective on plane geometry and covers non-neutral Euclidean geometry, circles and regular polygons, projective geometry, symmetries, inversions, informal topology, and more. Includes 1,000 practice problems. Solutions available. 2003 edition. 480pp. 6 1/8 x 9 1/4. 0-486-47459-3

TOPOLOGICAL VECTOR SPACES, DISTRIBUTIONS AND KERNELS, François Trèves. Extending beyond the boundaries of Hilbert and Banach space theory, this text focuses on key aspects of functional analysis, particularly in regard to solving partial differential equations. 1967 edition. 592pp. 5 3/8 x 8 1/2. 0-486-45352-9

INTRODUCTION TO PROJECTIVE GEOMETRY, C. R. Wylie, Jr. This introductory volume offers strong reinforcement for its teachings, with detailed examples and numerous theorems, proofs, and exercises, plus complete answers to all odd-numbered end-of-chapter problems. 1970 edition. 576pp. 6 1/8 x 9 1/4. 0-486-46895-X

FOUNDATIONS OF GEOMETRY, C. R. Wylie, Jr. Geared toward students preparing to teach high school mathematics, this text explores the principles of Euclidean and non-Euclidean geometry and covers both generalities and specifics of the axiomatic method. 1964 edition. 352pp. 6 x 9. 0-486-47214-0

Mathematics–History

THE WORKS OF ARCHIMEDES, Archimedes. Translated by Sir Thomas Heath. Complete works of ancient geometer feature such topics as the famous problems of the ratio of the areas of a cylinder and an inscribed sphere; the properties of conoids, spheroids, and spirals; more. 326pp. 5 3/8 x 8 1/2. 0-486-42084-1

THE HISTORICAL ROOTS OF ELEMENTARY MATHEMATICS, Lucas N. H. Bunt, Phillip S. Jones, and Jack D. Bedient. Exciting, hands-on approach to understanding fundamental underpinnings of modern arithmetic, algebra, geometry and number systems examines their origins in early Egyptian, Babylonian, and Greek sources. 336pp. 5 3/8 x 8 1/2. 0-486-25563-8

THE THIRTEEN BOOKS OF EUCLID'S ELEMENTS, Euclid. Contains complete English text of all 13 books of the Elements plus critical apparatus analyzing each definition, postulate, and proposition in great detail. Covers textual and linguistic matters; mathematical analyses of Euclid's ideas; classical, medieval, Renaissance and modern commentators; refutations, supports, extrapolations, reinterpretations and historical notes. 995 figures. Total of 1,425pp. All books 5 3/8 x 8 1/2.
Vol. I: 443pp. 0-486-60088-2
Vol. II: 464pp. 0-486-60089-0
Vol. III: 546pp. 0-486-60090-4

A HISTORY OF GREEK MATHEMATICS, Sir Thomas Heath. This authoritative two-volume set that covers the essentials of mathematics and features every landmark innovation and every important figure, including Euclid, Apollonius, and others. 5 3/8 x 8 1/2.
Vol. I: 461pp. 0-486-24073-8
Vol. II: 597pp. 0-486-24074-6

A MANUAL OF GREEK MATHEMATICS, Sir Thomas L. Heath. This concise but thorough history encompasses the enduring contributions of the ancient Greek mathematicians whose works form the basis of most modern mathematics. Discusses Pythagorean arithmetic, Plato, Euclid, more. 1931 edition. 576pp. 5 3/8 x 8 1/2. 0-486-43231-9

CHINESE MATHEMATICS IN THE THIRTEENTH CENTURY, Ulrich Libbrecht. An exploration of the 13th-century mathematician Ch'in, this fascinating book combines what is known of the mathematician's life with a history of his only extant work, the Shu-shu chiu-chang. 1973 edition. 592pp. 5 3/8 x 8 1/2. 0-486-44619-0

PHILOSOPHY OF MATHEMATICS AND DEDUCTIVE STRUCTURE IN EUCLID'S ELEMENTS, Ian Mueller. This text provides an understanding of the classical Greek conception of mathematics as expressed in Euclid's Elements. It focuses on philosophical, foundational, and logical questions and features helpful appendixes. 400pp. 6 1/2 x 9 1/4. 0-486-45300-6

BEYOND GEOMETRY: Classic Papers from Riemann to Einstein, Edited with an Introduction and Notes by Peter Pesic. This is the only English-language collection of these 8 accessible essays. They trace seminal ideas about the foundations of geometry that led to Einstein's general theory of relativity. 224pp. 6 1/8 x 9 1/4. 0-486-45350-2

HISTORY OF MATHEMATICS, David E. Smith. Two-volume history – from Egyptian papyri and medieval maps to modern graphs and diagrams. Non-technical chronological survey with thousands of biographical notes, critical evaluations, and contemporary opinions on over 1,100 mathematicians. 5 3/8 x 8 1/2.
Vol. I: 618pp. 0-486-20429-4
Vol. II: 736pp. 0-486-20430-8

Physics

THEORETICAL NUCLEAR PHYSICS, John M. Blatt and Victor F. Weisskopf. An uncommonly clear and cogent investigation and correlation of key aspects of theoretical nuclear physics by leading experts: the nucleus, nuclear forces, nuclear spectroscopy, two-, three- and four-body problems, nuclear reactions, beta-decay and nuclear shell structure. 896pp. 5 3/8 x 8 1/2. 0-486-66827-4

QUANTUM THEORY, David Bohm. This advanced undergraduate-level text presents the quantum theory in terms of qualitative and imaginative concepts, followed by specific applications worked out in mathematical detail. 655pp. 5 3/8 x 8 1/2. 0-486-65969-0

ATOMIC PHYSICS AND HUMAN KNOWLEDGE, Niels Bohr. Articles and speeches by the Nobel Prize–winning physicist, dating from 1934 to 1958, offer philosophical explorations of the relevance of atomic physics to many areas of human endeavor. 1961 edition. 112pp. 5 3/8 x 8 1/2. 0-486-47928-5

COSMOLOGY, Hermann Bondi. A co-developer of the steady-state theory explores his conception of the expanding universe. This historic book was among the first to present cosmology as a separate branch of physics. 1961 edition. 192pp. 5 3/8 x 8 1/2. 0-486-47483-6

LECTURES ON QUANTUM MECHANICS, Paul A. M. Dirac. Four concise, brilliant lectures on mathematical methods in quantum mechanics from Nobel Prize-winning quantum pioneer build on idea of visualizing quantum theory through the use of classical mechanics. 96pp. 5 3/8 x 8 1/2. 0-486-41713-1

THE PRINCIPLE OF RELATIVITY, Albert Einstein and Frances A. Davis. Eleven papers that forged the general and special theories of relativity include seven papers by Einstein, two by Lorentz, and one each by Minkowski and Weyl. 1923 edition. 240pp. 5 3/8 x 8 1/2. 0-486-60081-5

PHYSICS OF WAVES, William C. Elmore and Mark A. Heald. Ideal as a classroom text or for individual study, this unique one-volume overview of classical wave theory covers wave phenomena of acoustics, optics, electromagnetic radiations, and more. 477pp. 5 3/8 x 8 1/2. 0-486-64926-1

THERMODYNAMICS, Enrico Fermi. In this classic of modern science, the Nobel Laureate presents a clear treatment of systems, the First and Second Laws of Thermodynamics, entropy, thermodynamic potentials, and much more. Calculus required. 160pp. 5 3/8 x 8 1/2. 0-486-60361-X

QUANTUM THEORY OF MANY-PARTICLE SYSTEMS, Alexander L. Fetter and John Dirk Walecka. Self-contained treatment of nonrelativistic many-particle systems discusses both formalism and applications in terms of ground-state (zero-temperature) formalism, finite-temperature formalism, canonical transformations, and applications to physical systems. 1971 edition. 640pp. 5 3/8 x 8 1/2. 0-486-42827-3

QUANTUM MECHANICS AND PATH INTEGRALS: Emended Edition, Richard P. Feynman and Albert R. Hibbs. Emended by Daniel F. Styer. The Nobel Prize–winning physicist presents unique insights into his theory and its applications. Feynman starts with fundamentals and advances to the perturbation method, quantum electrodynamics, and statistical mechanics. 1965 edition, emended in 2005. 384pp. 6 1/8 x 9 1/4. 0-486-47722-3

Physics

INTRODUCTION TO MODERN OPTICS, Grant R. Fowles. A complete basic undergraduate course in modern optics for students in physics, technology, and engineering. The first half deals with classical physical optics; the second, quantum nature of light. Solutions. 336pp. 5 3/8 x 8 1/2. 0-486-65957-7

THE QUANTUM THEORY OF RADIATION: Third Edition, W. Heitler. The first comprehensive treatment of quantum physics in any language, this classic introduction to basic theory remains highly recommended and widely used, both as a text and as a reference. 1954 edition. 464pp. 5 3/8 x 8 1/2. 0-486-64558-4

QUANTUM FIELD THEORY, Claude Itzykson and Jean-Bernard Zuber. This comprehensive text begins with the standard quantization of electrodynamics and perturbative renormalization, advancing to functional methods, relativistic bound states, broken symmetries, nonabelian gauge fields, and asymptotic behavior. 1980 edition. 752pp. 6 1/2 x 9 1/4. 0-486-44568-2

FOUNDATIONS OF POTENTIAL THERY, Oliver D. Kellogg. Introduction to fundamentals of potential functions covers the force of gravity, fields of force, potentials, harmonic functions, electric images and Green's function, sequences of harmonic functions, fundamental existence theorems, and much more. 400pp. 5 3/8 x 8 1/2. 0-486-60144-7

FUNDAMENTALS OF MATHEMATICAL PHYSICS, Edgar A. Kraut. Indispensable for students of modern physics, this text provides the necessary background in mathematics to study the concepts of electromagnetic theory and quantum mechanics. 1967 edition. 480pp. 6 1/2 x 9 1/4. 0-486-45809-1

GEOMETRY AND LIGHT: The Science of Invisibility, Ulf Leonhardt and Thomas Philbin. Suitable for advanced undergraduate and graduate students of engineering, physics, and mathematics and scientific researchers of all types, this is the first authoritative text on invisibility and the science behind it. More than 100 full-color illustrations, plus exercises with solutions. 2010 edition. 288pp. 7 x 9 1/4. 0-486-47693-6

QUANTUM MECHANICS: New Approaches to Selected Topics, Harry J. Lipkin. Acclaimed as "excellent" (*Nature*) and "very original and refreshing" (*Physics Today*), these studies examine the Mössbauer effect, many-body quantum mechanics, scattering theory, Feynman diagrams, and relativistic quantum mechanics. 1973 edition. 480pp. 5 3/8 x 8 1/2. 0-486-45893-8

THEORY OF HEAT, James Clerk Maxwell. This classic sets forth the fundamentals of thermodynamics and kinetic theory simply enough to be understood by beginners, yet with enough subtlety to appeal to more advanced readers, too. 352pp. 5 3/8 x 8 1/2. 0-486-41735-2

QUANTUM MECHANICS, Albert Messiah. Subjects include formalism and its interpretation, analysis of simple systems, symmetries and invariance, methods of approximation, elements of relativistic quantum mechanics, much more. "Strongly recommended." – *American Journal of Physics.* 1152pp. 5 3/8 x 8 1/2. 0-486-40924-4

RELATIVISTIC QUANTUM FIELDS, Charles Nash. This graduate-level text contains techniques for performing calculations in quantum field theory. It focuses chiefly on the dimensional method and the renormalization group methods. Additional topics include functional integration and differentiation. 1978 edition. 240pp. 5 3/8 x 8 1/2. 0-486-47752-5

Physics

MATHEMATICAL TOOLS FOR PHYSICS, James Nearing. Encouraging students' development of intuition, this original work begins with a review of basic mathematics and advances to infinite series, complex algebra, differential equations, Fourier series, and more. 2010 edition. 496pp. 6 1/8 x 9 1/4. 0-486-48212-X

TREATISE ON THERMODYNAMICS, Max Planck. Great classic, still one of the best introductions to thermodynamics. Fundamentals, first and second principles of thermodynamics, applications to special states of equilibrium, more. Numerous worked examples. 1917 edition. 297pp. 5 3/8 x 8. 0-486-66371-X

AN INTRODUCTION TO RELATIVISTIC QUANTUM FIELD THEORY, Silvan S. Schweber. Complete, systematic, and self-contained, this text introduces modern quantum field theory. "Combines thorough knowledge with a high degree of didactic ability and a delightful style." – *Mathematical Reviews.* 1961 edition. 928pp. 5 3/8 x 8 1/2. 0-486-44228-4

THE ELECTROMAGNETIC FIELD, Albert Shadowitz. Comprehensive undergraduate text covers basics of electric and magnetic fields, building up to electromagnetic theory. Related topics include relativity theory. Over 900 problems, some with solutions. 1975 edition. 768pp. 5 5/8 x 8 1/4. 0-486-65660-8

THE PRINCIPLES OF STATISTICAL MECHANICS, Richard C. Tolman. Definitive treatise offers a concise exposition of classical statistical mechanics and a thorough elucidation of quantum statistical mechanics, plus applications of statistical mechanics to thermodynamic behavior. 1930 edition. 704pp. 5 5/8 x 8 1/4. 0-486-63896-0

INTRODUCTION TO THE PHYSICS OF FLUIDS AND SOLIDS, James S. Trefil. This interesting, informative survey by a well-known science author ranges from classical physics and geophysical topics, from the rings of Saturn and the rotation of the galaxy to underground nuclear tests. 1975 edition. 320pp. 5 3/8 x 8 1/2. 0-486-47437-2

STATISTICAL PHYSICS, Gregory H. Wannier. Classic text combines thermodynamics, statistical mechanics, and kinetic theory in one unified presentation. Topics include equilibrium statistics of special systems, kinetic theory, transport coefficients, and fluctuations. Problems with solutions. 1966 edition. 532pp. 5 3/8 x 8 1/2. 0-486-65401-X

SPACE, TIME, MATTER, Hermann Weyl. Excellent introduction probes deeply into Euclidean space, Riemann's space, Einstein's general relativity, gravitational waves and energy, and laws of conservation. "A classic of physics." – *British Journal for Philosophy and Science.* 330pp. 5 3/8 x 8 1/2. 0-486-60267-2

RANDOM VIBRATIONS: Theory and Practice, Paul H. Wirsching, Thomas L. Paez and Keith Ortiz. Comprehensive text and reference covers topics in probability, statistics, and random processes, plus methods for analyzing and controlling random vibrations. Suitable for graduate students and mechanical, structural, and aerospace engineers. 1995 edition. 464pp. 5 3/8 x 8 1/2. 0-486-45015-5

PHYSICS OF SHOCK WAVES AND HIGH-TEMPERATURE HYDRO DYNAMIC PHENOMENA, Ya B. Zel'dovich and Yu P. Raizer. Physical, chemical processes in gases at high temperatures are focus of outstanding text, which combines material from gas dynamics, shock-wave theory, thermodynamics and statistical physics, other fields. 284 illustrations. 1966–1967 edition. 944pp. 6 1/8 x 9 1/4. 0-486-42002-7